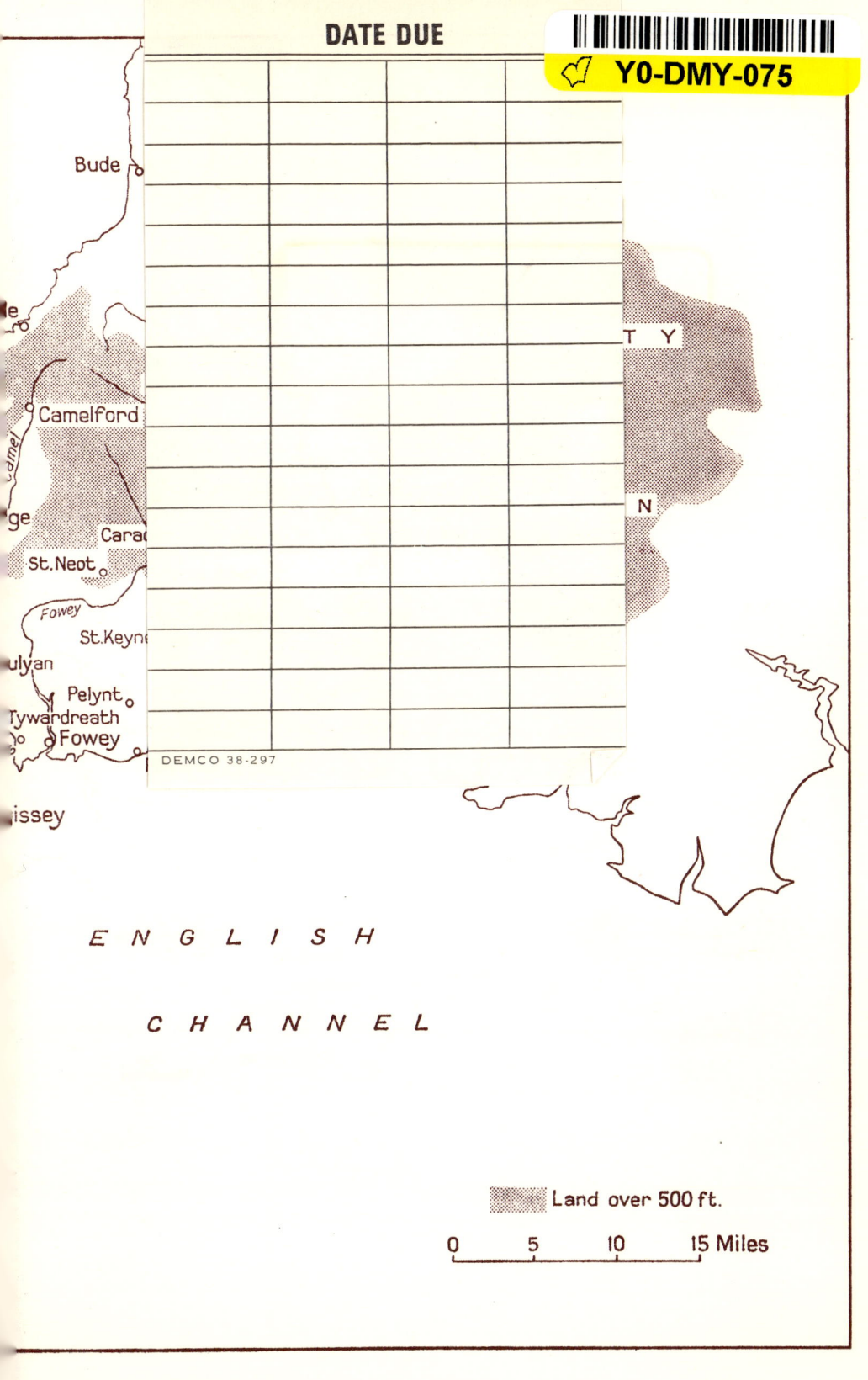

The Hard-Rock Men

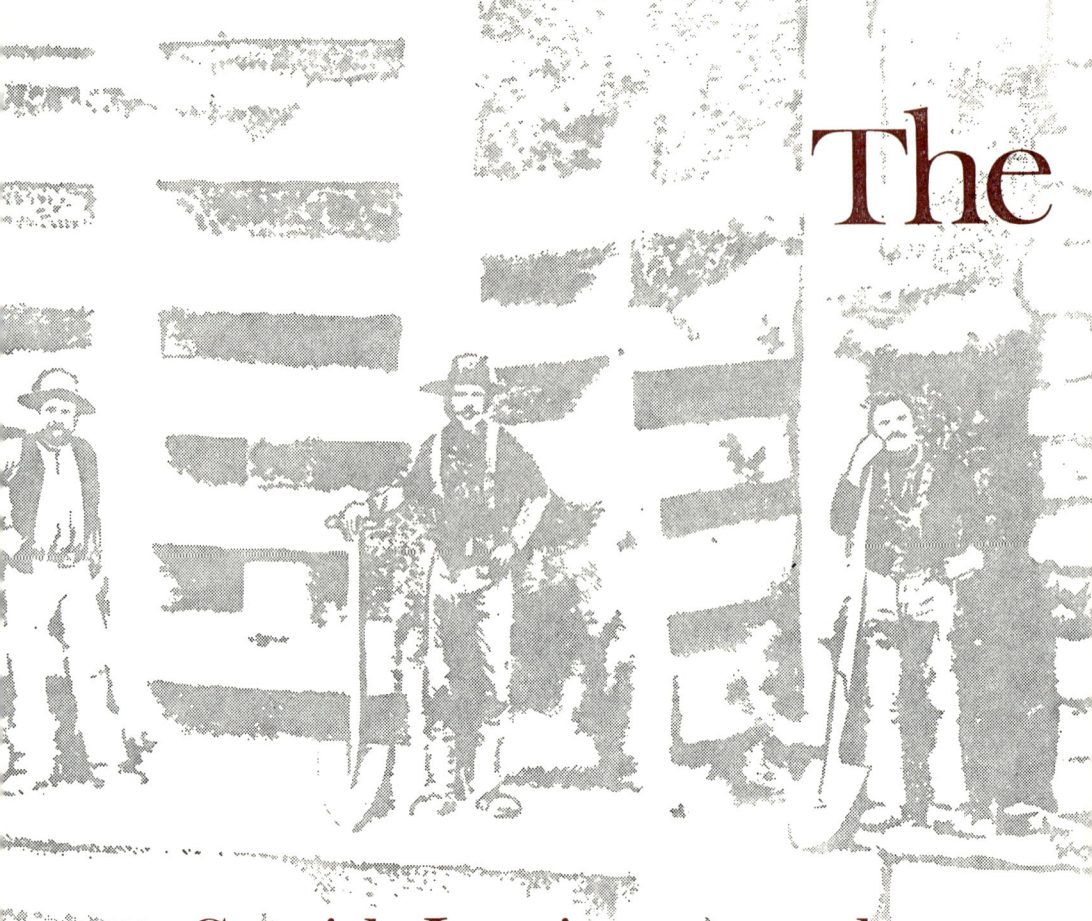

The

Cornish Immigrants and the North American Mining Frontier

Barnes & Noble Books
A division of Harper & Row Publishers, Inc.

Hard-Rock Men

John Rowe
Senior Lecturer in Modern History in the University of Liverpool

1974

Published in the U.S.A. 1974 by
HARPER & ROW PUBLISHERS, INC.
BARNES & NOBLE IMPORT DIVISION

Copyright © *John Rowe 1974*

All rights reserved. No part of this book may be reproduced in any form without permission in writing from the publishers, except by a reviewer who wishes to quote brief passages in connection with a review written for inclusion in a magazine or newspaper

ISBN 06 496014 5

First published 1974

Printed in Great Britain

Preface

This study of Cornish immigrants in America is the outcome of a variety of circumstances and interests. It is a theme which must attract any Cornishman, but the reason for the present writer attempting it demands fuller explanation.

Perhaps the obvious explanation is the right one—childhood and youthful memories of the region within a mile or so radius of his first home. This being largely an account of western mining frontiers, so that home in a way had something of the characteristics usually attributed to a frontier environment. Lying about the meeting point of granite and killas soils, it was also near the contour line which, in Cornwall, marks the border between mixed farming and rough grazing. There, on the col between the headwaters of the River Looe and the upper reaches of the Fowey, were to be found traces and traditions going far back into the historic past, and even into the often nearer-than-thought prehistoric past.

Of the miners that have many times been referred to as 'the old men' by later workers in that calling, there were many relics. Along the valley of the Fowey, between Trekeive and Treverbyn, there were not only pits and hummocks left by tin-streamers, but adits driven into the steep hillsides, most likely by prospectors in search of tin and copper. There was Larkholes, which may have been worked in the 1840s, but hardly ever successfully or profitably—unless to professional mine-promoters of the ilk found in Trollope's *The Three Clerks*.

Although the Caradon 'bals' had been 'knacked' not so many years before, the more active connections of the people of that district had for some time been with foreign mining camps; more particularly, it is true, with the Rand—whence many had returned home dying of the 'African complaint', another name like 'rocks on the chest', for Miner's 'con', or phthisis—but also with the American camps of Cripple Creek, Grass Valley, the 'Lakes', and above all, Butte City, of which it seems possible that more people in Commonmoor had some sort of knowledge, or connection, than with Plymouth or Exeter. Just as down along in westernmost Penwith, Pendeeners were wont to refer to the States as 'the next parish', so in this more eastern part of Cornwall

PREFACE

America was looked upon as neighbouring mining land, and as neighbours always seem to be just that bit better off than oneself, very much a land of material economic promise.

Even more close was direct family connection. The writer's own father was a 'bird of passage' who had sojourned twice in the northland in the Edmonton district of Alberta, and in between had a spell in Western Australia. Other relations had taken themselves overseas, contact being completely lost in some instances after they arrived in Canada, New Zealand, Upper Michigan, and other places far from the Old Country. Many of them went not as miners but just in search of better economic opportunities. That had been the reason why so many others had emigrated from Cornwall to North America. That so many of them were miners was due to the prevalence and vicissitudes of that industry in late eighteenth- and nineteenth-century Cornwall.

It was in Wisconsin, the first American mining region to which they went in considerable numbers, that they first became generally known as 'hard-rock miners'. Their skill in dealing with hard rock, and in underground mining, became proverbial in the New World. More colloquially they were called 'Cousin Jacks', but no one really knows the origin of the term. The most common explanation is that when a job fell vacant there would be a Cornish worker ready to tell the boss or foreman that he would send home for his 'Cousin Jack' to fill it. Some in Northern Michigan, however, did not give much credit to this tradition and, aware of the profanity of many Cornish workmates, averred that it was just a corruption of *cussin' Jack*. But the term may have originated even before the mound-building Indians of Iowa had been chipping away at the masses of native copper up in Ontonagan and Keweenaw: Richard Carew made a reference to all Cornish gentry being cousins or kin in the casual manner in which an old proverb or adage is repeated.

But this is only surmise about a mystery that is unlikely ever to be solved. The historian can leave it thus; his task is to record facts, and tentatively to suggest conclusions, judgments, and opinions, that may be deduced from them. He cannot hope to find all the facts, but he can discover and record what he trusts are the most significant fragments of the past and of the people who then lived. The Cornish emigrants, 'birds of passage', and even stay-at-homes, who appear in the following pages are but few of a vast throng, but it is hoped they are in some ways typical and representative.

J.R.

Irby, Cheshire.

Contents

	PREFACE	v
	LIST OF ILLUSTRATIONS	ix
	ACKNOWLEDGEMENTS	xi
1	The land whence they came	1
2	To go or not to go	18
3	The first western mining frontier	37
4	Copper frontier in Upper Michigan	62
5	California—here we come	96
6	Toil, trouble, and tribulation	127
7	Go they must—but where?	155
8	South-western desert empire	177
9	Gold in them thar hills—perhaps	229
10	A home from home	258
	APPENDICES	
	I Public sales of copper ores in Cornwall and Devon 1845–75	297
	II Shipments of rough copper from Lake Superior 1845–61	298
	III Production of fine copper, Michigan and Cornwall 1862–72	299
	BIBLIOGRAPHY	301
	INDEX	305

Illustrations

PLATES *between pages 76–7*
1 A page from Richard Wearne's journal, 4–7 July 1848
2 Cliff Mine, in the Keweenaw district of Upper Michigan
3 The wreck of the emigrant ship 'John' on the Manacles, May 1855

between pages 124–5
4 Hydraulic mining at the Malakoff Diggins in 1870, and the same site today
5 Grass Valley, Nevada County, California, in 1858
6 Bringing water to the mines. A flume near Quaker Hill, Nevada County, California, and Treffry Viaduct and Aqueduct, Luxulyan, Cornwall

7 'Billy' Barker of Barkerville, British Columbia *facing page* 148
8 Barkerville, B.C., before and after the great fire of September 1868   149
9 On the way to Last Chance Gulch, via the Missouri, about 1875   164
10 Miners at Speculator Mine, Montana, in 1890   164
11 The Comstock country, near Virginia City, Nevada   165

between pages 220–1
12 Two Cornishmen who were successful in North America: Thomas V. Keam (top) and Richard Pearce
13 Caribou, Colorado, in its bonanza days, and the same site today
14 The Tamarack Location, Calumet, about 1880
15 Copper Queen Mine glory-hole, Bisbee, Arizona

ILLUSTRATIONS

16 Cornish miners' homes, in the Copper Country of Upper Michigan (top), and at Mineral Point, Wisconsin
17 Miners' Union Day parade in Butte, Montana, and the Boston and Montana Band of Butte, 1888   *facing page* 276
18 A rock-drilling contest and a drinking party at Butte, about 1890   277
19 The graves of Cornish immigrants at Mineral Point, Wisconsin (top left) and at Virginia City, Nevada (top right and bottom)   292
20 'Trails for gold lead to the grave'—memorial verses on William Wearne, who died at Panama, 19 December 1852   293

MAPS
Cornwall *front endpaper*   North America *back endpaper*

Grateful acknowledgement for permission to reproduce illustrations is made to the following: Marquette County Historical Society; Arizona Pioneers' Historical Society; Bancroft Library, University of California; Charles Woolf, Newquay, Cornwall; Denver Public Library Western Collection; C. Owen Smithers, Butte, Montana; and Montana Historical Society, Helena. The wreck of the emigrant ship *John* is reproduced from *The Cornish Magazine*, vol. i (Truro, 1898).

x

Acknowledgements

Although the people who helped me in many different ways in the preparation and completion of this work are far too many to be enumerated here, some of my debts are too great to pass without acknowledgement. I would take this opportunity of expressing my gratitude to the indefatigably helpful and hospitable staffs of the Bancroft Library at Berkeley, California, to the Arizona Pioneers' Library at Tucson, Arizona, to the Michigan State Historical Collections at Ann Arbor, the Wisconsin State Historical Records Library at Madison, the Library of the Montana Historical Society at Helena, the Marquette County Historical Society at Marquette, Michigan, the Library of the University of Colorado at Boulder, the Butte Public Library and the National Archives at Washington, D.C. In the Old Country I owe a very great debt of gratitude to the staffs of the Harold Cohen Library of the University of Liverpool, of the Picton Reference Library, Liverpool, of the Morrab Library, Penzance, of the Royal Cornwall Polytechnic Society at Falmouth, and of the Royal Institution of Cornwall at Truro.

I would also express my most sincere thanks to many active newspaper staffs who, despite unremitting pressure of their daily calling, sacrificed time to help out a researcher whose main interest could hardly be described as news. To the *West Briton* at Truro I am particularly indebted, and it is a lasting grief to me that its former editor, Claude Berry, did not live to see the book published although he read and offered most helpful comments on an early draft. And I would add a word of gratitude to the staffs of the *Grass Valley Union*, the *Boulder Daily Camera*, and the *Marquette Daily Miner*.

The individuals who have helped in many ways have been numerous indeed; to one and all of them I would extend my most grateful thanks. The full list of them, however, would be far too long to record here, although I must express my especial gratitude to the late Mrs. Carroll Paul and the late Mr. Kenyon Boyer of Marquette, to Howard and Lois Keast of Dover, New Jersey, to Frank and Elsie Corin of Grass Valley to whom, among other debts, I am grateful for the Richard Wearne manuscript journal, to University staffs at Berkeley, Boulder, Ann Arbor, Tucson, Missoula, Brown, Lexington, Madison, and

ACKNOWLEDGEMENTS

Liverpool, who have all encouraged me in my work, and to the long-suffering typists who had the unenviable task of dealing with my rough drafts which were as erratic as the more faulted and twisted lodes of Caradon or Carn Menallis, Mrs. Irene Griffith, Mrs. Rosemary Gentleman, and Miss Lesley Roberts.

To the British Association for American Studies and the Rockefeller Foundation I would here take the opportunity to express my gratitude for the most generous grant that made it possible for me to spend a year engaged upon historical research in the United States, and to my own University, Liverpool, I wish to extend my thanks for allowing me a year's leave of absence to carry out those researches. I also wish to extend my sincere thanks for a travel grant to the Fulbright Commission.

Above all I would thank Professor David B. Quinn for his continued encouragement in the completion of the work and for reading the drafts of earlier versions and giving me much helpful criticism and advice. Any faults that still remain are my responsibility alone. I would add a specially personal word of thanks to Bob Athearn, Wilbur Sheppardson, Harry Allen, Herbert Nicholas, Leslie Rowse, and Fred and Gladys Harris for their continuing interest in my work. The final and certainly far from least expression of thanks is to my wife who went along with me on a twelve-month Odyssey, following the trails of the Cousin Jacks, from Atlantic to Pacific twice over, and ranging from two miles high altitudes in the Rockies to the twelve-hundred level of a Butte mine where, not all that surprisingly, we found a Cousin Jack, not, however, engaged in digging for red metal but in showing tourist parties round the works, a Truronian who still remembered the Delectable Duchy and its treacle and cream.

J.R.

1 The land whence they came

In the Upper Peninsula of Michigan, on the desert hills of south-eastern Arizona, among the forested western foothills of the Sierra Nevada along California's Mother Lode, near the Continental Divide in Western Montana, by the tributaries running into the Mississippi in southern Wisconsin and Illinois, beneath the shadows of Pike's Peak, on the Comstock Lode, and in countless other places they were to be found. The Cousin Jacks! Sometimes they said they were English, but just as often when asked what country they had come from they answered 'Cornwall'. Most of them were miners, and it was in the settlement of the mining frontier that they contributed most to America's history. Several, however, followed other trades while the goad of stark necessity in the 'Old Country' had made the miners themselves used to other callings. There were farmers and quarrymen, blacksmiths and carpenters, ministers, teachers, and journalists. Indeed, all they had in common was the land whence they came seeking new homes and a living in the Western Continent for most of the nineteenth and early twentieth century.

In the homeland, environmental and hereditary factors contributed to the growth of a provincialism which marked the Cornish as a race apart from those who lived east of the Tamar. They were both clannish and individualistic, cautious in some respects and reckless in others, romantic and yet practical and realistic. Their motto 'One and All' indicated how they clung together, yet brothers' quarrels and family feuds among them were bitter and lasting. At home traditional rivalries between adjoining parishes and townships were notorious such as those between Penzance and Newlyn, Looe and Polperro, Pelynt and Duloe, and, above all, between Camborne and Redruth. Nor was there much affection at times between 'Church' and 'Chapel'. Typically enough their traditional sports were the individualistic combats of the wrestling ring and the free-for-all hurling matches.

'About thirteen hundred square miles of land and thrice that number of years of history' might be taken as the essential description of Cornwall and its people. To it might be added a geological history going back to primeval chaos that saw granitic and porphyritic dykes thrusting up through crumpling sedimentary beds of rock, fusion and fission of elements, titanic earth movements and, by contrast, long slow geological eras of denudation when even the hardest rocks weathered away and decomposed. So the stage was set for man where he could not only hunt and fish and cultivate the soil but delve for metal beneath the surface of the earth—for tin and copper, some lead, clay, and slate, here and there a speck or so of silver and gold, even a trace of platinum and rather more than a trace of uranium.

The site of this rich stage was the south-westernmost county of England, a peninsula jutting out into the Atlantic in the cooler temperate zone of the northern hemisphere, swept by south-westerly winds, but warmed to some degree by the Gulf Stream. The sea had as great an influence as the land on the course of local history and upon the life of the people who lived in Cornwall. Ages before the galleys of halfmythical Phoenician peoples ventured from the eastern into the western basin of the Mediterranean, and made their way into the Atlantic, were erected the megalithic tombs of Lanyon, Trethevy, and Zennor which have unquestionable affinities with the chambered burial places of eastern Mediterranean peoples as old as, if not older than, the earliest builders of Troy. The Celts, who left Cornwall a language which only died out less than three centuries ago, were comparatively late arrivals. Centuries later came the wandering Celtic saints of Ireland, Wales, and Brittany, Petroc and Keverne, Just and Keyne, and a host of others to give their names to so many Cornish parishes. The sea and those who dared travel on it, in truth, left an indelible impression on Cornwall in times that historian and archaeologist can only dimly and doubtfully descry. Despite the hazards of the sea, land travel long after these times was slower, costlier, just as dangerous and almost as uncomfortable, and the sea continued to be the main link between Cornwall and the rest of the British Isles. It was to be especially important in maintaining a close connection with South Wales, particularly towards the end of the seventeenth century when deforestation in Cornwall had left next to no fuel for smelting purposes, and still less for the steam pumping engines that had to be brought into use in the next century to cope with drainage difficulties as the mines increased in depth.

Long before that time the Cornish had gained wide repute as miners. Although the early tin workings were alluvial, by the sixteenth century techniques of following lodes underground had developed from deep opencast workings to methods of sinking shafts and driving underground levels and adits. The unsuccessful gold-mining expeditions of Martin Frobisher in Baffin Land and of Walter Ralegh and other early Virginian adventurers almost certainly included Cornish 'tinners' among their personnel. Possibly a few early immigrants from Cornwall helped to start the copper mines in New Jersey whose owners induced Josiah Hornblower to come out from England to install a pumping engine in 1752.[1] In 1771 a small band of Cousin Jack 'prospectors' came to Upper Michigan on behalf of a British mining company to investigate mineral deposits that might help to provide some recompense for the costly French and Indian Wars. Till then Detroit and Michilimackinac had been the farthest frontier, and although the Cornish located copper on the Ontonagon River, they concluded that it was too far beyond the borders of civilization to be worked profitably, and vanished back whence they came.[2]

At the time of this early venture to Upper Michigan the copper mines of Cornwall were facing extremely severe competition from the copper deposits recently found, with some Cornish assistance, at Parys Mountain in Anglesey.[3] They had other problems, too. The miners had gone deep down, and the deeper they were the harder it was not only to hoist rock and orestuff to the surface but to keep the lower working levels reasonably clear of water. Man power, horse power, and, finally, steam power in turn were used, while hoisting and pumping machinery were devised and improved until, by the early nineteenth century, some of the Cornish mines were the most modern and mechanised in the world, and the fame of Cornish engineers had gone far afield. Josiah Hornblower was but the first of many associated with Cornish mining engineering to cross the Atlantic. Sixty years after he settled in New Jersey his nephew's close associate, Richard Trevithick of Camborne came to Peru to help the owners of long-worked silver mines modernise their methods and revive past prosperity.[4] Not long after that Trevithick's own Camborne partners, the Vivians, set up as engineers in Pittsburgh.[5]

These inventors and engineers and the machines which they devised

1. L. P. Lorre: 'The First Steam Engine in America', *Trans. Newcomen Soc.*, Vol. x (1930), pp. 15 ff.
2. W. Pryce: *Mineralogia Cornubiensis*, (1778), p. 61; J. B. Martin: *Call It North Country*, p. 38.
3. J. Rowe: *Cornwall in the Age of the Industrial Revolution*, pp. 90–1.
4. H. W. Dickinson and A. Titley: *Richard Trevithick, The Engineer and the Man*, pp. 159 ff.
5. *West Briton:* (henceforth cited as *W.B.*), 24 February 1865.

and improved were the more striking examples of a dogged and ingenious mining people who, for generations, wrested wealth from the depths of the earth. It was a hard way of making a living. Quick agility, strong hearts, and tough lungs were needed to descend fathoms deep by ricketty ladders into dark shafts, twist along narrow levels and adits, get out of the way quickly when overhanging rocks cracked and fell away. It demanded nerve and endurance to go down deep into places darker than midnight and work long hours. That work was breaking rock, and killas and granite were hard; dykes of spar and porphyry harder still. Tools were poor and primitive—iron picks, shovels, drills, and sledge-hammers. Until gunpowder was introduced in the eighteenth century, and even, in places, many years after, they used fire to break the harder rocks, piling up faggots of furze against the rock face at the end of a level and then setting fire to them. The quick, fierce blaze usually split the rocks but not to any great depth. When gunpowder came in nearly a century went by with numerous blasting accidents killing, maiming, or, very frequently, blinding miners using it before a Cornishman, William Bickford, patented a safety fuse. The only light to set a fuse was flint and steel or a tinder box or the flickering homemade candles of rush-pith dipped in fat. Matches were a late innovation, and when one St Agnes miner sold off all his belongings to go to Australia in 1865, and there was a box left lying on the mantel shelf an old lady told him to take them with him, 'for thee'st might never get another box out there'.[6]

Working 'at grass' on the ores brought up from underground was just as hard if less hazardous. Only in the eighteenth century did mechanical means of crushing supersede—in part—stone-breaking by hand. Copper ore was taken off to the coalfields of South Wales for smelting,[7] but practically all the tin raised in Cornwall until well on into the nineteenth century was smelted in the county, in the local 'blowing houses', mainly to enable the Duchy of Cornwall to collect its medieval 'coinage' dues on tin. The Duchy was not the only 'sleeping partner' exacting tolls from the mining industry. Landowners usually leased mining rights on their property to 'adventurers' in return for about a sixth or an eighth of the mineral mined. Smelters charged as high as they could for their services, and before the nineteenth century there had been several outcries against copper smelters' 'rings'. With all these

---

6. *W.B.* 9 May 1889; letter of William Mitchell.

7. In the latter half of the eighteenth century some copper was smelted at Hayle, but the cost of importing coal and Welsh competition made it unprofitable.

and more living on the mining industry, working miners were left with lean pickings indeed.

They might have been even less had it not been for the survival of the systems of working by 'tutwork' and 'tribute'. Tutworkers were men who contracted to do a certain amount of work in a period of, generally, one or two months, usually working in 'pares', or small groups, one man making the contract 'bargain' for all of them. Tributers were similarly grouped in small partnerships of two, three, or perhaps half a dozen men, who contracted to work a 'pitch' in a mine for an agreed percentage of the proceeds over monthly periods. Tutwork was usually exploratory or development work, sinking shafts, driving levels, and so forth. The tribute pitches were let out at monthly or longer intervals to the 'pares' who offered the lowest tender. The system had its origins in times when, generally through increased depth of working, 'free' mining without the assistance of 'outside' capital became impossible, and was a method of sharing the proceeds of mining between labour and capital. Naturally enough a number of abuses crept in; men working on a good pitch would conceal part of the mineral they extracted, work slow, or, on occasions, 'pick out the eyes of the lode' in order to get what in transatlantic idiom would have been termed a 'bonanza' month's working. On the other hand the labouring miners were dependent upon the assayers and agents of the smelters for a fair appraisal of the ores they brought to 'grass' and payment was, at best, a month and more frequently two months in arrears. It was the calendar month, too, which meant that the labourer had to wait five or nine weeks before he received full remuneration for his toil; although advances on pay, known as 'subsist', were commonly made when any miners and their families ran deep into debt and, when they got paid, bills and charges for credit swept away most of their earnings. Mining in Cornwall was a hazardous and a hand-to-mouth occupation: some it made thrifty, others improvident.

To these difficulties must be added the vagaries of metal markets, prices slumping and booming with rumours of war, changes of government, and the like. In the eighteenth century, possibly much earlier, the copper- and tin-mining industries had been affected by speculative booms and slumps, and these became more marked with the growing specialisation of the industry. In earlier times, however, the miner could, when metal prices dropped, turn his attention elsewhere, to patches of land on which he could raise crops and stock enough to save his family from starvation if not from 'short commons'. Those living near the coasts had the shoals of pilchards in late summer and through the fall to help stock their larders, and there was a fair amount of seasonal work on the farms at harvest-tide and in the spring. The Cousin Jack was a man of several trades, and a tolerable enough master of them in most instances, and although there were invariably the shiftless and incompetent ones in every parish the authorities who doled out relief did all they could to ensure that they did not become a great burden upon the community. Nevertheless, until the transformations of social and economic life by the technological developments of the 'Industrial Revolution', the entire emphasis was upon individual, family, and local self-sufficiency.

Cornwall, perhaps more than many other parts of England, became vulnerably dependent on essential supplies from outside its own borders towards the end of the eighteenth century. A century earlier it had started 'importing' charcoal from the Isle of Wight and south-eastern England to smelt tin, a staple commodity which had to be marketed far outside the country in which it was produced. Commercial contacts with other regions, in Britain and abroad, fostered travel and migration although on a very small scale, and it was limited to the richer and more adventurous individuals, to some who found it expedient to leave the county for not always reputable private or social reasons, and to seafarers.

The soils and climate of Cornwall could support within its boundaries a fairly large number of people, but the development of mining in the eighteenth century not only diverted labour from producing the primary necessities of life but contributed to an increase in population which outran local means of subsistence. Along with this went the growth of an economy in which technological progress and the existing structure of society caused periods of underemployment when labouring men were hard put to it to earn even the shadow of a living wage. Among the factors which led to the migration of many Cornish people to Wisconsin, Illinois, Pennsylvania, and Michigan before 1850 must be reckoned persistence of potato blight in the Old Country year after year from the early 'forties', besides the better prospects of reasonably remunerative employment in those American regions.

Out of Cornwall's total acreage of roughly 870,000 rather less than a half could be described as being under cultivation or as moderately good pastureland at the beginning of the nineteenth century. Nearly a quarter was rough moorland tracts providing scanty grazing for tough but scrubby cattle, sheep, and ponies, while hardly a quarter was reasonably highly cultivated. The highly mineralised zones were mostly in the higher and more hilly districts which afforded few facilities for the husbandman. However, many mining areas contained tracts of land which could be made moderately productive by men who had enough time to cultivate them as gardens rather than as fields and be content with crops that would supply a fair proportion of their own family needs. Large-scale reclamation of such lands was not possible until the middle of the nineteenth century when iron ploughs were introduced and, in some places, steam power was used to work them. Until then the most successful improvements of wastelands were those wrought by sheer back-breaking, muscle-racking, manual labour. The working miners, as a rule, had a fair amount of free time to devote to this, and landowners were quite willing to allow them to reclaim waste lands and use them on long leases at merely nominal rents. In this way, much land was reclaimed around Camborne and Redruth in the eighteenth century by tribute workers who sometimes forewent taking pitches in the mines for a month or two, or who, when working in the mines felt that six or seven hours underground was enough for a single day; tut-workers who performed their contracted stint in less than the agreed-upon month, too, often spent the rest of that time working a plot of ground. In times of unemployment, of course, many chose to work as small-holding husbandmen rather than do nothing.

Such men did not only grow food crops but often kept livestock as well, as there was still much open waste ground that anyone could use to depasture stock. Many miners kept some poultry and a pig; more than a few, especially in the western mining districts, rented cows from farmers. For about £3 a year in 1750, rising to £6 by 1800, and to £10 or £12 by the last decades of the nineteenth century, a man rented a cow undertaking to rear its calf and return it at the end of the year; not only did he get the milk but the farmer provided winter fodder for the animal and bore the loss if it died. The farmers who rented out cows were content to maintain herds of beef and draught oxen this way without the labour of milking and tending cows and young calves; generally the man who rented the cow had more milk, butter, and even cream than his family required and sold such produce to his neighbours, and there would also be skimmed milk to help feed a pig or two. Often a 'dairyman' rented more than one cow, and there is no doubt that this system made some miners inclined to turn to farming instead of spending all their days in the dark, deep, and dangerous mines. Desiring a 'real farm' of their own, many miners hoped they might one day make a lucky tribute bargain that could enable them to realise their dream. But such lucky bargains were rare and seemed to get even rarer as time went on.

If some miners yearned for a small holding or farm, tenant farmers had their worries and troubles too. There were rent, rates, and tithes to be paid and other bills to be met, and these from an occupation which, in Cornwall, involved accommodation to natural forces which were often very unfavourable. Many districts were very stony, and there were few relatively extensive tracts of moderately level ground. Soils off the granite worked fairly easily, but some of them were cold and damp. Under the grass roots of the growan, or granite, soils there were often beds of quartz and gravel. The horses bred in Cornwall to work these lands were of a notoriously inferior breed, while Cornish oxen were certainly no better than those reared elsewhere. Hardly any lime was available in a county whose soils were predominantly acid, and the use of sea-sand in its stead was limited to districts within fairly short distance of tidal water if anything like an adequate amount was to be applied to the cultivations.

The climate, however, was, perhaps, the greatest natural force hampering farming. The saying that it has no climate, only weather, is more true of Cornwall than it is of many places to which that remark has jestingly been ascribed. Averages of yearly rainfall and of mean January and July temperatures are misleading; frost has rarely if ever been recorded in July, but it has occurred in practically every other month; furious gales and thunderstorms can come in any month; hail has lashed down and shattered proud-looking fields of wheat and barley and oats in August; snow has whitened the ground in May; some July days have been as cold as New Years' Eves and January suns have shone warmly from cloudless skies. In some years 'February fill-dyke' has more than earned its appellation, but in others it has been one of the driest months. April showers in many years are rare. Crops mature in sheltered valleys a month or six weeks earlier than they do a couple of hundred feet higher up on an exposed upland that may be less than half a mile away. Thick fogs often shroud the coast between Gurnard's Head and Cape Cornwall when at Penzance, barely half a dozen miles away, the sun is shining from a cloudless sky.

From the farmer's point of view the climate of Cornwall is very unreliable; the experience of years would suggest that in the normal seven years' lease term of a tenant farm, the occupier could, perhaps a little optimistically, gamble upon getting four 'average' years, the others being either abnormally good or bad; the hopeful counted on there being two very good seasons in the seven, the pessimist on at least two 'washouts'. It was hard to assess what did most damage—rain or drought. Heavy rains in July and August sometimes smashed standing crops to the ground, or left reaped crops rotting in the sheaves. Or, again, there were years when a cold, dry spring was followed by a drought in June, and by the normal reaping times in late July and early August farmers declared that, from the field gateway, they could see fieldmice running across the centre of the fields. Such seasons meant short supplies of fodder for farm livestock, and if there followed—as more often than not seemed to be the case—long, hard winters, there was difficulty in keeping animals fed well enough to withstand disease. Such problems were eased somewhat when improved facilities of transportation made foreign feed stuffs available—provided the farmers had money or credit enough to obtain them—or, alternatively, to ship the stock which they could not feed adequately to markets outside the county for, naturally, no-one wanted to buy more stock in Cornwall itself at such times.

Cornish fishermen, too, had to adapt themselves to hard times, when the shoals of pilchards, always waywardly unpredictable in their migratory movements, failed to come to the Cornish coasts. The fortunes of all three—miner, farmer, and fisherman—affected those who performed ancillary services for them. When mines closed down, blacksmiths and carpenters who did much work for them were left wondering whence their next penny or loaf of bread might come. A bad harvest left the village wheelwrights and town harness-makers whistling for the money they were owed. A dearth of pilchards, and it was not only the fisherfolk who went hungry but the poor labourers and cottagers who for miles around relied on a store of salted fish to save outgoings of scanty wages on dearer foodstuffs. Clergymen whose tithes, till Victorian times, were paid in kind went short, while landowning squires often had to forego part of their rents to be sure of any money being paid by their tenants to them at all. Many, whatever their trade or social standing, may have saved up against the coming of the proverbial rainy day, but not all people were thrifty or far-seeing; it seems, however, that Cornish farmers were somewhat more economical and rather less improvident than the miners and fishermen.

Such generalisations, however, are of only partial validity, and all sorts and conditions made up the social community of Cornwall. It was an intensely provincial community as was shown by the survival of the old Celtic language until early Hanoverian times. On the south-east the Tamar was a formidable barrier to travel, no really reliable ferry service existing until well into the nineteenth century. If people wished to travel, they went on horseback or afoot; more often they stayed at home, with the exception of the seafarers, who, however, were only a small percentage of the population. The development of copper mining in the eighteenth century slowly furthered the improvement of necessary commercial communications with other industrialised regions of Britain.

Cornwall gained a reputation for lawlessness through later converts to Methodist sects exaggerating the unregeneracy of their own early days or, more often, those of their forebears, but there is no reason to believe that crime was any more prevalent in Cornwall than it was in other English counties. There were a few sensational crimes and acts of violence which would have made news in any age and be long remembered. In 1734 Henry Rogers claimed and seized a tenement in Crowan parish and 'shot it out' with the sheriff's men sent to arrest him, while half a century later 'King' John Carter engaged in a shooting battle with the excisemen who sought to check his smuggling activities at Prussia Cove, but such incidents were rare. In times of acute food shortages mobs of hungry labouring men raided the warehouses of grain merchants in some Cornish ports; but if violent blows were dealt and received in these affairs there is no record of many fatalities or of much more serious injuries than those often sustained in hurling games. Two types of crime associated with the sea coast have appealed to romantic imaginations and gained overmuch in successive retellings. Undoubtedly there were many fights between smugglers and preventive men, but neither the somewhat indifferent zeal and the small numbers of the latter would suggest that such conflicts were either frequent or very violent, even when it is remembered that the Cornish 'free-traders' deprived George III annually of about as much revenue as he and George Grenville hoped to get from the American Stamp Act. As for the wreckers, enough ships came to grief on rocky Cornish shores to make it unnecessary for a man to risk neck and limb on wild stormy nights, scrambling about the cliffs with a horn-lantern simply on the chance that it might lure some vessel to its doom. When a ship ran ashore it was certainly regarded as a godsend by those bold and rash enough to risk their lives in the raging breakers attempting to salvage the cargoes for their own use as zealously, if not more zealously, as they attempted to save shipwrecked mariners' lives. Yet it is almost impossible to find the name of any Cornish wrecker in old records of criminal courts; although some smugglers, like the Carters, gained a rather pallid immortality—mainly because one of them, Henry, became a Methodist and left an autobiographical sketch which, however, did not suggest that he was ever convinced by John Wesley's arguments that defrauding the king's revenues was a heinous sin. It may be regarded as significant that in Cornish tradition the outstandingly evil man was merely an unjust steward, Tregeagle, who extorted all he could from the Robartses' tenantry in the seventeenth century, and the ballad of his posthumous sufferings hardly compares with that about

>Bill Brewer, Jan Stewer, Peter Gurney, Peter Davy,
>Dan'l Whidden, Harry Hawke,
>Ole Uncle Tom Cobleigh and all

who, in the neighbouring county pursued the 'horrid career' of highwaymen.

Devonians, however, might well retort that if their county song commemorated highwaymen, that of their Cornish neighbours was as defiantly rebellious as the Marseillaise. Whether or not 'twenty thousand Cornishmen' were prepared to march on London when James II imprisoned Jonathan Trelawney and six other bishops in the Tower is very uncertain, but after 1688 there were a few dissident Jacobites in the county, while it is ironical that the early Methodists in a locality in which they later became so strong were resented as 'hidden Papists', and John Wesley himself narrowly escaped lynching as an agent of the Pretender.

The main disturbances after 1745 were food riots in times of scarcity and high prices, some of the most widespread occurring as late as 1847. Chartism, however, never attracted much of a following in Cornwall, and little in the nature of 'modern' industrial labour disputes occurred until after the collapse of the copper-mining industry in 1866. In fact, the great clay workers' strike in 1913 might well be associated with the wave of industrial disputes in American mining regions at that time, and men in the Old Country were encouraged to strike by Cousin Jacks who, in America, had joined the Western Federation of Miners and other militant unions. It is possible, though not very likely, that returned emigrants brought back with them from Pacific Coast mining camps the anti-Irish feeling which erupted in the Camborne riots of 1882.

Generally, however, in these later times, things were quiet and tranquil in Cornwall, a state of affairs that has often been attributed to the hold Methodism got in the county through the numerous visits of John Wesley himself and the proselytising zeal of his followers and successors. Wesleyanism did contribute much to Cornish life and, in some ways, radically transformed it, yet if it acted as a stabilising influence it also brought social antagonisms and conflicts to the Cornish community.

While Cornish society in the eighteenth century had marked class divisions, actual class conflict was rare if it existed at all. In early medieval times a Celtic clan system had been modified by a type of feudalism imposed by invaders from outside, and then there had been further modifications caused by the freedom of natural opportunity afforded by what can truly be regarded as a pioneering mining frontier. When the medieval crown transferred the mineral rights it claimed on tin-mines to the direct male heir-apparent to the throne, it made concessions designed to encourage mining in order to enhance the revenues accruing to it from the active pursuit of that occupation. The coinage duties on all tin mined in Britain—which meant the south-western peninsula almost exclusively—and the development of Stannary Parliaments and Courts to regulate tin-mining originated from the need to accommodate the rights and privileges of those who claimed immediate and ultimate title to the ownership of mineral lands with those who, to gain a livelihood, sought to exploit such lands. The Stannary organisation, which survived into the nineteenth century, can be regarded as an adaptation and development of a simple mine camp meeting, similar to those which were so prevalent in early Californian gold-rush days, to a feudal and royal landowning milieu which already had had to make concessions to Celtic clan conventions and usages. Further modifications were necessary as time went by, notably to fit in capitalists whose support became increasingly necessary to keep the industry alive. As always, no matter how much conditions of working changed, claimants of traditional rights clung desperately to their privileges and perquisites. The last Stannary Parliament met in 1750, by which time Hanoverian administrative centralisation had made it obsolete; coinage dues and the body of officials which had been necessary to collect them, but which had proliferated like all administrative bodies, survived until the capitalist middle classes forced through the great political reforms that ushered in the Victorian age; the Stannary Court of jurisdiction survived another two generations, its competence limited to disputed claims of mine boundaries, ownership, and similar civil causes, a far cry indeed from the days when one Stannary Court held its sessions at Lydford, just over the Tamar boundary, and made all and sundry tremble at the name of

> Lydford law,
> Where first they hang and draw
> And hold the trial after.

Old-time Stannary Courts in south-western England, it would seem, did not differ greatly from vigilante tribunals on the mining frontier of the Pacific Coast.

Under the Stannary organisation there developed a race of workmen who were very conscious both of their rights and of their obligations and of the rights of others. It did not always work well, but the fact that the poor communications of the time minimised administrative efficiency was not invariably harmful. In theory and law it was far from free enterprise; in practice and fact it was nearly so for those who were ready to follow counsels of expediency rather than adhere to time-wasting conventions. Furthermore, the division of Duchy mineral rights from landowning rights made the owners of the latter less exacting and less disposed to active participation in, or interference with, mining than otherwise they might have been; many landowners, however, did not only encourage mining to get their own 'lord's dish' therefrom,[8] but were paid officials of the Duchy stannary organisation as well. In short, the higher and lower levels of 'central' and 'local' authority were quite closely interconnected, and their economic interests were complementary not conflicting.

Of inferior status were many small landowners who themselves worked the minerals underlying their own acres. Such landowners had, perforce, to deal directly with contractors who, might, from a social point of view, be regarded as their inferiors, with merchants and tradesmen who supplied the necessary materials for mining operations, and even with actual tributers and tutworkers. Such direct dealings, however, were sometimes delegated to stewards and agents, often lawyers and attorneys who, in turn, used their agencies to 'feather their own nests' and advance a few rungs on the social ladder.

There was a similar mingling of the classes connected with agriculture. Smaller squires and younger sons of landowners became virtually yeomen farmers, owning lands they farmed themselves; in hard times such men were economically far worse off than tenant farmers who ex-

8. i.e. the proportion of minerals raised which was allotted to the landowner who leased the working of mineral rights to the actual mining 'adventurers'.

pected their landlords to provide and maintain farm buildings and the like. Lower still in the social scale were the miners and fishermen who spent their spare time cultivating small tracts of land rented for low rents from farmers and squires, often renting cows and other livestock, too; by careful husbandry and thrift, helped now and again by a good tribute pitch, such men sometimes began the upward social ascent, becoming small tenant or even yeomen farmers. On the other hand a poor harvest or two, or stock losses, might relegate tenants or their sons down to the 'labouring' classes. Briefly, there was a social hierarchy in Cornwall, closely associated with economic life, but a man could rise—or fall—within it.

Until the later eighteenth century such social movements in the Cornish class structure were neither considerable nor sensational. Then came the rapid expansion of a more specialised mining industry closely associated with engineering technological progress. At the same time, in part necessitated by this development, communications were improved, linking the industry of the county with outside economic developments. All this hastened the transformation of agriculture from a local subsistence basis to a commercial one. Social fluidity increased in consequence, but these economic changes also created and aggravated class and even caste tensions since they not only offered vaster opportunities to the enterprising but left the unenterprising and unfortunate faced with the threat of losing social status. Some of the latter left the county to conceal from their neighbours their loss of wealth and standing, while others emigrated in the hope of gaining both socially and economically elsewhere.

Before these developments occurred, or rather assumed an aggravated form, people had, for the most part, quietly accepted the position they occupied in local society and their means of livelihood. Whatever the tenant farmer said about his landlord behind the latter's back, he was courteous and deferential when meeting him face to face. The working miner might call the mine purser all things obscene under the sun when talking to his fellows, but, when tribute-setting time came round was polite enough to him. As always, trades people 'on the make' exacted all the outward signs of respectful deference they could from their less affluent neighbours although by so doing they lost real affection and esteem.

Such a society was both stable and dynamic, although the word 'organic' might be more aptly applied. The sense or belief that it was divinely ordained was general, but only as long as there was a generally accepted church and religious organisation. The Cornwall of the eighteenth century was far removed from the Cornwall of the Age of the Saints, and it was to undergo a radical transformation when the advent of Wesleyan Methodism brought in a revival of both religious individualism and religious itinerancy. This development was associated with the changing economic and social conditions of the time which, in part, contributed to its appearance and growth.

The increasing pace of social and economic change stimulated a demand for a more dynamic religious faith. The regime of parson and squire was out of place on a mining frontier where fortunes might be made—or lost—almost overnight; it only required a few men to make fortunes by 'lucky strikes' or by financial speculation in mining ventures to rouse social discontent. A fair number of men by providing supplies and services to the growing mining industry achieved wealth greater than that which many landowning gentry could count on from rents or parsons from tithes. With economic opportunity open not to all but certainly to more than the few score individuals and their families from whom the most of Cornwall's landowning gentry and clergy were drawn, unsatisfied social ambitions were born and antagonisms to the old order developed. These moderately wealthy but socially 'unrecognised' people were naturally attracted by the Arminian theology of the Wesleys, challenging the narrow group of the elect who, hitherto, had taken its heritage of the best of this world and the next for granted. Once this process began class and caste barriers crumbled. John Wesley, fundamentally a conservative to the core of his being, had initiated a movement he could not control and whose ultimate consequences would have left him aghast.

Periodic evangelistic missions and mass revivalism can only be regarded as egalitarian and democratic; the organisation of the ministry as the servant of the whole community of believers rather than its directors tended the same way; the employment of local preachers and class leaders to maintain the loyalty of converts, initiated at first through the sheer lack of adequate 'professional' ministers, emphasised the breach with the old hierarchic order. In Cornwall, John Wesley found a fertile field. The old parochial system had hardly been changed since pre-Reformation times, but the mining industry had caused concentration of population in districts remote from the old parish churches. No new churches had been built to serve the needs of these mining communities, and, furthermore, even where there were churches in some mining areas there was often not an adequate number of fully ordained clergy to undertake the full cure of souls. Starting with open-air meetings and gatherings in private houses, it was not long before Wesley's followers built their own preaching-houses and chapels. Wesley himself had no desire to secede from the Established Church, but once his conservative hand was removed, the 'people called Methodists' soon had their chapels in practically every parish and in many more than one. In very early days, too, some mining and commercial men, genuine enough converts to Methodism, thought that they had a natural right to occupy such a position in the local chapels as land-owning squires had for centuries arrogated to themselves in the parish churches, and secessions within Methodism itself early in the nineteenth century hastened this development, while a few local secessions came about through conflicting personal social ambitions of this nature. The rift between Methodism and Anglicanism became deep and wide; the unity of parochial society in religious matters was destroyed. New causes of tension and conflict had appeared within Cornish communal life, and long traditional social allegiances were shattered; all sharpened the desire for change, progress, and freedom, a desire which led several to leave the homes their families had occupied for centuries and migrate overseas.

Yet those who left the 'Old Country' generally departed with mixed feelings and not a few misgivings. The Cornwall they knew had many disadvantages and shortcomings, yet it was home and a pleasant land. The moors and the sea; the hills and valleys; the woodlands and fields; deep sunken lanes running down to the rivers and coasts, shadowed by overhanging trees, and the rough tracks cutting across the barren heaths; grey granite and whitewashed cob cottages, snug farm-houses, and stately mansions; tall-towered churches, minestacks, and the mysterious cromlechs and stone circles of an older, pagan time—all these struck responsive chords in the hearts of the men and women who had lived among them. When they left they took with them and cherished memories of the 'Old Country'.

Of that, perhaps, the greatest living proof is in the number of people in the United States who, today, say that they would love sometime to visit Cornwall, not because of travel propaganda disseminated by press, radio, television, or film, but because of tales or descriptions handed down to them by grandparents or great-grandparents. Americans whose ancestors came from other lands sometimes, too, express an inclination to visit those countries, but few of them with such intensity as the Cornish-Americans. First-generation immigrants have sometimes, perhaps mainly through the desire to conform completely to what they regard as the 100 per cent American way of life, been indifferent and even hostile to the country from which their parents came; it should be added, too, that many of that generation could hardly be expected to yearn for a land from which hard economic necessity had driven their parents to seek a living in a new land. But with the second, third, and even the fourth generations there has grown a romantic interest in the old ancestral homeland; there has been more leisure time to think and dream of it, and the passage of time has tended to blur and efface many of the harsh realities of life and the struggle to make a living in the Old Country in the far from 'good old days'. If these grandchildren and great-grandchildren visit Cornwall they find a country differing in so many ways from their own that it has the enchantment of novelty, but these very differences tend to obscure the great changes which have taken place in the Old Country since their forebears left. Some of those changes have been for the good, but not all equally so. Be that as it may, Americans visiting Cornwall come as foreigners, but many of those with Cornish origins retain, consciously or more likely unconsciously, so many of the Old Country ways of manners and speech that they are given the welcome of neighbours, perhaps somewhat distant neighbours. Such a statement, however, cannot be easily or even truly made of the thousands of holiday-makers coming into Cornwall every year from that traditional 'foreign' land across the Tamar. In fact, though seemingly fantastic, it is true that America to many Cornish people has never been a foreign land but, rather, the next parish to St Just and Sennen, just across the water surging against the rocks of Cape Cornwall and Tol-pedn-penwith.

2 To go or not to go

About the time the United States broke away from Britain, migration from Cornwall on a small scale began. The few prospectors on Lake Superior in 1771 were of little account, though their long journey proved interesting in the light of developments three generations later. What was immediately significant was the state of Cornish mining in the last quarter of the eighteenth century, more especially copper mining. Many Cornish copper-mines had, by this time, been worked for well over fifty years and the more accessible lodes had been worked out; generally more mineral veins existed at greater depths which could not be worked, or not profitably, with the crude machinery then available. Even when James Watt and Matthew Boulton introduced the separate condenser the problem was not solved. Steam engines fitted with Watt's patent and used to pump water from the mines were, in the main, only superior to the Newcomen engines which had been operating in Cornish mines for the past thirty years or more in the fuel saving they effected; in fact Watt and Boulton merely enabled a few more 'marginal' mining ventures to work and pay dividends. For doing this, however, the Soho partners, till their patent rights expired in 1800, exacted a percentage of the savings in fuel effected by the engines they supplied or supplied parts for, or from any engine that could be regarded as incorporating any feature described in their patent specifications. Their monopolistic patent had been so drafted that it could apply to almost any pumping engine then in operation. In the later period of their partnership Boulton and Watt began supplying machines from their Soho foundry besides technical and supervisory advice, and greatly improved the standards of machine production and maintenance, yet it cannot be claimed that they brought a new era of prosperity to the mining industry; at most, they helped to keep copper mining going in Cornwall.

The Lake Superior prospecting venture and the far more successful exploitation of the rich, shallow, Anglesey deposits represented efforts to exploit copper deposits with the limited technical equipment then devised. Anglesey was so rich and easy to work that it glutted the markets, forced prices down, led to the 'knacking' of several Cornish mines which could not be so cheaply operated, and caused economic depression and unemployment in Cornwall. At the same time the older tin-mining industry was encountering lean times, and until the great French wars both Cornish mining interests found it hard to sell their product. This prolonged period of depression came at the time when a marked increase in population had become obvious and even alarming. In April 1788, distress and food riots in western mining districts led to Boulton's suggestion that Cornish miners seek work at the Combe Martin lead mines in Devonshire and in collieries still further up country,[1] while some Cousin Jacks found employment in the Denbighshire lead-mines of North Wales. Later, when discoveries of copper were made in mid Cornwall, near St Austell in 1812, on Caradon Hill in East Cornwall in 1835, and in the Tavistock district just over the Devon border eight or nine years after, miners came to these new areas from the older western mining centres of St Just, Camborne, and Redruth. To make a living from mining, men had to go where prospects seemed best, and a new area always seemed to have a greater attraction than an old one; it only required a few men to set the example of moving away to start movements of migration and emigration.

In the early nineteenth century, wealth could be made from mining in Cornwall, although only by well-directed labour and endeavour. Furthermore, the developments of technology and financial organisation seemed to be progressively reducing the amount going to ordinary working miners as wages or tribute. To the 'adventurer' shareholders, who, in earlier times, had advanced the capital and supplies needed in initial exploratory and discovery work, were added men who had capital and

1. The firm of Boulton and Watt wrote to their Cornish agent, Thomas Wilson on 18 April 1788, the letter being in Boulton's handwriting:

'As the quantity of Copper and Tin lately raised in Cornwall, is much too great for all the Markets to take off, it is proper the miners and Adventurers should turn their Eyes to Lead as the price of that Metal is almost double what it was some years ago, it being now from 21 to 23£ per ton. The Mine at Combemartin and some others in Devonshire have always been talked of as rational undertakings and as a principal part of their riches arises from the silver extracted from the lead there is no danger of overstocking the country with that commodity.

'Pray learn all you can about the (lead) mines at Penrose near Helston as well as those near Wadebridge. Setting aside the lead, mines now working in England and Scotland are desirous of putting into the Mines all the Miners they can get and consequently most of your young men may find employment in them if they are disposed to be active. There is also a want of miners in the Shropshire and Staffordshire collieries. It appears to me that it would be much better to provide for the Miners in either of the ways aforesaid than to open new Tin or Copper Mines and thereby endanger the total ruin of the County by overloading the markets...'

who were able to wait for returns when prices slumped. Mining operations, however, had to be carried on despite the state of the metal markets to ensure that when better times returned full advantage could be taken of them. The maintenance of mines in workable order was growing increasingly expensive; the deeper and more extensive a mine the greater the amount of timberwork to be kept in safe condition; if a mine was 'wet', it was now necessary to maintain steam-driven pumps, and that, until 1800, meant regular premiums to Boulton and Watt besides the regular payment of competent engineers commanding high salaries to keep machinery in running order. The whole economy had become fiercely competitive, and had practically eliminated the individual miner working on his own account with two or three labouring partners. Although tutwork and tribute continued, the increase in population meant savage undercutting at contract-making times by labourers themselves competing against each other for a meagre chance of a livelihood. 'Bonanza' pioneer days for the individual miner in Cornwall were over by the middle of the eighteenth century; those who made fortunes from mining after that time were either merchants who supplied mine materials, smelters, and men engaged in transportation. Small investors also received fair dividends on occasion, but nothing like the gains made by unscrupulous mine company promoters whose activities were a feature of every feverish period of economic boom until the great collapse of 1866. Such mine promoters, however, outdid themselves so far as Cornish mining was concerned in the early eighteen-twenties. After the panic of 1825 it was extremely difficult to attract outside investment in Cornish mines and the industry, whatever the state of the metal market and however great the continuing richness of west country mines, found it difficult to command adequate capital to maintain the optimum level of productive activity.

Agriculturalists in Cornwall were also facing difficult times when the inflated price levels of the Napoleonic war years collapsed. The long wars together with increasing mining activity had accelerated the change from local subsistence to commercial agriculture, and even before the wars a fashion among the landed gentry, set by George III himself, for 'improved' farming had caused rural tensions and upheavals. In all likelihood these changes and 'improvements' were inevitable, but in many cases the king's example encouraged landowners to overhasty adoption of new methods without due consideration of all the economic and social implications involved. The demands of the increasing population for foodstuffs together with feverish wartime price fluctuations encouraged and accelerated the process of change. Just as elsewhere the 'agricultural revolution' resulted in the elimination of the cottager with rights of pasture, turbary, and so forth on common lands, while small-holders and labourers were forced out of husbandry by the new methods and improved agricultural techniques.

The older system of agriculture had been concerned mainly with producing stock and crops for family and local consumption. The new commercial agriculture specialised in supplying certain staple commodities to more distant markets in which large consuming populations had appeared. This had two dangerous consequences for the farmer. The first was dependence on transportation and marketing agents to dispose of his produce, and the payment they demanded for their services was often disproportionately high. The second was excessive reliance on returns made from one or two products most in demand in the markets; this was particularly dangerous to the farmers whose holdings were not large enough to raise sufficient in average seasons to build up a reserve of capital against bad years.

Landowning, too, had become a commercial business besides conferring local social prestige. The growing affluence of their mine-owning neighbours encouraged the landed gentry to look on their estates primarily as sources of income which should be managed far more economically than they had been. It was much more profitable to have tenancy agreements with ten or a dozen large farmers occupying two thousand acres than to have a hundred small tenants occupying the same area of land; not only did it reduce legal and administrative costs of running the estate but it meant a great reduction in the number of houses and outbuildings that the landowner had to maintain in good repair. At the same time a tenant farmer, especially if he had a sizeable family of his own to work for him, was eager to get a larger acreage so as to be able to produce more for the markets; costs of marketing did not rise in proportion to quantity sold, for a couple of drovers could drive thirty cattle as easily as ten and it cost little more time to send off a wagon-load of grain as half a load; when prices ruled low, too, a few more bushels of wheat or barley or half a dozen more fat beasts to sell might mean all the difference between a profitable and a losing season. Few early nineteenth-century farmers, in any case, had much knowledge of economic laws and, like many Cornish mine adventurers, were apt to believe that financial difficulties could be overcome simply by raising and selling more produce.

The prevalent leasehold systems facilitated change. Most farms came to be held by tenants for periods of years, usually seven or fourteen years, sometimes for twenty-one. At the end of the term the landlord could refuse to renew the lease of a farm and incorporate it with a neighbouring holding. The miner–small holder usually holding under the uncertain three-life system was in an even more insecure position than the tenant farmer who knew that he might have to seek another farm at the end of seven or fourteen years and could prepare for such an eventuality.

Under this system a miner took over a dwelling house and an acre or two of ground; frequently he took the ground only and built the house himself, on condition that he and his heirs paid a nominal rent for it during the term of three lives named on the deed of lease, besides usually paying an initial fee for the holding. The practice was to name three young lives, usually children over four or five years of age owing to the then prevalent high infant mortality. When the last life expired the entire holding—buildings and lands and all improvements the holders had made on them—reverted to the landowner or his heirs. Sometimes an additional life was added for a monetary consideration if one or two of the originally named lives expired within a few years of the first agreement; occasionally this was done by some landowners in any event on the death of the second life. Leniency, or the reverse, in such tenancy arrangements depended absolutely on the charity or fair-mindedness of the landowner or his agent. Normally after fifty or sixty years these holdings would fall back into the hands of the landowners, and a fair number of such reversions occurred fairly regularly on estates where mineral discoveries several years back had led to the creation of several holdings to house and help maintain a mining population. On reversions landlords could, if they wished, re-lease to the occupiers or to another man on similar terms, or they could merge the holdings into larger farms letting for periods of years. The temptation to do the latter was strong in the early nineteenth century: from the landlord's point of view such a course reduced pettifogging details of estate management and ensured more regular income from his estates. Many such small holdings, too, had been so improved by men who often had reclaimed them from wasteland that they were Naboth's vineyards to neighbouring tenant farmers. Such small holdings, too, were often extremely irregular in size and shape and at times formed extremely inconvenient enclaves in neighbouring farms.

Despite this, the three-life system persisted through the nineteenth century, and emigration actually aggravated tenurial insecurity. Many of those who emigrated overseas did not keep up any correspondence with their relatives and friends at home. Among them were some whose lives had been named in leases. When they vanished without trace, no-one in Cornwall knowing whether they were alive or dead, there was no telling whether a leasehold on lives should have reverted to the landowner or not. The tendency to profit by the normal greater female expectancy of life by the lessee, who suggested the lives on which he should hold his tenement, did not help, for women emigrated as well, and if they married or remarried after going overseas might be even more difficult to trace than men. Children named in leases were often unaware of their connection with a tenement in the occupancy of one who might not even be related to them, and if they emigrated certainly felt no obligation to keep in touch with that particular person and even less with the landowner. Nevertheless, a death from fever in the lead-mining region of Wisconsin, a cholera fatality on the overland trail to California, a drowning on Fraser River, or a 'shooting scrape' in Last Chance Gulch or Virginia City could leave a worn-out miner or an aged widow in Cornwall with no right to live in the cottage that they had regarded as their own, perhaps for the whole of their adult lives.

These considerations must be regarded as among the uprooting forces driving both miners and husbandmen to emigrate from Cornwall in the early nineteenth century. To them should be added, with emphasis, cycles of economic boom and depression in both mining and farming and in their ancillary trades. A period of marked economic instability followed the end of the French wars in 1815, aggravated by some abnormally bad harvests while in some mines the fact had to be faced that once mineral had been dug out of the ground it was gone for ever. With their crude methods the 'old men' had removed an inordinately large proportion of the better ores, and this had, doubtless, been encouraged by heavy demands for copper during the war years, especially as the Anglesey deposits had, by 1800, been almost exhausted. Furthermore, men had been living under the stress and excitement of war time conditions so long that the return of peace left them with feelings of restlessness, and with an inclination to seek out opportunities of making a living elsewhere, particularly since prosperity did not accompany peace.

Times were unsettled. All that seemed at all certain to many people were the prospects of hard times and little reward if they stayed where they were and tried to carry on the trades to which they had been accustomed and trained. Undoubtedly the majority were inclined to heed the counsels of caution—to stay and endure the evils they knew as they and their fathers had so often done before. To others, of more sanguine and restless temperaments, the last thing to do was to stay where they were and as they were. That, too, posed great questions—the question where to go and the question could they, all things considered, go.

The first of these questions posed two others. Was it possible to find a new home and a better livelihood elsewhere in Britain or should they emigrate overseas? Practically every mining area that had been, or was to be, developed in the British Isles attracted at least a few Cornish miners. Although they seemed to prefer mining non-ferrous metals to working in collieries and iron-mines, a considerable number of Cornish miners in the course of the nineteenth century settled and worked in the coalfields of South Wales, Tyneside, Lancashire, Yorkshire, and Lanarkshire. Non-ferrous minerals, expecially copper and lead drew others to North Wales, Cumberland, the Isle of Man and Ireland. Commercial connections like those which existed between Cornwall and the metal-smelting districts of South Wales, the manufacturing and engineering regions of the 'Black Country', and the clay-using Staffordshire Potteries brought some Cornish settlers into those areas. As the years passed and the metropolitan capital rapidly grew in population, more and more Cornish people found a living in London. Towards the close of the nineteenth century, the blighting depression of English agriculture which bankrupted so many 'up country' tenant farmers brought in to take their place the harder living West Country farmers as well as Scots, particularly in Essex, Sussex, Oxfordshire, Warwickshire, and Northamptonshire. A factor which stimulated migration from Cornwall to other parts of Britain was the itinerant Methodist ministry, beginning with John Wesley himself, which brought to the remote south-western peninsula first-hand information about the regions east of the Tamar. As the Methodist Connection rapidly grew in numbers in the nineteenth century and became a new focus of social life, it afforded, both in Britain and overseas, personal and social contacts for the migrant; sometimes a man coming to a new home already knew the Methodist minister then temporarily stationed there, and in any case, the growing Methodist societies were always ready, in Britain and America, to welcome newcomers to their membership and this provided them with an immediate entry into the community life of their new homes.

Outside Britain the places that offered most to those who did not want to expatriate themselves completely were the scattered colonial possessions of the Empire, but in the early nineteenth century there were great drawbacks associated with practically every one of them. South Africa which drew so many Cornish immigrants after the discovery of diamonds and gold had very little to offer before 1869. India, the West Indies, and a number of other tropical British possessions offered opportunities to a limited number of individuals, but tropical climates were widely regarded as deleterious to white men if not positively lethal. Australia suffered from its identification with 'Botany Bay'—the jail settlement on the other side of the world to which, since 1789, Britain had been despatching its social misfits and undesirables along with a semi-military force of guards and warders who, morally and socially, were hardly distinguishable from the criminals they guarded and ruled. New Zealand was the resort of whalers whose long voyages beyond the limits of civilisation had, with more than a little justice, gained them a reputation hardly more savoury than that of pirates; its natives had, through missionary propaganda, become known as repulsive cannibals; so the distant Antipodean islands were hardly likely to arouse the active interest of men whose foremost object was a more secure and adequate living than that which they had had at home in recent years.

Some did go from Cornwall to these distant British possessions, but the less distant American continent attracted far more. Latin America drew a number, principally adventurers like Richard Trevithick, the Camborne engineer, after the collapse of the Spanish Empire in the first decades of the nineteenth century. In the early eighteen-twenties more Cornish miners and engineers were sent out by speculative British capitalists who had obtained mining concessions in various parts of South America. This mining boom collapsed in 1824–5, and thereafter the British capitalists were more wary about South American mines. Later, however, considerable Cornish colonies settled at the Morro de Velho gold-mines in Brazil, at the Real del Monte silver-mines in Mexico and at Tocapillo on the copper–nitrate coast which Chile seized from Bolivia in the Pacific War of 1880.

Latin America, however, for Cornishmen, had two grave disadvantages—political instability and a foreign language. The same drawbacks were not apparent in North America, although political motives were prominent in British schemes to establish more settlers in Upper Canada in the years immediately after the War of 1812. On the side of the American Republic, too, there was awareness of the need for more people to settle in western lands, not only to get some return for the millions Jefferson had spent in buying 'Louisiana' but also to guard the frontiers of existing settlements which had been threatened, first by Tecumseh and his Indian confederacy, and later by the British on the northern shores of the Great Lakes. Fur-trading companies and shipping interests had already initiated Anglo-American rivalry in transcontinental drives to the Pacific Coast, and beyond the Pacific were the reputedly rich markets of China.

These grandiose but vague political and commercial considerations were, however, of little significance to prospective emigrants; their interest was in the material opportunities the vast unpopulated regions of North America offered. As yet there was little inkling of the existence of much in the way of mineral wealth, especially in the British possessions; the copper of Upper Michigan was still regarded as inaccessible, and the lead prospects near the old French fur-trading post of Prairie du Chien had not attracted much interest. But land was cheap and there were forests, and it was the latter that first gave the opportunity of large-scale emigration from Britain. During the Napoleonic Wars, with her own forests gravely depleted by the demands of building, industry, shipping and commerce, and with Baltic supplies liable to be cut off by the enemy, Britain had begun exploiting the forests of her North American colonies; the ships that brought timber over to Britain came to be used for carrying emigrants cheaply to North America. Once there, however, there was nothing to keep such emigrants under the British flag, and many of them drifted southwards into the United States, the attraction being that many Americans were leaving farmsteads on the Atlantic coastlands for regions further west. This latter phenomenon is not easy to explain. In the years of the War of Independence and the War of 1812, a number had left the coast because it had been too exposed to attack by British naval power. As settlement progressed westward, there were those who moved on sooner than be in sight of smoke from a neighbour's cabin, but their number has been exaggerated by romantic writers on the rugged individualism of the American frontiersman, who seem to forget that on many frontiers menaced by Indian attack such hermitical and anti-social habits were positively suicidal. Defeats inflicted on hostile Indian tribes brought considerable numbers of American settlers on to their former hunting grounds, but it seems more reasonable to suggest that Americans had become used to a rather extensive system of agriculture through the very plenitude of land in pioneering times and that, as population increased in the older settlements, they wanted to carry on in the same way, leaving the eastern regions for people who were more willing and able to follow more intensive farming methods. The immigrant from the Old Country was used to such a type of agriculture, besides being less adept at forest clearing and lumbering than the Americans. A further and, possibly, the most important consideration was simply climate—the Canadian winters were too extreme, and the attraction of a more southerly clime was almost irresistible.

Some notion of the countries overseas and some ideas—true or false though they might be—of the prospects they offered were necessary to start a migration of people. About the United States the ordinary working miner and small farmer knew comparatively little in the early nineteenth century. A few British veterans of the Revolutionary War and of the War of 1812 brought back accounts of the country where they had campaigned, but such accounts, being based on the hardships of military campaigns that had been crowned with little glory, honour or success were hardly likely to be favourable. With so much of the land between the Alleghenies and the Mississippi Valley sparsely peopled and hardly explored, let alone adequately surveyed, little was known of America's mineral resources, although an incentive to discovering and working them had been provided by the extremely strained Anglo-American relations after 1803, culminating in war; after the Peace of Ghent, too, many Americans felt their national security depended on locating and mining iron, lead, copper, and other metals within their own territorial limits if at all possible. The few metallic objects in Indian possession, some copper and lead that had come into the hands of French fur-traders, gave no real indication of the extent of copper deposits in Upper Michigan or of galena lead ores in south-west Wisconsin. Until the latter came within fairly easy reach of frontier settlements extending up the higher reaches of the Mississippi, no mineral strikes made were big enough to get much publicity outside the reports of surveyors and geologists, certainly nothing like Cerro de Pasco and Real del Monte.

Yet if there was little exact knowledge of conditions and prospects in America there was some, and, in many instances, its very indefiniteness led to exaggerated hopes and surmises. Some men and their families emigrated. A few of them came back, and their accounts of privations and ill success almost certainly deterred their friends, who had had thoughts of emigrating, from leaving the Old Country. This was not so in every case, however, since many of those who had failed as emigrants were known to be incompetent ne'er-do-wells who had done nothing much at home and who were thought, justly, to be incapable of making anything of their opportunities anywhere. Their tales, too, could not be reconciled with some letters sent home by those who had made a success of life in the New World, written by people who were known by their friends to be energetic, capable, and industrious. It is impossible to calculate the influence of such letters, but for an ordinary working class or small-holding family to receive a letter in the first half of the nineteenth century, when postage was dear, postal services unreliable, and so many could hardly read or write at all, was very much of an event in their lives; letters that came from distant countries were proudly shown and discussed with all the friends of the recipients and senders in local taverns and Methodist class meetings alike. Sometimes such letters were given to local newspapers, and so came to be read by many.[2]

One such letter came from Peter Davey who, about 1830, left Treverbyn in St Neot for Upper Canada. In recent years life had not been easy in that district, predominantly a farming one characterised by only moderately fertile soils, steep hillsides and deep valleys, and more liable than many places in Cornwall to suffer from late frosts and wet summers; the nearest market town, Liskeard, was some five miles away, and the roads to it were rutted lanes twisting and turning a tortuous way up hill and down dale; probably too much wheat had been grown there during the Napoleonic war years, and it was hard and costly to get sufficient lime in those days to counter the increasing acidity and sourness of the arable lands. All the time rent, rates, tithes, and other charges had to be met and other bills paid.

At Coburg in Upper Canada, Davey found conditions that were more or less typical of the interior region of North America lying between the southern flanks of the Laurentian massif and the Ohio river. To a friend he wrote in November 1831:

I should have written to you before, but not having seen sufficient of this country I thought it best to delay my letter till I had, that I might be more certain of stating the truth. Soon after my arrival, I bought a lot of land (200 acres) for which I gave £275; the wood will nearly pay for the land and clearing. We can make 6/3d per cord for wood and 6 dollars per hundred

2. Many letters published in newspapers, however, were communicated to local editors by persons desirous of promoting or deterring emigration for personal reasons; agents of shipping firms and colonial speculators certainly gave very different accounts from those sent by mine-owners fearing a labour shortage at home or by those in the Colonies fearful lest an increase in immigration depress wages in the new lands.

bushels of coals,[3] 16 gallons; I have burned one pit, and there is a good sale, as the smiths all work with them. Everything grows well; the soil is rich and will bear many crops without manure. Cucumbers, pumpkins and watermelons grow in natural soil here, in the season better than in England. Wheat, Indian corn, pease and potatoes also produce a fine crop. Wheat sells for 5s a bushel, 8 gallons; Indian corn 3/6; peas 3s; potatoes 1/3d; oxen £15 and £20 a pair; cows at £4 to £6 each; fat cattle 15s to 20s per hundred(weight) and make from 3½d to 4d a pound; butter 7½d; cheese 5d to 6d: Land is getting up and the country is improving very fast. I shall have forty acres of wheat next year, and having no rent to pay, no poor rates, no tithes, no Church rates, no land tax and only about five shillings a year to Government, I may fairly hope to do well: but it is useless for idlers or drunkards to come here, as they will be sure to starve. Industrious labourers can support themselves and families well; wages are from 3/9d to 5s a day. Tell Mallett and Keast if they could get here their families would soon cease to be a trouble to them.—We live in great harmony, so much that we care little about locking our doors at night; in truth, I would not return to England if I could have the land of the estate I rented in St Neot given to me.[4]

It is reasonable to assume that for a farm holding of the size which he bought outright in Ontario, Peter Davey would have, in Cornwall, been called upon to pay a rent of about £200 annually. In some years he would have made more than five shillings a bushel for wheat in the Old Country, but the prices for dairy produce and beef cattle were about the same at home as in Canada; cows were cheap and working oxen decidedly dear in comparison with Cornish market prices. Just how much wheat Davey raised on the virgin soils he reclaimed from forest in Canada can only be a matter of pure surmise, but it is hardly probable that forty acres of arable land in St Neot would have yielded more than a hundred and twenty quarters of wheat, worth on his reckoning £240 in Ontario. A comparable crop in Cornwall a few years before, when prices ruled high, would have been worth somewhere around £500, but during the ten years prior to his departure Davey would have been lucky if he had made as much as £400. With the rent deducted from that sum, a tithe for the parson, rather less than a tithe for seed corn, and much the same for poor rates, church rates, land taxes, and other charges, hardly £100 would have been left. This calculation would leave nothing for labour costs, and on a holding of this size and with such an acreage under cultivation, these, even ignoring family help, must have been approximately £1 a week, leaving

3. Charcoals.   4. W.B., 10 February 1832.

Davey and his family the same far from princely sum on which to live. Some profits had, doubtless, been realised from other crops and from livestock, but hardly enough to give the farming family a really high standard of living or, for that matter, even to maintain the standards to which they had become used in times when the price of wheat had, on the whole, been easily a third above the average levels of the eighteen-twenties. By the end of 1834, too, Cornish wheat prices were to fall below the five shillings per bushel of Ontario, but no-one could have foreseen this at the time Davey's letter reached his friends in St Neot.

In Canada and the United States, the higher cost of labour did not take up anything like half the total value of the wheat crop, and other expenses left the farmer with, in a case like Davey's, the equivalent of at least double his Old Country income, quite apart from what he was making from lumber and charcoal off his forest clearances. In this rough analysis of farm costs in Cornwall, too, no allowance has been made for manures; while the stand-by was farmyard dung, burnt stubbles and hedgerow parings, and sweepings from the roads and lanes there were, perforce, especially on the more acid growan or granite soils, outgoings for lime and sea-sand; farm covenants or leases, too, drawn up by landlords or their agents, prescribed the manuring that had to be carried out by the tenant to ensure that the land remained in 'good heart' at the expiration of the lease. High rents and such restrictive covenants accounted for bitter relations between landlords and tenants in those times of falling wheat prices, and hence the attraction of 'no rent to pay, no poor rates, no tithes, no Church rates' and so forth. Under these circumstances, and with Cornish wheat prices down to 7s. 6d. per bushel—and still falling—it was not surprising that two months later the local paper which had published Davey's letter informed its readers that:

The rage for emigration that now prevails in the North of this County is wholly unprecedented in Cornwall. In different parishes from two hundred to three hundred persons each have either departed or are preparing to leave for Canada or the United States. Last week a vessel named the *Springflower*, sailed from Padstow, (for Quebec) having on board a hundred and eighty persons, and another—the *Economist*—is now ready to sail with two hundred more. The recent regulations of the Privy Council, requiring that a regular medical practitioner be engaged for the voyage, in every vessel taking more than a hundred and fifty passengers, has caused some delay.[5]

5. *W.B.*, 6 April 1832.

The region of North Cornwall which so many were leaving was one of rather poor, cold soils: apart from the local limited opportunities of employment afforded by the Delabole slate quarries and the fisheries of Port Isaac and one or two smaller coves, it was wholly agricultural; much of it was marginal land that had been brought under cultivation during the Napoleonic war years and for that reason its farmers were affected all the more severely by falling wheat prices; there is good reason to surmise that in recent years rents had gone up while actual crop yields had fallen owing to the difficulty and costs of keeping such mediocre soils in even moderately 'good heart'.

The allusion to the provision of surgeons on the emigrant ships indicates a factor that may have made many intending emigrants of those times pause and think again. Sickness and mortality on board ships engaged in the emigrant traffic, along with the intense alarm caused by the appearance of epidemic Asiatic cholera in Britain, had led the British Government to issue the Order in Council compelling passenger vessels of moderate size to carry a surgeon. It might have been asked how good and well qualified were the doctors carried by the Padstow ships, *Springflower* and *Economist;* still the fact that there was a medical practitioner on board may have decided persons with infant families or of no very robust constitution in favour of going; a few even went in the hope of regaining health in a new country.

Other things had to be considered as well. Few of those who emigrated overseas in the first half of the nineteenth century had, previous to their going, travelled more than ten or a dozen miles from the places where they had been born. The cost, too, was an even weightier consideration. Tenant farmers who were able to get away before the lean years swallowed the surpluses they had been able to save during the fat, had a fair amount of capital to defray the expenses of travel first to a port of embarkation, then across the ocean, and finally to the place where they decided to settle. They had to allow for not only the cost of subsistence during their travels but for a considerable period that might elapse before they could begin to make some sort of living in their new homes. It was wise to be prepared for an ocean voyage across the Atlantic by sailing ship lasting five or six weeks, but, even before a ship sailed, three weeks might pass through contrary winds and tides, a captain refusing to sail until he had more than a full complement of passengers or found a surgeon willing to go on the voyage. Once in the New World, two or three months might slip by before the emigrant finally got 'located' and was able to start work, and then there would be a further wait before that work yielded returns and a regular living. Miners, naturally, had less reserves of capital, although in many instances the disposal of their leasehold cottage holdings, the sale of a few goods and chattels, and, now and again, the windfall given by a lucky tribute pitch they had been fortunate enough to strike was enough to see them through without excessive discomfort and privation. Miners, too, tended to emigrate as individuals, only sending back for their families to join them when they had worked long enough in the new country to save the passage money for their wives and children; if they failed to 'make good', they often managed somehow or other to make their way home again, although the number of those who, once overseas, vanished without trace so far as their families were concerned was by no means small. Several miners and engineers went overseas on contracts arranged by British commercial connections of American mining speculators. As time went by and more emigrants were successful in the new country, they encouraged and helped their relatives and friends to come out and join them.

All those who emigrated or thought of emigrating, however, had hard decisions to make and risks to face. Knowledge or lack of knowledge of conditions and prospects abroad cut both ways. Many emigrants would never have left home had they been fully informed of all that lay ahead of them when they set off. Conditions on many of the emigrant ships were incredibly bad. In the early days most of the shipping companies only provided the passage and water, and emigrants had to bring along their own provisions and other necessities, although some supplies could be bought on board the ships. A letter from a later emigrant to Australia described typical emigrant ship conditions to his Redruth friends in 1849 thus:

We had not been out many weeks when we were told that the assorted soups and preserved meats were done, and there would be served out, instead, soup or *bouille*, which they (i.e. the shipping agents) say, is stewed beef and vegetables, but which is far more like old smiths' aprons and rotten carrots. The biscuit is of the lowest quality to be had, and as for the salt beef and pork, why I would make a bet that you would never guess what it was—the bones are such as no-one here ever saw before; I think that they are from animals that existed previous to the Flood. You, perhaps, are not aware that on ships of this kind the agents send out for sale every day large quantities of spirits, wine, ale, cheese, ham, flour, preserved fruits, pickles, etc., and there are stated hours each day for selling them; so that you can see it is in their interest to give us bad provisions, we being then compelled to buy many things to make anything like a living. One family on board has spent nearly £80, and several families it has cost from £10 to £15 ...[6]

Costs on the much shorter Atlantic crossing, where, too, there was far more competition between shipping companies than on the Australian voyages, were much less than these, although 'overland' travel after arrival in America drastically reduced this saving.

Many of the early emigrant sailing ships were fearfully overcrowded. Live pigs and poultry kept on and between decks were conducive to neither cleanliness nor hygiene. Cooking facilities were utterly inadequate. Inefficient surgeons aggravated rather than alleviated sickness and mortality. Furthermore, there were the normal hazards of storm, shipwreck, and, more rarely yet worst of all when it did occur, fire at sea. Many disasters occurred on the emigrant sea routes, but probably that which attracted most attention in Cornwall and brought home to Cornish people the dangers of emigration happened on the night of 3 May 1855, when the sailing ship *John* struck the notorious Manacles reef off St Keverne a few hours after putting out of Plymouth. Seventy-two of the two hundred and sixty-two passengers, were saved, and the entire crew of nineteen. A local newspaper commented that:

In the long history of disasters at sea, the loss of the *John* on the Manacles is without parallel. Only a few hours out of port, she was wrecked on a weather shore in fine weather, and out of four boats only one was safely launched, but without oars, and she broke adrift with only five persons on board although she could have saved twelve or fourteen. Another boat was stove in in the attempt to get her over the bulwark, a third broke adrift, and the fourth the captain would not consent to launch till daybreak, but the tide meanwhile swept it from the deck. The only remaining expedient for escape was taking to the rigging as the rising tide had filled the decks, and the crew together with some of the passengers got aloft, but others, who had not the activity or the nerve to climb, were washed from the poop or knocked overboard by the boom which got adrift. And thus miserably two hundred perished. All this happened within hail of the land, but the ship was not provided with lights or any means of signalling her distress.[7]

Other emigrant ships were to be wrecked though few with such heavy loss of life and none so near the Cornish homeland. On the western side of the Atlantic were the additional hazards of fog and icebergs off Newfoundland, and on the St Lawrence and Great Lakes perils from ice and blizzard storms in the fall and spring. And even

6. *W.B.*, 22 June 1849.   7. *W.B.*, 10 August 1855.

when safely ashore, the emigrant had to encounter much that was unknown and unexpected before he could make a living.

The cost of emigrating and the perils of travel were not the only deterrents. Ties of home and kindred, too, were very strong. Circumstances had to be really desperate for many to decide to quit the Old Country and leave parents, relations, sweethearts and friends not knowing when, if ever, they would see them again. Over the years there were to be scores and hundreds of leave-takings in Cornwall like that of fifteen members of the Wearne family, who, largely through stress of difficult times and the doubtful future the Old Country offered them and their children, left West Cornwall which had been their home for close on eight centuries. Richard Wearne in after years recalled how they left for the distant lands beyond the 'Western Sea'.

The Day was very fine. Thousands of People was on the warfs to witness our departure. Some, that I had taken sweet Counsel with for many Years others that I had given Employment to, but among the Multitude was my Poor Aged Father and Mother them that had Promised themselves we should be a comfort to them in their old age Bath'd in tears to see no more.

At 6 O'clock in the Evening, Tuesday the 4th July 1848, we was towd out by the Brilliant steamer to the Bay of St. Ives, and we could see the sand Banks covered with our Friends waving their anchifts, etc. untill we gragurly disapared. Tho' I join to sing Hyms with our Party on bord, it would not drown that great Affection that was so deeply stampt within my hart, it was as the Devishing the sinewrs from the Marrow, thinking on those that with Pain and sorrow brought us to Manhood to see, Props, no more now in their decline of Life. The Evening came on a Calm sea sickness soon came on some very ill, all of our Famly was very well Except two of our little Girls they were very bad. The Next Morning at 4 O'clock little Breeze move on the water at 12 next day Cape Cornwall graguly Disapared one thing greatly comforted me to know I had left some behind that would not forget to Pray for us . . .[8]

The Wearnes had plenty of time for sad thoughts on a far from comfortable voyage that took over seven weeks. They reached Montreal on 23 August and finding business depressed there, determined to go to south-west Wisconsin where they had relatives by marriage. They reached Linden eleven weeks after they had left Hayle. Then they had to find a way to make a living and to get some sort of a roof over their heads before winter set in, and on their arrival, too, they found out that those who had in letters home, in all probability, urged them to follow

[8]. Richard Wearne MS Journal. Wearne was fifty-two when he left England; and the aged parents, Richard and Biddy were seventy-five and seventy-three respectively; both died in 1851.

them overseas were not so ready to give them a hospitable welcome when they arrived.

The Wearnes and many others uprooted themselves; in the long voyage they had plenty of time to think of those they left behind. There were many instances of young men sending for their girls to come out and marry them, while several came back to Cornwall for a brief holiday, married, and sailed away again. Many family ties and betrothals, however, were broken, and in numerous cases emigration broke up a family as surely and as irrevocably as death.

Yet there were some who left on the spur of the moment, taking thought neither of the morrow nor of those they left behind, although few were as irresponsible and feckless as one old St Just miner who, towards the end of the century loafing about the quay at Penzance saw a ship ready to leave for America, and, with no further thought and no luggage save a pocket handkerchief and the clothes he stood in, went on board and sailed to the States. Another St Juster is reported to have offered to see a young friend off on a ship bound for Australia from Plymouth; when they arrived at the dock he said to the youth, 'Well, I may as well see 'ee all the way there!' and with no further thought went off to the Antipodes. Men involved in poaching scrapes, Benedicks with shrewish wives, and unwed fathers comprised the most of fly-by-night emigrants who slipped away without announcing their departure.

To go or not to go, that was the question which many in Cornwall had to answer in the early part of the nineteenth century. If they decided to go away they left home and those nearest and dearest to them most likely for ever, but they had a chance of a better living than they had, or were likely to have, had they chosen to stay. In some districts, too, economic conditions had reached a state in which accustomed living standards could not be maintained. A kind of compromise solution might have seemed to be possible by going 'up country' to the rapidly growing industrial towns and cities, but this did not appeal to men wedded to a rural way of life, while the metal miners of Cornwall seem to have had little liking for coal and not much for iron mining. In any case, London, Lancashire, Tyneside, and Lanarkshire were as remote to people who looked on the Tamar river as the boundary of a 'foreign' land, who even regarded the inhabitants of the next parish or the next parish but one as 'foreigners', as places on the further shore of the ocean that lapped against the Cornish coast. If one had to leave home, it was just as well to go three or four thousand miles as three or four hundred. Something, too, could be said for quitting the Old Country entirely with all its parsons and squires, taxes and tithes, poor rates and church rates, its limited area and limited opportunities, for a new, freer, vaster country. Already the demand for political and social reform in Britain had evoked some talk of American republican institutions and freedom at a time when, through Spanish competition, the lead and copper mines of western England had fallen on difficult days. Just at that time news came that in far-off Wisconsin, the Americans had come upon rich lead deposits but hardly knew how to work them; the response from Cornwall initiated the exodus of miners from that county which was to play a leading part in the saga of the American mining frontier.

3 The first western mining frontier

The region of the Upper Mississippi which first attracted a considerable immigration from Cornwall, was on the south-western boundaries of the original North-west Territory, some fifty miles below the confluence of the Mississippi and Wisconsin rivers. In earlier times French fur traders and missionaries had heard of the existence of lead deposits in the region from Indians, but little was done about it until 1788, when Julian Dubuque secured some sort of concessionary rights from the natives to mine lead. As a mining pioneer, however, Dubuque did comparatively little, getting out just enough lead to indicate the existence of galena ores in the region round the trading post which was destined to become the city bearing his name on the Iowa side of the upper Mississippi,[1] and when he died in 1810, north-western Illinois, south-western Wisconsin, and eastern Iowa were hardly more than names to the American people and still less to the people of the Old World. Americans who had crossed the Alleghenies by the Cumberland and other gaps to the Ohio Valley which led to the Mississippi, then tended to go down rather than up river, more particularly after the Louisiana Purchase of 1803. Some, however, did look upstream at the confluence of the Ohio and Mississippi towards the old French settlements of Kaskasia and St Louis above which again was the meeting of the Mississippi flowing down from the north and the Missouri coming from the west. Along the latter the Lewis and Clark expedition went to find a hazardous way to the Western Ocean, while at the same time Zebulon Pike explored the northern Mississippi without discovering any easy route to the Great Lakes. These expeditions, however, were far in advance of the frontier of settlement; although Ohio had been admitted

---

1. When Zebulon Pike was engaged in exploring the Upper Mississippi in 1805, he visited Dubuque, but not the actual mines; Dubuque was very guarded in answers to Pike's queries, saying he mined between twenty and forty thousand pounds weight of seventy-five per cent ores per annum, and that copper deposits were associated with the lead. Z. M. Pike, *Expeditions*, p. 10 and Appendix to Part I, p. 5.

to statehood as early as 1803, the Indians were formidable north and west of that river, and even after the defeat of Tecumseh's confederacy at Tippecanoe Creek in 1811 the entire country between Lake Michigan and the Upper Mississippi was still a wilderness and still peopled by tribes on whose friendship no American could rely.

The wilderness was a deterrent as well as a challenge. The adventurous have always made a greater mark in history than the common man who only sought a livelihood and security. Politicians might coin a phrase like 'Manifest Destiny', but farmers, labourers, and tradesmen living, according to their lights, in moderate comfort, were far less enthusiastic about dreams of continental expansion and the like which seemed, sooner or later, invariably to involve them in troubles and wars. Every twenty years or so there seemed to be a major Indian crisis, and when the young Republic emerged from the war of 1812 with its head unbowed but decidedly bloody, many of its citizens felt that further encroachments on Indian hunting grounds and quarrels with the British in Canada should, if possible, be avoided. The Ohio and Wabash valleys were now safe for settlers and extensive enough without venturing further west to the valley of the Illinois and its tributaries; vast areas to the southward, too, seemed to offer better and less risky opportunities for those ambitious to bring wildernesses under the plough. Furthermore, the richest section of the Union, New England, had shown a marked disinclination to finance the establishment of potential new states whose entry into the Federal Republic would reduce its own comparative political weight and influence in national affairs and counsels.

On the other hand the war of 1812 had emphasised the Republic's need of achieving greater national economic self-sufficiency. Industries had started in Pennsylvania and New England to provide some of the manufactured goods previously obtained from Britain. The war had made obvious the need of developing American supplies of strategic minerals, especially lead and copper, and after the end of the war both considerations of national security and the furtherance of the infant manufacturing industries gave an impetus to American mining. Scattered iron and copper deposits in the Atlantic States had been already worked for some time. Iron was plentiful enough, but copper was scarce and lead even scarcer. Some newer finds of copper in Maryland and Kentucky supplemented the older workings in Connecticut and New Jersey whose richer and more accessible ores had been taken by the earlier miners. These, however, were inadequate to meet the whole needs of the country, whether in peace or war, and the tales of fur traders and explorers about the metals worked by Indians in the western wildernesses came to be discussed as possibly indicating the means of supplying the country's needs. Julian Dubuque's lead-mines were accessible by the Mississippi, despite hazards of river navigation and the falls at St Louis, but the copper ores of Lake Superior were beyond the falls of St Mary; river and lake navigation alike were closed for at least four months of the year, and on the river, though the thaw came earlier, there were also risks from flood; at other times, through drought, only vessels of very shallow draught could reach Dubuque.

In any case lead was more valuable and scarcer than copper, and within a decade of the war of 1812 the first American pioneer miners came into the lead district round Dubuque's former trading post. First of the 'Cousin Jacks' to follow in their wake was Francis Clymo, who may well have had experience of both lead and copper mining in the Old Country before emigrating, at the age of twenty-seven, to Maryland in 1819 from the parish of Perranzabuloe. After a few years' mining or prospecting for copper in Maryland and Virginia, he moved with his family to Kentucky and thence to Wisconsin about 1830.[2] By that date several American pioneer miners had settled in the region, but they seem to have been more adept at locating metallic ores than in mining them. Most of the early American settlers in this area, too, came up the Mississippi from the southern border States and were far more accustomed to the long rifle and axe than they were to pick and gad, while many were of the volatile, adventurous type apt to be quickly disheartened after the first romantic excitement of pioneering discovery died away, leaving them faced not with fortune but by the burdensome, stark necessity of dull, monotonous labour to get a mere living.

In south-west Wisconsin, north-west Illinois, and eastern Iowa it was the same as it was later to be in Michigan and California, in Montana, Nevada, Idaho, and Colorado, in fact everywhere on the mining frontiers of all the continents. Daring, sanguine, but feckless adventurers were often the first on the scene, but they were also the first to leave, their place being taken by men for whom mining was a job and a skill and whose rewards, for the generality of those engaged in it, were not affluence but modest living wages.

The early days in this lead-mining region of the Upper Mississippi

2. L. A. Copeland, 'The Cornish in South-West Wisconsin,' *Wisconsin Historical Collections*, Vol. XIV (1898), pp. 305–6n.

were stirring enough, but it is the simple passing of time that has been mainly responsible for tinting pioneering hardships with glamorous hues. Those who referred to the country as a 'howling wilderness' were as right as those who vaunted it as a land of promise. All the strange novelty and violence of pioneering and prospecting in Indian country were real enough, but the settlement of this and every other mining frontier called for sweat and tears as well as blood.

Blood was shed in the lead-mining region. When the first miners and prospectors came up the river in the early eighteen-twenties, it was Indian country, beyond the limits of existing state and territorial governments, and in theory under the control of the Federal War Department. A Congressional Act of 1807 had stipulated that mineral lands in the national domains were not to be sold but only leased to persons locating on them, and for thirty-nine years the development of the lead region was delayed and impeded by both the observance and breaches of this law. 'Finding's keeping!' was the prospector's first article of faith, and the attempt of officials of government to enforce the taking of leases for working mineral lands for terms of a single year or two provoked resentment, defiance, and contempt for the law. With federal military forces at extremely low strength and scattered all over the sparsely populated frontiers of settlement in the vast western domains of the United States, and with the administrative headquarters of both military and civilian authorities often long and difficult journeys from the district where lead discoveries were made, there was bound to be trouble from the outset, especially as there was the ever-present risk that a rich claim might be 'jumped' while its discoverer was away seeking to get legal title to work it. Later, the officials empowered to register, grant and confirm leases were suspected of corruption, of taking bribes, favouring personal cronies, and the like, nor were matters improved by the way in which the national elections of 1828, 1840 and 1844 resulted in changes of such administrative personnel. Until the law was changed in 1846, this system of leasing Federal mineral lands discouraged capital investment in the lead-mining industry, and practically restricted operations to one-man concerns or to small partnerships of labouring miners, besides having the effect of creating unleased and unoccupied tracts of confirmed or suspected mineral lands that otherwise might have been systematically worked.

The law, in brief, bred lawlessness, and America's first mining frontier had its full share of stabbing and shooting 'scrapes'. The Southern immigrant element brought its 'code' of duelling as a substitute for due but tardy processes of law to settle personal grievances and quarrels. Thus, in December 1839, Martin Nash and Patrick McKenney were killed by Thomas Ryan, Thomas Flanagan, and John Gardner in disputes over mining claims at Snake Hollow in Grant County, Wisconsin;[3] a year later, Mineral Point, which had already a number of Cornish settlers, was buzzing with the news of the affray between the 'two-gun packing' editor of the local *Miners' Free Press*, Henry B. Walsh, and 'Charley' Breeden, one of the more prominent public figures in that seven- or eight-year-old mining community. Walsh, who subsequently claimed that his pistols were 'harmless toys' incapable of injuring a man beyond the range of two paces, had apparently cocked his armament first at Breeden, who was armed with a musket fifty or sixty yards distant, but the latter was quicker on the draw and shot first, hitting the editor in the shoulder, and had then run away; Walsh had grabbed a pistol, presumably something more business-like than his own 'toys', and sent a parting shot after Breeden, but missed. Breeden had a grievance, for Walsh in editorials had called him, among other things, liar, coward, and scoundrel, and it seems certain that the feud between the two had its origins in scandals over leases of mining claims in Mineral Point which had occasioned the dismissal of Walsh's friend, John F. Sheldon, registrar of the Mineral Point Land Office. A later inquiry indicated that Sheldon had blocked bids being made for certain lands which he had then taken up himself, and that he had also, in contravention of the 1807 law, sold outright titles to some of the most valuable local mineral lands.[4] Disputes over mining claims, too, may have been the cause of the shooting in February 1842, of the young Territorial representative Charles Arndt by another representative, James R. Vineyard.[5] On the other side of the picture should be recorded the appearance in the lead-mining district of miners' meetings to establish their own local government, to prescribe rules and conditions about the size of claims laying down that if they were not worked for a certain time every year they lapsed and could be taken up by another party, and so forth.[6]

Despite the legal uncertainties about mining rights, it was apparent, before 1830, that the lead deposits of the Upper Mississippi offered reasonable opportunities for making a living to men who were prepared

3. *Madison Express*, 4 January 1840.
4. *Miners' Free Press*, 1 December 1840; *Madison Express*, 22 April 1841.
5. *Madison Express*, 12 February 1842.
6. Allard J. Smith, 'The Lead Region of Dubuque County, Iowa': *Trans. Wisconsin Soc. of Science, Arts, and Letters*, Vol. XII, Pt. I (1900), p. 223.

to stand the inconveniences of primitive frontier life, and by that year a few more Cousin Jacks had arrived in the region. Most of them, like Francis Clymo, had already been in America a few years and had been working in New Jersey and Pennsylvania iron-mines or in other small mines in the Atlantic States. There they had heard of the prospects, often much exaggerated, of the western lead-mines. Having once migrated across the Atlantic, a further move into the heart of the continent had no great terrors for them. Some, too, were disappointed and disillusioned with the first jobs and opportunities they had had in their new country. They came to the lead mines on the Mississippi, found prospects that were, if not as bright as they had been reported, reasonably good, and letters they wrote to kinsfolk and friends in the Old Country touched off direct emigration from Cornwall to south-west Wisconsin, north-west Illinois, and even to Iowa on the further side of the great river.

The legal prohibition of sales of mineral lands in this region, even when enforced, which was certainly not invariably the case, did not stop any man from working 'his own mine'. Even if he could only secure a short-term lease of mineral property, it was usually on condition that a proportion of the proceeds of smelted ores, a tenth or a sixteenth, went to the Federal Government, which was no intolerable exaction in the eyes of Cornish miners who had been used to working tribute pitches in which a third, a half, or even more of the ores they raised went to the mine-owners or adventurers who, in their turn, paid the owner of the land the 'lord's dish' of about an eighth of the mineral produced, while in the tin-mines additional 'coinage' dues had had to be paid to the Duchy of Cornwall. In place of these two or three interests exacting a toll from the actual mine operatives there was, in America, only one—and that a remoter, less efficient, and more evadeable one than those in the Old Country. In Cornwall, too, the more shrewd and enterprising miners had developed ways and means of avoiding the payment of all the dues claimed by investors, landlords, and Duchy officials, and these could be and were used to their own advantage in the United States.

From the earlier fur traders, from the Indians, and from evidences of early prospecting of previous red and white people in the region, the American pioneers knew just where to prospect for lead. Many outcroppings of metal-bearing rock strata could be easily identified by men with little previous mining experience and just enough knowledge of lead ore to recognise its presence in such rock outcrops. The hilly character of much of the district and a fair number of ravine-like gulleys gave the sharp observer a fair idea of the beds and dips of stratified rocks, while it did not demand much geological knowledge for a man to surmise—and often correctly—that mineral would be found in a locality presenting similar topographical features to one in which it had already been found. Whether the so-called 'masonic' plant or 'lead weed' was an infallible indication of the existence of lead ores at its roots was doubtful, just as in Cornwall elder and blackthorn were not trustworthy indications of either barren or 'keenly' lodes of tin. In many American mining districts, too, there were disputes about the efficacy of the divining or 'dowsing' rod in prospecting for mineral; just as in the Old Country some miners were sceptical about its virtues. In any event, in the eighteen-twenties many lead deposits had been found by American immigrants in the Upper Mississippi region and Federal restrictions on sales and leases of mineral lands had certainly encouraged some to travel far afield in the limestone districts in the hopes of locating rich deposits they could exploit before any government agent caught up with them.

Locating lodes, however, was one thing and successfully working them another. Actual outcroppings, though fairly numerous, were very limited both in area and richness. Experience seemed to suggest that the deeper the mineral-bearing lodes were traced, the richer the ore. A few workings were enough to show that the galena and Trenton limestones dipped below other rock strata, and also suggested that the richest ores were to be found in the Trenton beds. It was not long before it became obvious that only men skilled in deep mining techniques, acquainted with the arts of sinking shafts and driving levels through comparatively hard rock would get much out of the mineral deposits of this region, and the Cornish miners entered on the scene.

There were only a few of them at first, and only a handful of Cousin Jacks served in the Black Hawk War, among them being the Camborne miners Francis Vivian, Stephen Terrill, William Bennett, and the James brothers, Joseph and Edward, while two Prideaux brothers, James and William, came from the neighbouring parish of Illogan.[7] The battle of Bad Ax ended any threat of further trouble from the Indians in the lead region, and in the course of the next few years scores of Cousin Jacks came into the district; by 1850 there were nearly six thousand Cornish people in the three Wisconsin counties of Grant, Iowa, and Lafayette, and perhaps two or three thousand more in the neighbouring counties

7. Copeland, op. cit., pp. 308–10.

of Jo Daviess, Illinois and Dubuque, Iowa. Of these Cornish-Americans of 1850, it is reasonable to assume that a third were infants born after the arrival of their parents in the New World, but at least four or five thousand of the settlers in the region had emigrated thither more or less directly from Cornwall in the years between the end of the Black Hawk War in August 1832, and the discovery of gold in California.

Comparatively few of this really considerable multitude left much in the way of personal records. In State and Federal census returns one or two designated 'Cornwall' as the 'country' of their birth, but most wrote 'England'. Local county histories, which became popular towards the end of the nineteenth century, gave biographical details of some of the most successful of these men—probably because the compilers of these bulky and expensive compendiums had induced them to subscribe to these publications. Patient genealogical research even yet may unearth a few more facts about these early settlers from the Old Country. It is unlikely, however, that even the most microscopic techniques of historical research will alter the fundamental picture—that of a group of emigrants who came to a new country, faced many privations and hardships, made some sort of a living, and, above all, contributed to the settlement and the transformation of a howling wilderness of a mining frontier into civilisation. American adventurers had found the mineral wealth of the district, but might have left it and the region had not these skilled immigrant miners come in to show them the best modes then known of working it; these miners provided the market and some of the labour for the pioneer farmers who established themselves on the better agricultural tracts of the country. That much can be confidently asserted, but what is open to surmise, is what would have happened in the lead-mining regions of Wisconsin and Illinois had the Cornish not come in, as they did and when they did, in such numbers, and especially in relation to the crisis of the Union over slavery in the middle of the century. The first immigrants into the region were from the South, and several of them had brought slaves with them; the Cornish came as free labourers without slaves, and when their work and skill ensured the development of the lead mines, there followed the 'free-soilers' and anti-slavery immigrants from New England and the Northern States to engage in the ancillary trades and services required by the mining industry and to farm the land around the mines.

Most of the Cornish immigrants were young men. Hardly one of the Camborne and Illogan men who served in the Black Hawk War was over thirty years old and the later horde of migrants mostly were between eighteen and thirty-five years old. At first there was a preponderance of single men; several of these were later joined and married by girls they had courted in Cornwall, while some went back to the Old Country for a few months, married, and came out to the lead region again. Thus Philip Allen, later a prominent storekeeper in Mineral Point first came there in 1842 when twenty-six years old, returned to Cornwall four years later to marry Elizabeth James and bring her out to Wisconsin.[8] Joseph Gundry came to Mineral Point from Porkellis in Wendron parish when twenty-three in 1845; after two years he went home for a few months, and came back with his bride, Sarah Perry, a Wendron girl.[9] Simon Lanyon of St Allen married a girl of the same parish in 1838, when he was twenty-three, and almost immediately left her in Cornwall while he came out to find a new home for her in Linden where she joined him two years later, coming out, however, accompanied by her brother-in-law, William Lanyon and his wife and children. Another Lanyon, John, did not come out to Mineral Point until 1858, by which time his blacksmith brothers were well on their way to prosperity.[10]

The list could be extended almost indefinitely, but these few instances suggest the heartaching partings and separations suffered by many of the early Cornish settlers and their womenfolk in the lead region. Other hardships were encountered by William and Lavinia Rablin, who were married in Camborne in the spring of 1835, and straightway left for Wisconsin. They reached Mineral Point towards the end of June to find conditions that are more usually attributed to the later roaring mining camps of the Pacific Coast than they are to the lead towns of the Upper Mississippi. Perhaps the Americans in the district were rehearsing for Fourth of July celebrations the coming Saturday; certainly liquor was flowing freely, and while some men grow affable in their cups the tempers of others become frayed to breaking and fighting point. At home the social life of the Rablins had centred round the Methodist chapel, and they sought out kindred spirits in the 'far West'. William started working as a miner at Linden with his brother, and left his wife at Mineral Point for a time, but she insisted on coming out to housekeep for the brothers in a crude sod hut they had hastily built. The only social life they had for some years was Sunday preaching

---

8. *History of Iowa County* (Chicago, 1881), p. 853.

9. *Commemorative Biographical Record of the Counties of Rock, Green, Grant, Iowa, and Lafayette, Wisconsin* (Chicago, 1901), p. 605.

10. *History of Iowa County*, p. 865.

services held in the cabins of the miners before a church was built. In later years Rablin recalled going to these cabins of Methodist neighbours in the wilderness, often taking his wife up in his arms to carry her over the swampy places, and jumping from one log-stump or stepping-stone to another. Yet in less than ten years he was wealthy enough to start farming near Mineral Point and to build a fairly substantial residence in that town. When he retired from farming, in 1867, he had a six-hundred-acre farm that in Camborne parish might well have been worth £15,000, and it is doubtful whether in the Old Country he had ever made a wage averaging £2 10s or £3 a month.[11]

The sod cabin which had been the Rablins' first home in Wisconsin was typical of many of the early immigrant dwellings in the lead region. Men who came out alone to mine were disposed to 'rough it' and to make do with tiny, cramped shelters hastily cobbled together rather than built of sod, wood, or brush, some with no more than an entrance doorway and a hole in the roof for the smoke to go through. The womenfolk who came with or joined them later naturally enough demanded more comfort than such makeshift shelters afforded, especially in wintertime. Sure of reasonable returns from mines which were virtually, and certainly after 1846, their own property, they had the advantage, too, of abundant building material close to hand. There was plenty of timber in the early days, while the extensive surface beds of limestone provided a building stone that could be quarried and worked easily. Galena limestone, in particular, quarried from the earth could be cut and 'dressed' with little trouble, but it hardened quickly on exposure to the air. Near the first mining settlement at Mineral Point, which in early days was known as Shake-Rag-under-the-Hill,[12] quite a number of substantial houses were erected by Cornish immigrants; both limestone and timber were used in their construction, but it is noteworthy that while good, shapely, and well-dressed stone was used for the fronts, the sides and still more the backs, were of rough stone plastered together with lime mortar. Just as at home, in Cornwall, many houses and cottages with fairly good exterior walls were filled out with 'cob' or dried mud, a coat of limewash serving to hide the cheap and crude building materials. When, in Mineral Point, wood was used, it was green and unseasoned, and in the course of years many

11. *History of Iowa County*, p. 271.
12. The name is said to have originated from the practice of the Cornish women summoning their menfolk home to meals from the opencast lead workings on the opposite hill by waving a cloth or rag.

floors and ceilings became warped and twisted, and with listing floors and walls cracking it was not surprising that many of these early habitations scarcely survived their builders, though in recent years nearly a score of them have been skilfully restored, and these remain to this day as a living memorial of a typical Wisconsin Cornish mining settlement.[13]

For one reason and another, and certainly not least from the natural disinclination to render to the Federal Government and its officials all that it claimed by the mineral lands leasing system up to 1846, there are only fragmentary records of the actual output of lead from this region in the early days, and only surmises can be made about the earnings and profits of individual miners and small partnerships. Early on, several miners began acquiring farm land, and for many years it was usual for a man to spend most of the summer about his farm and the more open winter days mining lead; in fact, this region was one of both family-operated farms and family-operated mines, and many combined the two callings. Comparatively few of the mines went down more than a hundred feet, and many were less than sixty feet deep. Most of the ores were got out directly from shaft excavations and few subterranean levels or drifts were driven, but many, if not most, of the shafts were inclined, not horizontal ones. The nature of the 'country' rock, as it would have been called at home in Cornwall, and the comparatively shallow depths to which the lead veins were worked, meant that most drainage problems could be solved by buckets and windlasses, or by simple pumps that could be worked by a horse. Geology and climate co-operated to make this a country of modest opportunity for the farmer-miner, and it was not the actual mining so much as smelting and transporting ores and metals out of the region that afforded much scope for capital and engineering enterprise.

Conditions changed somewhat when zinc ores began to be worked round about the time of the Civil War, but the lead region of the Upper Mississippi never reached the stage where the maintenance of its mining industry came to depend upon elaborate and costly mechanisation and investment of capital. At first there seemed to be ample opportunities for all who came; then, even before the most accessible and cheaply worked lead deposits were exhausted, what seemed to be richer mining prospects beckoned the more enterprising and adventurous miners away,

13. The above paragraph is based on Robert M. Neal's account of his work of restoring the old Cornish homes in the *Wisconsin Magazine of History*, 1946, pp. 3–13, and on Mary E. Rowe's article 'Little Bit of Cornwall Lives in Houses on Shake Rag Street', *Milwaukee Journal*, 9 October 1948.

first to the copper country of Michigan, then, more dramatically, and in far greater numbers, to the gold-fields of California. All the time, too, there was the immediate neighbouring attraction of cheap, potentially rich, farming land: quite a number of the early Cornish immigrant miners by 1845 owned holdings of some forty, eighty or even a hundred acres in Wisconsin. Ten or twenty years later there were many more who, having spent a few successful years in California and other western gold camps came back, not to Cornwall but to Wisconsin, where they bought up holdings that rivalled in size the estates of many Old Country squires.

Three or four acres held, like the roof over his head, on an uncertain tenure of lives, a pig or two to keep his family in pork and bacon, now and again a 'rented' cow, a patch of land to raise sufficient potatoes—for his family's needs—provided the dread blight or disease did not appear—were all a labouring miner could hope to achieve those days in Cornwall by hard and dangerous work. To that consideration must be added the increasing poverty of the older mines of the far western Cornish districts round St Just, Helston, Camborne, and Redruth. This should be, and by many was, contrasted to the opportunities offered by Wisconsin, Illinois, and Iowa. The Federal Census of 1850 shows that John Rule of Mineral Point, who had come out from Cornwall about ten or twelve years before, owned a two-hundred-acre farm, and had improved sixty acres of it; his livestock included seven horses, two cows, four bullocks, and two pigs; in the last season he had raised over a thousand bushels of cereal crops,[14] besides a hundred and fifty bushels of potatoes and twenty-five tons of hay; his butter production, however, was only fifty pounds, but that may have been a surplus left after the needs of his own household had been supplied. Another return, made by Sarah Jenkins, gave details of a forty-acre holding; she may have been widowed or her husband absent in California, but even with a family of six, the eldest sixteen the youngest five years of age, she was far better off than she would have been in Cornwall, for she had two horses, four cows, four other head of cattle, and seven pigs, and her farm had yielded a hundred bushels of Indian corn, three hundred bushels of oats, a hundred bushels of potatoes and twelve tons of hay in the previous season.[15]

14. The exact return was 50 bushels of wheat, 375 of Indian corn, 700 of oats, and 30 of barley.

15. Sarah Jenkins gave her age as 37; all her children were returned as born in Wisconsin, suggesting that she left Cornwall before 1834. There were other Jenkins families returned in the 1850 Iowa County returns to the Federal Census, viz, Samuel aged 58, a stonemason, with

Bare statistical records in census returns, however, only suggest that within a few years of their arrival and settlement in the lead region a number of Cornish families had done fairly well for themselves by emigrating. Letters written home confirm this, and that written by William Retallick to his brother from Lockport in Illinois on 5 June 1846, summarises the experiences of many who went out from Cornwall in those days. Retallick, however, came from a Cornish farming family and went to Illinois as a farmer with some capital to begin with, though not enough to meet all his requirements in a new country right away. Unlike the Cousin Jacks in the lead region who turned from mining to part- or whole-time farming, he apparently took over a holding on which the pioneer clearing land reclamation work had already been partly done, and on which there were quite adequate housing and outbuildings when he arrived. He wrote home:

My farm is a hundred acres of land, between sixty and seventy acres cleared, with a frame house, barn, and other outbuildings and a good orchard of apple, plum, peach, and cherry trees. Raspberries and gooseberries grow wild in the woods. There was no wheat sown on my farm last year. I sowed three acres of oats, two acres of Indian corn, and some potatoes. I shall cut eighteen acres of hay. It is surprising to see how fast things grow in this country. What I intend to cut for hay has been cut for many years past, and never had any manure. I never cut as good in England. I have bought two horses, they cost 175 dollars, that is £36.9.7d. We turn all our cattle in the woods; they cost nothing for keep in the summer. My farm cost two thousand dollars, eight hundred dollars to be paid in May, but I was not able to do so, so I have paid five hundred dollars for the farm, horses, cows, pigs, sheep, wagons, ploughs, and harrow, and everything except the oxen paid for. The oxen cost seventy dollars or £14.1.3d. Blacksmith's work is very dear; they charge 12s a shoe for horses, and 6/4d for removing. Labour here is very

---

wife aged 56, and their English-born 'children' aged 20, 18, and 16; Mary Jenkins aged 34, with children of 5 and 9 months born in England, and 'lodging' with her Thomas Jenkins, an 18-year-old miner, who may have been her brother-in-law; John Jenkins aged 23, with wife and an infant child of three months. The ages given suggest that they may all have belonged to a single family, had it not been that Mary Jenkins' children were born in England. The ages of both Sarah and Mary make it possible that their husbands had gone off to the Californian gold diggings. Samuel Jenkins, later engaged in a mercantile business in Mineral Point, who was born in Cornwall in 1824, and came first to Wisconsin in 1841, was certainly in California at the time of the census (*History of Iowa County*, p. 864). Another Jenkins family, that of Benjamin and Martha, had come to Wisconsin in 1848, the wife arriving first, he being at the time in Mexico; Benjamin's father had died in Guatemala, and his own eldest son was to lose his life in the Fraser River gold rush after some years in California; the second son, Thomas, was to work in Californian and Montanan mines, before settling at Plattville (*Commemorative Biographical Record of the Counties of Rock, Green, Grant, Iowa, and Lafayette, Wisconsin*, p. 460).

dear; I have paid 4/2d a day for some days, so we cannot afford to employ much. Land sells from £3 to £6 an acre. Horses are nearly as dear here as in England, but all other kinds of cattle are cheap; you can get a good cow for £3 or £4. I bought two in April for twenty-eight dollars; they have been in the woods ever since; one of them got calf this week. I have been offered twenty dollars for her, but I intent to keep her. They are the most friendly people here I ever saw; as they persuaded me last Saturday I asked five or six, and I had more than twenty (some of them I never saw before) to clear the land and make the fences. I wish I could induce you all to come to America—you would get a hundred acres of land for each of your children—you would have no need to work, but ride from farm to farm to enjoy yourself, and see them on their own estates enjoying the fruits of their labour. I like the States better than Canada; they have plenty of soldiers there to keep them safe, but they must be maintained by someone. Interest is very high; seven per cent on land security; some would give more if they could get it. I have never wanted friends since I left England; if a man is poor, let him tell the truth; and if he is industrious, he can have what he wants, at least I can. . . .[16]

The passing allusion to Canada suggests that Retallick, like the Wearnes of Hayle, a few years before,[17] had had thoughts of settling in British territory but that it had not proved to his liking and, so, he had moved down south into Illinois; the forthright expression that British troops were costly does not come unexpectedly from a tax-paying agriculturalist who had had more than enough taxes to pay in the country of his birth. Nor was it surprising that a farm-labourer's wage of a dollar a day struck him as excessive—it was at least twice what farm-hands got in Cornwall, but it should have been set against the neighbourly land-clearing and fencing 'bee' when a score of his neighbours turned up to give Retallick a helping hand. When he left Cornwall he had been given a letter of introduction and commendation by the clergyman of his parish church where he had been a regular communicant; this may have helped him to find friends though it may have been originally intended to pave his way into Anglican church life in British Canada rather than in Illinois. At all events his neighbours trooped in in full force to help him, and he found that honesty and industry were the qualities held in highest esteem in the farming community where he settled.

16. *Royal Cornwall Gazette*, 31 July 1846. Retallick's figures are somewhat inconsistent when he attempts to change dollars into pounds; ignoring fluctuations in the rate of exchange, caused by the Anglo-American Oregon crisis, which was settled by April 1846, and the outbreak of war between the United States and Mexico, in May 1846, the normal sterling value of the dollar was roughly 4s. 2d.   17. See above, p. 34

The Wearnes who too had had initial plans for settling in Canada, had hardly stopped in Montreal a day before they decided to move down into Wisconsin. They had been fifty-one days on the ocean and St Lawrence, and it took another eleven days for them to reach Milwaukee, encountering yet more rough 'seas' on Lake Huron. At Milwaukee they found boarding-house charges excessive, but soon managed to hire an ox-team to take them and their possessions to the lead-mining region. By this time they had been travelling so long that they were becoming more than impatient to reach a final destination, but the pace of the ox was, at best, a mere fifteen miles a day. The roads were bad, if they could be called roads, and many times they were afraid that the wagon, lurching and creaking along, would capsize. It was not easy to get provisions along the way through a sparsely settled region; at one place they got a little sour bread from an Indian for which they had to pay half a dollar, which Richard Wearne reckoned to be ten times what its equivalent would have cost them in the Old Country. For the most part, however, they had to rely on their muskets to provide them with food as they journeyed across southern Wisconsin. It is unlikely that any of the party had had any previous experience of 'camping out', and they were unlucky in the way that bad weather dogged their tracks; the women and children complained of chills and sickness, but after nine days' ordeal they reached Dodgeville.

Their troubles did not end there, for they then had to find friends and relations scattered about the mining districts; when they found them they did not get quite the welcome they anticipated, and they were chagrined to find that they were not expected—letters may have gone astray or the earlier settlers had hardly expected the Wearnes to emigrate. Richard Wearne's journal indicates that he had thought his 'Friend', William Goldsworthy, who had married his cousin, would have been more hospitable at Linden; Goldsworthy gave part of the Wearne family a roof for a single night, and then all he could or would provide them with was a log hut which belonged to his son. It had no door; the light and the cold came in 'through a thousand holes', and Goldsworthy charged two dollars a month rent for it, but

we was to make it Tenetable and he to allow it in the Rent. I put much cost about it. But he was not to his word, however, we took it, and I and John my Brother sleep in it that Night. Sooner Tenant to Stringers than your own People.

Although one might feel some sympathy for Goldsworthy at the unexpected arrival of fifteen relations, most of them adults, yet Richard

Wearne had naturally expected more kindness from a man to whom and to whose family he had in former days given 'House Room ... for one year without cost to him'. Still they worked hard and fast to get the hut into a more habitable state, but it is hardly surprising that some, if not all, the party wished themselves home again. Tempers became frayed, and quarrels nearly blazed out, but on the first Sunday when some of their old friends called on them they felt less homesick, even felt almost at home in Hayle chapel when they all joined to sing 'a few of Wesley old Hymns'.

Only two of the seven men in the family, Richard and his brother Henry, had definite plans what they intended to do in Wisconsin, although his son William Wearne, fated to die four years later of fever at Panama on his way to California, thought he might make a living at his tailoring trade, while John Wearne hoped to find plentiful scope as a stonemason. Richard and Henry meant to start a small iron foundry, but such a scheme was not immediately practicable, so they and three of their sons set about prospecting and mining. At the same time Richard bought a quarter section of land which he promptly called Wearne's Creek. In his journal he recorded that he had

> bought it on account of water to erect a small Wheel for Iron Foundry as I and my Brother have been talking about it for Years. The land is not adapted for a good teeling[18] spot but a good Hay Bottom and good for Raising Cattle. My sons did not like the Place but John my Brother spoke highly of it.

Winter came on early and hard that year, 1848, and the average gains the Wearnes made from their mining activities to the end of the year did not come altogether to more than fifty dollars for the five of them, and to add to their troubles Richard Wearne himself was laid up for a fortnight when, in the course of work, he was hit in the eye with a flying stone. By Christmas some members of the family were feeling very homesick and asking themselves if they had been wise to come out to this winter-ridden land. Although bad times and bleak prospects had led them to emigrate, their circumstances in Cornwall had not been all that bad and comfortless—it was the threat of poverty and of declining social status rather than poverty itself which had led to their leaving the Old Country, and the same could have been said about many other Cornish families in the Mississippi lead region.

Sober analysis of the sparse statistics that are available suggest that the lead region offered fair opportunities of a decent living to working

18. i.e. ploughing and cultivating.

miners, but that it never held out much hopes of vast fortune-making; there was always the chance of silver bonanzas being associated with galena ores, but this region was not destined to prove a Leadville. When, in 1848, the Wearnes arrived in Linden, mining had been going on in the district for over twenty years, and for the past decade on quite an extensive scale; miners coming into the region had ranged far and wide in prospecting ventures, and had certainly got out the richest of the most easily worked deposits. Shipments of lead out of Galena reached their peak in the three years 1845–47, when they averaged about two million dollars in value annually;[19] some more lead was sent out through Milwaukee and other ports on Lake Michigan, but probably ninety per cent of all the lead raised was shipped out through the Fever River port, six miles above the confluence of that river with the Mississippi in Jo Daviess County, Illinois, just south of the Wisconsin line. In those years it seems probable that some twenty thousand men were full- or part-time miners in the lead region, and their average annual earnings from this can only have been about a hundred dollars each. In the next five years the output of lead declined by nearly a third, but this was partly, if not mainly, caused by the exodus of so many miners to California and elsewhere. From the time American miners first came into the region, in the eighteen-twenties, to the end of 1850, it is hardly likely that the aggregate quantity of lead raised in it exceeded 25,000,000 tons; the market price fluctuated, but it averaged around eighty dollars per ton in the eighteen-forties, and the working miners certainly did not get anything like the whole proceeds for their labours. What they did get was a larger share of the monetary returns from their toils than ever they had done in Cornwall, making an average wage just that much higher than in the Old Country to make the difference between a comfortable standard of living and straitened circumstances or even privation. For many it was actually true that they gained economic security by coming to this western mining frontier. There, in the Upper Mississippi Valley they were not so utterly dependent upon their single main calling. Land was there, cheap and in abundance, for those with any inclination towards husbandry, while on the wild lands there was game in abundance, free for all who wanted a meal and had a gun or snare, and with no gamekeepers or squires to say them nay.

This last, indeed, provided one of the most glaring contrasts to life in the Old Country. On their journey from Milwaukee to Linden the

19. *Wisconsin Express*, 13 May 1852.

Wearnes had relied on their muskets to provide them with food. Anywhere in the region of the Upper Mississippi in those times a man could go out into the prairies or woods at the right time of the year and whatever he shot or trapped was his own. Pigeons, wild duck, geese, deer were all there for the taking, while fish abounded in the lakes and streams. In Cornwall there were places where a labouring man hardly dared to look at a rabbit or hare, while the lives of pheasants and partridges were, so far as he was concerned, sacrosanct. If a cottager snared a rabbit to feed his hungry family and was caught by the squire's gamekeeper he was liable to imprisonment; if he was caught a second time and brought before Quarter Sessions, that court of landowning gentry could sentence him to transportation to Van Diemen's Land. Some magistrates and justices were kindly and lenient, but the savage penalties prescribed by English law in early Victorian times did increase the charms of the New World for many; among the causes of emigration was the by no means uncommon necessity of many young, enterprising, and sometimes reckless and even desperate, men to get out of the Old Country one jump ahead of the law. No small number of the Cousin Jacks who came to America came rather than stay at home till magistrates or judges sentenced them to what was, in fact, a prolonged term of slavery in the Antipodes or a shorter 'hard labour' term in a British jail. America, to them, was 'the land of freedom', no matter what those philanthropically minded individuals, including a fair sprinkling of land-owning justices of the peace, said about Negro slavery in the Southern states.

The mining frontier did not offer much of the glamour romantic writers later attributed to it, and it offered far less opportunities of wealth than contemporary emigration agents, shipping firms, and the like declared in the propaganda they put forth. But men who were prepared to adapt themselves to the new environment of America, if nothing untoward happened, made a fair living. The Wearnes, like scores and hundreds of others, had to adapt themselves to their new home; there were bound to be initial disappointments and frustrations; many unexpected little things went wrong, but in time, more especially as the memory of home grew a little more blurred, matters seemed to improve. In any case to return home, if they could, would have been a confession of failure and of mistake, and such confessions for many people are harder to make than perseverance in the courses upon which they had embarked.

In those early days in the lead region, too, they were all pioneers together, and this fostered a community spirit of good neighbourliness. Family disputes and quarrels there might be, as there had been at home, but all of them in the new land had the common experience and bond of emigration and settlement. For lack of churches in the early days they met in each other's homes, conducted services, sang hymns and then gossiped as they had done after Sunday church services and Methodist meetings at home, and as Richard Wearne wrote:

Folks in this Contery likes frequently to visit Each other. It is good if the fare Sects did not like to talk about their Nighbours and after bring forth strife.[20]

Within two months of their arrival in Linden, too, the family had celebrated the wedding of Richard's daughter, Selina, to a Cornish wheelwright, Thomas Tamblyn, and in the spring of 1849 mining prospects improved, and they were able to set about building better dwelling houses. Even letters from Hayle reporting that business conditions had not improved at all since they left and, further, that more of their relations and friends were coming to America, encouraged them by indicating that they had been well advised to emigrate. Tragedy came too, for in May 1849, Wearne's son-in-law, the engineer, Nicholas Phillips, died of cholera at Galena while on his way from Cornwall to Wisconsin.

Tragedies of that nature were far too frequent in those pioneering days and seemed all the more bitter when victims had only just come out from the Old Country. Yet if fever and cholera at times ravaged some of the mining districts, mortality was no greater than it was in similar epidemics in the Old Country. The great Dodgeville cholera epidemic of 1851 caused 136 deaths in a population of only 900 but at Mevagissey, in Cornwall, a few years before, nearly as many had died of the same disease, in a town of about the same population. Cholera was deadly wherever it struck, and the high mortality in Dodgeville should be qualified by the fact that the epidemic broke out at a time when, with so many men gone to California, there was a disproportionately high number of women and children in the Wisconsin town; although cholera swept away many in the prime of their life, it tended to exact its heaviest tolls from the youngest and oldest age groups, and women left with family cares, worrying about their absent menfolk, were often more prone to fall victims to any epidemic that appeared. In the Mississippi lead region, too, infant mortality was high, although no higher than in many Cornish mining parishes. Again, however, it must

20. Richard Wearne MS. Journal, entry of 7 January 1849.

be remembered that in a newly settled frontier region there was invariably a very high proportion of the population in the age group between twenty and forty-five years of age with another large proportion taken by their children mostly under five years old, and it was this last age-group that died in such tragic and startling numbers.

Comparisons between the mortality statistics of the Mississippi lead region and Cornwall indicate two significant differences. The proportion of men dying in the prime of life was smaller in the lead district. Mining accidents did occur, but the much less deep and more easily worked American mines were rather safer and less unhealthy than Cornish mines. There was less strain on heart and lungs caused by long climbs up and down ladders; it was far easier in the lead region, where men worked their own mines, to ease off when feeling slightly indisposed than it was in Cornwall—particularly when, in the latter in bad times, men taking a tribute pitch for two or three months had to slave at it to get something like a mere subsistence for themselves and their families. On the other side, pioneer conditions were hard on women. Deaths in childbirth were frequent in both countries, but on the mining frontier pioneer hardships, both physical and psychological, aged and killed the less robust women in far greater proportion than was prevalent in Cornwall. On the women fell the cares of bearing and rearing children, all the household chores to which they had been used to at home, and a great many more; many had to spin the yarn and weave the cloth for the clothes the family needed, to knit, darn and sew, cook the meals and tend the sick; where, too, the family had acquired land and stock, the women were often left to look after the animals while the men went to the mines, or decided to go hunting or, as in 1849 took themselves off to California. In Cornwall it was the miner who was old before he was thirty or thirty-five; on the early American mining frontier it was his wife.

It has been suggested that the hazards of mining led the womenfolk of Wisconsin lead miners to insist that their husbands leave the mines for the fields. While many Cousin Jacks were unwilling to give up their dangerous and far from healthy calling for farming or some other trade, yet it is doubtful if their wives preferred living on isolated farms to village mining settlements, which, after all, had a few stores or shops, churches, and schools. When the system of mixed farming and mining was adopted, the women and children had to work even harder about agricultural tasks, and it was to be still harder for them when many of their menfolk fell victims of 'California fever' in 1849 and 1850. Even before then, some men had left families behind to go to the copper country of Upper Michigan, and it was the men, too, who got the enjoyment of hunting and trapping in the wilderness around the mines.

If life was hard on the mining frontier of the Upper Mississippi, it had its compensations, although some of the amusements of those times would not be regarded with favour today. Settlers from Missouri and the South were usually given the credit—or discredit—for the introduction of wolf-baiting; a trapped grey wolf was turned loose to fight for its life or lose it at the fangs of three or four hound dogs; at the time, however, cock-fighting 'mains' were still held in the Old Country. Mining camps in Wisconsin doubtless staged wrestling matches like those in which so many Cousin Jacks were to participate in later years in Upper Michigan, California, Montana and elsewhere. But, as in Cornwall, most social life centered around the Methodist chapels, varying from emotional revival meetings and temperance rallies to Christmas and Sunday school anniversary 'tea treats' and sports, while the feature that probably drew more to the chapels than ministers and class-leaders would readily admit was the hymn-singing, which ofttimes made up in enthusiasm defects in melody and tunefulness.

The chief centre of the mining district in south-west Wisconsin was Mineral Point. Within half a decade of the first considerable discoveries of lead there, a Methodist church was organised in 1834 and among the leading spirits in its development were the Cornish settlers, William Phillips and his wife, Mrs S. Thomas, Andrew Rumphrey, and James Nancarrow.[21] Before another five years had passed the first chapel in Mineral Point was no longer large enough to accommodate all the Methodists in the vicinity, and the editor of the *Miners' Free Press*, advanced what today would seem an incredible reason for building a larger church—the existing building was so small that many people did not attend services because they felt that the building would be so crowded that they would be unable to get inside its doors.[22] Years later, Cornish-born local preachers often occupied the pulpits of many Methodist churches in the lead region. Philip Allen, who came out to Mineral Point from Cornwall in 1842 as a stonemason, and who in a few years became one of the most prosperous store keepers in the town, was one of the best-known Primitive Methodist local preachers of the neighbourhood;[23] another member of the same sect, John Toay, who also came out in 1842, was not only a noted local preacher but sufficiently

21. Copeland, op. cit. pp. 309–10. 22. *Miners' Free Press*, 29 January 1839.
23. *History of Iowa County*, p. 853.

active in public affairs to gain election to the State Legislature before his death in 1867.[24] The Cambornian, Joseph Bennett, who came to Dodgeville in 1845, when twenty-three years old, and who became a smelter and businessman, devoted much time to the duties of a Sunday School Superintendent in the Episcopal Methodist Connexion, and was also to secure a place in the State Legislature in 1871.[25]

These men were but three out of a veritable host of immigrants from the Old Country, and it is certain that they not only served their chapels and local communities well, but that they also helped to impart distinctively Cornish features to this American mining frontier. Even at the beginning of the present century round and about Mineral Point, Hazel Green, Plattville, and Dodgeville, the descendants of the early Cornish settlers spoke in a brogue and used terms and expressions that were neither 'standard' English nor American, but which would have been as familiar as day itself to the folk of Redruth and St Just, of St Austell and Liskeard—would, for that matter, be recognised as Cousin Jack talk in Butte City, Grass Valley, Calumet, and Johannesburg. They still called a mine a *bal*, referred to their *crib* or luncheon snack, spoke of going to the *kiddleywink* when they wanted stronger drink than a *dish o' tay*, talked of *clunkin' down* their food or *vittles*, had no compunction in telling a storekeeper who tried to sell them inferior goods that they wanted none of *thiccy passel o' traade*, called the wooden floors of their houses *planchins* and when feeling unwell admitted that they were *braave an' whisht*. Somewhat surprisingly, however, few Cornish names came to be attached to any of the more considerable mining settlements in the region; Redruth Hollow and British Hollow were about the only ones, although the original name of Mineral Point— Shake-Rag-under-the-Hill—and Hard-scrabble are of dubious origin. New Diggings was a name that might appear in a mining district of any English-speaking country. Hazel Green, White Oak Springs were purely descriptive place names, while Snake Hollow had a real frontier twang. Dodgeville, Mifflin, Penton, Shullsburg and possibly Ridgway, merely commemorated individuals, while a widely travelled miner may have given Potosi its name. Diamond Grove Diggings and Lost Grove Diggings have a romantic and nostalgic undertone, unlike Biddick Diggings, which may well have got its name from the Cornish name for pickaxe. The run of mines known as the Terrill Range were, in all likelihood, named after a member of that Redruth family, who emigrated to Wisconsin in the eighteen-thirties.

24. Ibid., p. 876.   25. Ibid., p. 880.

All things considered, the Cornish immigrants left their mark on the Upper Mississippi lead-mining region. Had they not come in when they did, with their hard-rock mining techniques, the permanent settlement of this mining frontier would have been much slower. In their wake, and to supply their needs, came the farmers, but in the early days the mineral region was the most thickly settled and populous part of the Territory, and but for the lead deposits it is hardly possible that Wisconsin would have sought and obtained admission to statehood in the Federal Union as early as 1846. The earlier admission of Missouri and the later admission of California as states provoked major crises in American history, but the sound and fury that raged in Congress in 1819–21 and in 1850 have obscured the circumstances attending and implications flowing from the admission of other states. The admission of Wisconsin was significant, because it was the first mining frontier area to come into the Union as a state directly on account of its mining industry. A new factor had appeared to affect the uneasy balance of economic sections in the Union; up till then the slave-holding commercial agriculture of the Southern states had been able to hold some sort of balance with the Northern free states of mixed agricultural, industrial and commercial interests, just as decisively, though less precipitously and sensationally than California, did Wisconsin's mining interests and progress tilt the balance against slavery; in the years of deepening crisis before the Civil War the Badger State became notorious for its 'Free Soil' sentiments, which culminated in turning a Southern political theory against the slave-owning states by the nullification of the Federal Fugitive Slave Law that had, after 1850, come to be regarded by Southern extremists as the *sine qua non* of their continued loyalty to, and membership of, the Federal Union. It might even be claimed that the arrival of scores and hundreds of free lead miners in the Upper Mississippi doomed slavery in the lower part of the Great Valley; at least, it contributed to that development in American history.

Without exaggeration it can be said that the arrival of so many Cornish immigrants in the lead region in the eighteen-thirties and forties accelerated the economic process, which transformed Wisconsin from frontier Indian country into one of the co-equal states of the Union; equal, if Southern political theories of State Rights were conceded to be the foundation of American political organisation, to South Carolina, then the extreme protagonist of State Rights theories and defender of the 'peculiar institution' of slavery; equal to Virginia who claimed to be the first and 'dominion' state; equal to Lone Star Texas admitted to the Union barely a twelve-month before the Badger State. As the crisis between North and South worsened, most of the Cornish in Wisconsin threw in their lot with the new Republican party; a few joined the Democrats, but not all members of that party were proslavery or disunionists. It was, however, not unadulterated altruistic idealism that determined their political affiliations, for, in the eighteen-forties, the lead-mining interests were convinced that they would be beneficiaries of the protectionist policies advocated by the Henry Clay Whigs, to whom Abraham Lincoln always claimed he belonged; when the later Republican party seemed to fall into the control of the moneyed power that Jacksonian Democrats had so furiously assailed and when, in hard times, working miners reckoned that capitalist interests were extorting an undue proportion of the results of their labours, there were some who went over to the Democratic party; others, however, stayed in the Republican fold, but worked hard to transform it towards the pattern of that later progressive Republicanism which was to make Wisconsin politics an unpredictable departure from American party norms. The teetotal Methodists, to whom many of the Cornish immigrants and their children belonged, went even further and helped to build up the Prohibitionist party in the latter years of the century.

As a rule, however, the Cornish immigrants were not very interested in politics. Some, of course, regarded themselves as strangers in a foreign land and aliens to its political life. More, however, having once uprooted themselves, were prepared to move on elsewhere in quest of a living if prospects for so doing seemed good. Rolling stones gather neither moss nor very settled political views or convictions. Several Cornish immigrants had spent some time on the Atlantic Coast, particularly in the anthracite coal mining region round Pottsville, in Pennsylvania, before coming to the lead country; they were not keen on coal mining, especially in a district where the claims of landed proprietors exacted tolls very like the lord's dish in the Old Country, and where, in any case, there was far less opportunity of any working man becoming the owner of the mine he worked. From Wisconsin a few moved up to the Copper Country of Michigan in the mid forties, while with the news of the Californian gold discoveries Cousin Jacks seemed to be leaving the Upper Mississippi lead region as quickly and in almost as great numbers as, a few years before, they had come to it. Yet six or seven out of every ten Cousin Jacks who went to California from Wisconsin returned to the Badger State, some of them after having called the gold diggings they worked Wisconsin Mine and after having lodged in boarding-houses in some of the more populous camps of the Mother Lode country which went by the name of the Wisconsin House or the Wisconsin Hotel. And there were Cousin Jacks who went even further afield than California and who finally returned to Wisconsin. Charles Dunn, who came as a lad of fifteen direct from Cornwall to Mineral Point in 1845, went off to California seven years later, stayed in the Golden State barely two years and then spent eight years in Australia before returning to Iowa County to settle down on a sizeable farm of a hundred and eighty acres.[26] The younger Richard Wearne, who had only been seven years old when his family reached Linden, in 1848, made his way to California ten years later, then, in 1862, went to the Fraser River gold diggings in British Columbia, staying there for four years before coming back to the machine-shop his family had established in the Wisconsin mining region.[27] Another Cousin Jack who toiled, and, unlike many, successfully, on the Fraser River was Christopher Clemens, who finally settled down as a farmer in Grant County in 1860; he had first come to Wisconsin in 1842, but after two years went off to Cuba for five years; he had then gone back to Cornwall for a visit, married, came out to Grant County, but then almost straightway went to the Californian diggings; after a few months he returned to Wisconsin, but later sojourned longer in California before he finally 'struck it rich' in the British Columbian gold-fields.[28]

The wanderings of these adventurous Cornish miners is amazing, and there were many other cases of men, Cornish-born, who finally settled down, farming or in business, in Wisconsin. There is no escaping the conclusion that they, after having seen 'a good bit of the world', had decided that the best place for them to establish a permanent home was in the lead-mining region of the Upper Mississippi. Others did not return either to the Old Country or to the Badger State. They were to play their parts on other American mining frontiers where many new arrivals from Cornwall were to join them, particularly in neighbouring Upper Michigan and in far-distant California.

26. *History of Iowa County*, p. 858.
27. Ibid., p. 879; another Richard Wearne, his cousin, went to California in 1852, returned to Hayle in 1854, married there, and subsequently emigrated to Australia.
28. *Commemorative Biographical Record of the Counties of Rock, Green, Grant, Iowa, and Lafayette, Wisconsin*, p. 394.

4 Copper frontier in Upper Michigan

The first Cornish immigrants in Wisconsin, coming mainly from the Camborne–Redruth area, knew more of hard-rock copper than lead mining, although the Lanyons of St Allen and others from the north-eastern fringes of the Carn Menellis mining zone must have been acquainted with galena ores and rock formations. For that reason they 'searched' in Wisconsin for copper ores, spurred on by the fact that in the Old Country discoveries of lead and copper had been made close to each other, notably in the Perranzabuloe and Caradon districts. Scientific or professional geologists, like Owen in Wisconsin and Houghton in Michigan, took cues from, rather than gave them to, practical mining prospectors, for the former's 'official' report on the copper ores of the former Territory only appeared some time after William Alford and Philo W. Thomas had set up a copper-smelting furnace at New Baltimore in the Mineral Point district.[1] As early as 1838 a group of speculative adventurers shipped fifty thousand pounds weight of copper ores, taken from a claim within a mile of Mineral Point, to England for smelting, and it yielded over twenty per cent pure metal as against the average Cornish produce of barely eight per cent at that time.[2] Owen, who made a geological survey of some of the Wisconsin mineral lands in 1840, claimed that

an analysis of a selected specimen of the *BEST* working Cornwall ore and of three AVERAGE specimens of Wisconsin ore showed that the latter contains from a *FIFTEENTH* to a *THIRD* more of copper than the former

and emphasised twenty per cent Wisconsin produce against the Cornish eight per cent average.

Owen's report, however, must be regarded with some suspicion. At the time it was published the Territorial Government of Wisconsin was trying to attract immigrant capital as well as population, and it is

1. *Madison Express*, 24 October 1840; *Miners' Free Press*, 12 January 1841.
2. Owen's report in *Miners' Free Press*, 12 January 1841.

likely that the recent acute financial slump had checked the Territory's growth towards a population large enough to make its claims for admission to statehood irresistible. America's financial troubles in the late eighteen-thirties were mainly caused by over-speculation in land, and possibly many now thought that the exploitation of mineral resources might provide a sounder economic basis on which to found a new commonwealth. Just as the lead discoveries in the Upper Mississippi Valley had practically made the United States independent of foreign supplies of that metal, so it was hoped that copper and iron would be found in the north-western frontier zone in sufficient quantities and richness for the Federal Republic to export instead of import those metals.[3] In the older mining districts of Europe some mines were being profitably worked more than two thousand feet deep and yielding dividends from copper ores of three per cent produce; on the American western frontier hunters and backwoodsmen had stumbled upon mineral outcroppings at grass roots that promised to be five or even ten times that rich. The mining frontier's problems were to obtain miners to work the deposits and to transport the ores to shipping points for dispatch to domestic and foreign markets. Cousin Jacks who had already come to America and had done moderately well there were prepared to write home to relatives and friends to come and join them in the new country. Those in the Old Country, feeling the pinch of hard times and aware that their longer-worked mines were declining in produce, were prepared to follow them overseas, while the yearning for greater freedom and independence, in the decade of Chartist agitation, along with considerable nonconformist religious ferment within the ranks of Cornish Methodism, made the republican 'land of the free' even more alluring.

All the glowing reports and sanguine theories of geologists, however, could not create bonanza lodes of copper in Wisconsin where some of the early finds of cupriferous veins had been made in limestone rock formations similar to lodes in Staffordshire which had not proved very remunerative in England. In Colonial times copper had been worked in a few places near the Atlantic coast, notably in Connecticut and New Jersey, but not on an extensive scale. French adventurers, fur traders, and missionaries in the 'old North West' had found traces of Indian copper mining in the Lake Superior region, but the Anglo-American prospecting party of 1771 in the Ontonagon country reported that the extensive copper deposits of that area were far too remote from the

3. *Madison Express* (quoting *Monroe Democrat*, n.d.), 9 October 1845.

Atlantic settlements to be profitably worked, situated as they were on the further side of a vast wilderness of forest and swamp and locked in the grip of ice and snow full five months of the year. Every generation after the collapse of Pontiac's confederation, saw an eruption of Indian hostility in this wilderness frontier against white encroachments under such redoubtable war leaders as Little Turtle, Tecumseh, and Black Hawk, and not until the last of these had been defeated and, a few years later, the Chippewa Indians ceded their claims on the Upper Peninsula of Michigan to the Federal Government was the copper frontier accessible to mining adventurers; thereafter those who came into the region had still to face natural obstacles which could not easily be overcome.

Deprived of the Toledo region for the benefit of Ohio and given the Upper Peninsula lying west of the Straits of Mackinac, the young State of Michigan took up the challenge to make the most of a tract of territory which, forgetful of frontier prowess in his younger and more adventurous days, the now elderly statesman Henry Clay dismissed disparagingly in Congress as being about as useful as the dark side of the moon. The young scientist, Dr Douglas Houghton, appointed State Geologist by influential political connections, soon showed that the post was no mere sinecure but made his way into the country whence had come tales of Indian workings for copper and of huge boulders of native copper lying by the banks of the Ontonagon River. Coming through the Mackinac Straits soon after the navigation season opened in 1840, Houghton made a rapid survey of a wide area of 'howling wilderness', and, in the following winter, drafted a report which was submitted to the Michigan State Legislature early in 1841. He reported that he had found various minerals—lead, iron, zinc, copper, manganese, and some silver, but concluded that 'the only ore of metallic minerals which can reasonably be hoped to turn, at present, to practical account is that of copper'.[4] He brought back four or five tons of copper ore specimens to Lower Michigan, and announced that on one occasion he had, with a single blast, blown out of a vein of copper-bearing rock nearly a couple of tons of ore containing numerous masses of pure native copper, some of which were forty pounds in weight. Such statements naturally evoked a lively response from ambitious provincial politicians who were eager to develop the resources of a State that had only gained admission to the Union in 1837, especially since at that time the United States was importing about half a million dollars worth of copper yearly and the price of the metal was in the region of

4. *Madison Express*, 23 June 1841.

three hundred dollars a ton.[5] Forty pound 'nuggets' of native copper were, it was true, worth only two or three dollars apiece, but if they existed in abundance—and Houghton's report implied that they did—there was reason to anticipate and speculate upon rich mining developments in Upper Michigan.

In his report, Houghton emphasised the fact that the mineral of the Upper Peninsula was native copper in strings, specks, and bunches associated with oxides and carbonates, but there was comparatively little pyritous copper. He made lengthy comparisons with the copper districts of Cornwall in a manner leaving no doubt that he was convinced that Upper Michigan would easily surpass the richest lodes ever worked in south-west England. He told the Michigan legislators in phrases still carrying the inflections of a young lecturer in a New York technical institute:

In the main, the resemblance between the character and contents of the copper veins of Cornwall and Michigan, so far as can be determined, is close; the veinstones are essentially the same; but in initiating this comparison, it should be borne in mind that the mineral veins of Cornwall have been in progress of exploitation for centuries, and the shafts and galleries have been carried to greater depths, while of those of Michigan simply superficial examinations have as yet been made, and those in a wilderness country, under circumstances of the utmost embarrassment and attended with the most excessive labour, privation and suffering.

An archaeologist, however, could have suggested that the primitive Indian workings on Lake Superior antedated those of the Old Country where copper had only been consistently mined for the past century and a half. Cornish tin-workings, of course, were far older, and the legendary Phoenicians and Jews and 'Saracens' of pre- and early Christian times afforded points of comparison with the copper-using 'Mound Builders' of Iowa who had come up to the Ontonagon country for copper, but who never settled in the region, merely spending brief summer mining seasons obtaining such supplies of the tough but malleable metal which they required for ornamental purposes. Since Houghton's time archaeological investigation has not thrown much light upon these early Indian copper miners, but both in the Upper Peninsula of Michigan and in Cornwall traces of 'old men's workings' more than once facilitated the discovery of rich lodes. One such excavation made in some remote time was right on the back of the rich Calumet conglomerate lode, but it is somewhat ironical that the credit

5. *Hunt's Merchants' Magazine and Commercial Review*, Vol. XIV (1846), p. 442.

for its discovery—or rediscovery—by Edwin James Hulbert was to be disputed by those who, perhaps out of Cornish clannishness, asserted that it was located by the Cousin Jack miner, Richard Tregaskis, possibly a member of one of the first families from the Old Country to settle in the Mineral Point district in the eighteen-thirties.[6]

Houghton's report, in scientifically guarded style, went on to suggest that the copper lodes of Upper Michigan were far richer and more extensive than those of Cornwall stating that:

The worked copper veins of Cornwall are stated by Mr Carne, to average from three to four feet in width and to have a length as yet undetermined. But few have been traced for a greater distance than one to one and a half miles, and but one has been traced for a distance of three miles.

The veins which I have examined in the mineral district of Michigan exceed the average of those last mentioned, but the imperfect examinations which have been made render it difficult to determine this with certainty. I have traced no one vein for a further distance than one mile, and usually for distances considerably less. It was not, however, supposed that those veins terminated at the points where they were left, but the fuller examinations were abandoned at these points, in consequence of physical difficulties connected with the present condition of the country. . . .

Of those (veins) which have been examined, embracing nearly the whole (and not including the native copper) the percentage of pure metal ranges from 9·5 to 51·72 and the average may be stated at 21·10.

In Cornwall, the average produce of the ores, since 1771, has never exceeded 12 per cent of the metal, and from 1818 to 1822 it was only 8·2. This shows the aggregate, and it is well known that while many of the productive veins are considerably below this, the largest average percentage of any single vein, in that district (Cornwall), it is believed has never been over 20 per cent, and it should be borne in mind that this average is taken after the ores have been carefully freed from all earthy and other impurities, which can be separated by breaking and picking.[7]

As for the famed Copper Rock of the Ontonagon River, about which tales had been circulating since late Colonial times and which, a few years later, was to be brought to the Smithsonian, Houghton thought that it had been moved to the site where he saw it by either natural or human agency from its original location in a vein. Despite bits chipped off by Indians and by travellers as souvenirs, it still weighed two tons

---

6. *Detroit Evening News*, 16 October 1899. Another account states that some stray pigs unearthed the lode for Hulbert (J. B. Martin: *Call It North Country*, p. 100), but even this account specifically stated that the conglomerate rocks were found in a pit excavated by Indians ages before.

7. *Madison Express*, 23 June 1841, quoting from *Detroit Western Farmer*, n.d.

when it reached the East and served further to propagandise the riches of the Michigan copper country. Houghton who estimated its weight as between three and four tons, reckoned that it was ninety-five per cent pure copper. Yet, compared to some of the masses of native copper that were, in the course of the next few years, to be excavated in the Upper Peninsula, it was a mere pebble; the greatest rock of native copper from the Minesota Mine weighed no less than five hundred tons.[8]

Geologists' reports, travellers' tales, traces of mysterious Indian workings of unknown antiquity, and specimens of native copper brought back to the East, all attracted the attention of mining men to the Michigan northland. Like later adventurers in Dakota they were hardly inclined to wait for the picturesque but tardy processes of Federal Indian diplomacy to clear land titles and throw the country open to settlement and mineral exploitation. So far as the Cornish were concerned, there were two routes of approach; and there is little doubt that the first Cousin Jacks in Ontonagon and Keweenaw came up from the lead-mining camps in Wisconsin and not direct from the Old Country, while a number came across the Lakes from Canada where, a comparatively short time before, some Cornish miners had found employment in the Bruce Iron Mines on the shores of Lake Huron.

The 'overland' route up from south-western Wisconsin in 1843 would have been remembered as one of romantic hazard and of human fortitude in the face of formidable natural obstacles had it not been for the Californian hegira half a dozen years later. Alfred Brunson, the Agent of the Chippewa Indians, who left Prairie du Chien for Keweenaw in mid May 1843, took fifty days to complete a trip through swamps and forests and over tracts of open prairie, misled by guides that went off the trails, suffering from the torrential downpours of one of the wettest summers ever recorded in that region, and plagued to frenzy like many other travellers by the persistent attacks of pestiferous mosquitoes and other insects. Others struck across southern Wisconsin to the shores of Lake Michigan to find shipping—if they were fortunate—to take them up through the Mackinac Straits to Luke Huron, on to the Falls of St Mary, then along Lake Superior to Keweenaw Point or on to the outfall of the Ontonagon. Time was lost by Brunson's party in seeking first for straying draught animals and then for those who went to search for them. Tamarack swamps were not only difficult to traverse with ox-wagons, but the water in them was most unpleasant to the taste.

8. J. B. Martin: *Call It North Country*, p. 74.

It was easy to mistake one hill or ridge for another in a very broken country and to go miles off the most direct course. It was a wild country in which to travel: in a humid summer the flies and mosquitoes were inescapable, while a wary eye had to be kept for poisonous snakes. On the other hand, abundance of passenger pigeons in those days afforded a welcome change of travellers' fare, while the lakes and rivers abounded in fish and presented no obstacles to travel.[9] Once the first travellers 'blazed' a trail, the overland route up from Wisconsin proved comparatively easy and safe except during the long winter months from December to April.

On the Lake route the Falls of the St Mary between Lake Superior and Lake Huron were not bypassed by the 'Soo' Canal until 1855. It was at this portage that the streams of migration from the Upper Mississippi and from the Old Country converged. The Erie Canal had been completed in 1825, while the old fur-trading post of Detroit became a main route terminal and depot point by the early eighteen-thirties, and soon became a port for vessels on Huron and Michigan. The copper country of Upper Michigan, however, lay beyond the falls of the St Mary's River, on the southern shores of Lake Superior, the remotest of the Great Lakes and the one fraught with the greatest navigational perils. In the days of sailing vessels—and even in later times—the Great Lakes were vast, treacherous inland seas subject to sudden squalls and storms, to fogs and to additional hazards of ice in winter time. From early December to late April navigation on Lake Superior was closed by the vast ice floes that packed up against the shores of that inland sea.[10]

The Lake route rather than the 'improved' Indian trails northwards from the Upper Mississippi was the main, almost the sole, link of the Keweenaw Peninsula with the outside world when Michigan copper mining began in the eighteen-forties. On the great 'inland seas' many vessels came to grief; gales drove some on to rocky uninhabited coasts; some vanished without trace; reckless captains and incompetent navigators piled their vessels up on reefs, collided with other ships, or blew their engines and ships to fragments in crazy attempts to race rivals or to reach a destination in record time. Many miners from the Old Country were lost in some of these disasters. A dozen or more

---

9. Brunson's account of his journey was published in the *Grant County Herald* and subsequently reprinted in the *Madison City Express* in its issues of 27 July and 3 and 24 August 1843.

10. Walter Havighurst's *The Long Ships Passing* (New York, 1942) is an excellent study of the history of the Great Lakes and of their navigational perils.

Cousin Jacks were among the seventy or eighty lives lost when the steamer *Pewabic*, coming down from Lake Superior, collided with the steamer *Meteor* about six miles from Thunder Bay Island, on Lake Huron, in August 1865. The *Pewabic* sank within four or five minutes of the collision, and as her books and papers sank with her it was impossible to draw up a casualty list. The letter which a morbid-minded religious sentimentalist wrote home after this disaster underlined a tragic aspect of Cornish emigration to the 'Lakes', besides giving a graphic picture of passenger life aboard the vessels which carried miners, their wives, families, and others to and fro along the Great Lakes in the boom days of the Michigan copper mines. Disasters and tragedies are the very stuff out of which news is made, yet 'W.P., junior', writing hastily to a Cornish newspaper from Kingston, Ontario, as soon as he heard of the loss of the *Pewabic*, caused painful anxiety and worry in many Cornish homes, when he described the disaster at considerable length, and then went on:

The passengers and officers of the *Pewabic* had just commenced a dance. This fascinating recreation had been indulged in every night of the way down, and this being the last night the gay and happy throng expected to be on board, they were determined to enjoy themselves. Alas! They little dreamed that instead of treading the giddy mazes of the dance, on that memorable night, they would be struggling for life amid the angry waters, many to go down unknelled and uncoffined ... The books and papers of the boat have all been lost, so that the names of those who are drowned will never be known, but it is thought there were from ten to fourteen Cornish miners lost. An uncertainty, terrible as death itself, will for a long time hang over the fate of many a loved one. The ear will anxiously await the sound of well-known footfalls, but they will not come, and slowly and terribly will the heartsickness of 'hope deferred' give way to wild despair. May He who 'tempers the wind to the shorn lamb', abide with both the living and the dead.[11]

The inexplicable feature about this collision was that it happened on a clear night, when the lights of each vessel must have been visible for miles around; furthermore, the *Pewabic* had exchanged signals with the *Meteor*. An even worse disaster had occurred on Lake Erie thirteen years before when a propeller from Cleveland, refusing, apparently, to alter its course had rammed the steamer *Atlantic* and sent it to the bottom in a matter of moments taking a toll of three hundred lives.[12] There were many other shipping disasters on the Lakes, some due to 'natural causes', others to human blunderings, on the route to the Upper

11. *W.B.*, 15 September 1865.   12. Ibid., 1 October 1852.

Peninsula; in the early days, particularly, those venturing thither from the Old Country, wittingly or unwittingly, faced as many perils as did later venturers to the more publicised 'diggings' of California and Nevada, of the Fraser River and the Black Hills.

Nevertheless, the reports of Houghton and others and the propaganda of mining company promoters drew immigrants to Upper Michigan. The more respectable of the latter, perhaps aware of the way in which Cornish hard-rock miners had been brought to the anthracite coal-fields of Pennsylvania and to the lead mines of the Upper Mississippi after the first American operatives had shown no great aptitude for deep mining, soon brought in Cousin Jacks from the Old Country. The 'booming' of Lake Superior copper in financial circles, especially in Boston and New York, began soon after the Federal Government concluded a treaty with the Chippewa Indians in 1843. Before the end of 1844 close on a thousand immigrants had poured into the region, but the writer in the *New York Tribune* who envisaged that the population of the Upper Peninsula would rise to ten thousand by the end of 1845 and reach a hundred thousand in 1846 was unduly sanguine.[13] Another newspaper man stated in the *Buffalo Commercial Advertiser*, early in 1846, that nearly six hundred claims or locations, covering between eight and nine hundred square miles, had been made for mining purposes, but he then candidly—and bluntly—went on to say that by no means would all these be soon developed since a large proportion of them were

> in control of men, who, caring little for the responsibilities of working the mines and trusting to their results, will endeavour to humbug the copper-fever-bitten man, who is ready to embark his capital in a business which promises to do well. But let no man imagine that without skill, economy, patience, and industry, fortunes are to be made in a day, even from an interest in the richest mine.[14]

Still more sober and cautious were statements made by Governor Alpheus Felch. On 6 January 1846, he informed the Senate and House of Representatives of Michigan that, 'from the best information to be obtained, there are remaining in the mining country during the present winter, some three hundred men', and a year later, in his next Annual Message to the State Legislature on 4 January 1847, estimated that at least a thousand men were wintering in the Copper Country. In this latter message, however, Felch optimistically commented that mining operations during the past year had

13. *Madison Express*, 11 September 1845.
14. Quoted in the *Madison Express*, 12 February 1846.

confirmed the general confidence in the belief that the region in question is one of the richest in the world in valuable ores, and it cannot be doubted that it will . . . become the theatre of extensive mining operations, the source of much wealth, and the residence of a numerous population.[15]

To some degree this optimistic view was justified. Just over seven years later, at the beginning of the mining season, a census was taken in May 1854. This revealed that the total population of the Copper Country was 6,492; of these, 4,695 were males and 1,797 females; 4,063 of the males were over ten years of age, and no less than 3,615 fell within the age group of twenty to forty-five years; the copper mines were then employing no less than 2,304 hands. Ten years later, in 1864, another State Census returned a population of 18,811 of which 11,279 were males, and the twenty to forty-five-year age group contained 6,218 of them; sixty-three copper-mines were then employing 5,447 hands.

As early as 1846 a dozen mining companies had commenced more or less active operations in the Copper Country of Michigan. Around Eagle River the Lake Superior Mining Company, the Eagle Harbor Company, the Albion Mining Company, the North American Company and the Chippewa Mining Company had established locations; the Pittsburgh Mining Company was located nearer Keweenaw Point, at Copper Harbor which had already assumed the rowdy hustling, bustling features of such later, far more publicised, mining camps as Nevada's Virginia City and Dakota's Deadwood; the Boston Company and the Bohemian Company had started operations near Agate Harbor, and other companies were beginning to start up in the district.[16] Nearly all these mining companies ran into difficulties which could have been avoided had their promoters known more about practical mining and less about financial speculation and gambling, and it would have been of the greatest advantage to all had they only taken more care beforehand to get more information about the actual conditions prevalent in the Keweenaw Peninsula. In 1843 and 1844, however, there was a veritable 'copper rush' to secure claims in the Copper Country, and capitalists in Boston, New York, Pittsburgh, and, for that matter, in London, paid little heed to such considerations as the costs and difficulties of transportation to Lake Superior, the hard long winters of the region, the necessity of clearing dense forest cover to get at the mineral lodes, the blazing of trails in a trackless wilderness, and, last but not least, the techniques required to exploit mass native copper which was essentially

15. *Messages of the Governors of Michigan*, Vol. I, pp. 38, 79.
16. *Hunt's Merchants' Magazine and Commercial Review*, Vol. XIV (May 1846), p. 440.

different from the pyritous deposits that had hitherto been the main source of the metal in the Old Country and elsewhere.

On Midsummer Day 1843, a Wisconsin adventurer wrote back to Mineral Point, after taking a month to reach Copper Harbor; he had taken a boat at Milwaukee, sailed up to the Falls of St Mary's, and then spent fifteen days through contrary winds on the last leg of his journey along Lake Superior, days rendered further miserable by torrential rains. He remarked that many had arrived in the Copper Country from Boston and New York, and that:

Several companies are this morning trying to organise to cut a road or a path to the highlands. Small parties have tried to make hills (from here) but have failed in consequence of the white cedar swamps they had to encounter in the vicinity. I cannot form a comparison for the lake shore so apt as the idea I have formed of the swamps in Florida; for not only the lowlands but the highest mountains are so densely covered with all the different species of pine, laurel, etc., that a single man cannot penetrate the country without a hatchet to cut the limbs from the trees. The country has never been burned within the recollection of the oldest French or half-breeds. I do not hazard the truth when I say that the decayed fallen timber and undergrowth, such as laurel, moss, etc., lies upon the ground from six to ten feet in depth in many places. I have in many places gone to my Knees in moss and decomposed vegetable matter on dry ground ... A country of this character will require a great deal of labour and privation as well as capital, and at least two years to determine its character as a copper region.[17]

Even when the forest and undergrowth were cleared, the lodes still had to be discovered and traced, and then might prove to be of little if any commercial value. Indeed, the writer of this letter was more than sceptical of the worth of one eighteen-foot lode Houghton had traced from Keweenaw Point, although in the early eighteen-forties, it should be remembered, influential Wisconsins tried to keep their miners at home if possible and lavished warnings against ventures into unknown mineral districts; many a miner in the lead region of the Upper Mississippi was reminded of the old proverb that a bird in the hand was worth two—at least—in the bush.[18]

Others were less sceptical of Upper Michigan, and during his survey of the region in the summer of 1844 Dr C. T. Jackson reported that on

17. Letter of H. Messersmith to the Mineral Point *Free Press*, dated Copper Harbor, June 24 1843, reproduced in the *Madison City Express*, 27 July 1843.

18. Wisconsin, too, still resented the way in which the Upper Peninsula had been taken from its own original territorial limits and given to Michigan as a 'sop' for the Toledo strip, *Madison Express*, 29 May 1845.

the Eagle River location of the Lake Superior Mining Company, which had only started operations that season, there was

one vein of copper, eleven feet wide and one mile long, that will repay all the outlay of the company. The Cornish miners there have sunk four shafts on the banks of the river, intending to work the mines under the river. One shaft is already sixty feet, another forty, another thirty feet deep—all done by hand power. The deeper they go the richer the mineral is, and it contains about one-fourth silver . . . And in working the single exploration shaft at the Eagle River Mine, the metallic contents brought out by hand are worth $30,000! The rock is amygdyloided, and blasts very easily; it does not take more than twenty minutes to make a hole for a blast. There is no water in the shafts at all, although they have been worked down twenty-five feet below the bed of the river, and instead of the water of the river troubling them, when the dam is built it will be their greatest friend. The water will raise the ore from the mine, pound it, blow the blast for the smelting furnaces, and saw the wood for the buildings. The prevailing ores there are the black oxide of copper and the silicate of copper; there are no sulphurets of copper found in the whole region. One valuable large vein contains, in the clean ore, twenty-five per cent of copper besides silver; and the deeper they go the better it becomes.[19]

Jackson described what he saw, but it was only a 'bonanza fluke', and within two or three years the company failed. The daring ingenuity of the Cornish miners in working beneath the river might be emphasised, and also the appearance in a geologist's sober report of the perennial article of the Cousin Jack's mining faith—the deeper the lode goes the richer it becomes. Incidentally on this mining property one of the causes of failure was the practice that became notorious in the gold-mines of Colorado and Nevada over half a century later—high grading—for it seems that the working miners pocketed the richest specimens of silver ores.[20]

The Boston and Lake Superior Mining Association also brought several Cornish miners to the Upper Peninsula in 1846. Through the agency of the Falmouth merchants, G. C. and R. W. Fox and Company, this New England mining venture obtained the services of 'Captain' Henry Clemo[21] and nine miners, who sailed from Falmouth to Dublin,

19. Report of Dr C. T. Jackson to the Convention of American Geologists, *Madison Express*, 29 May 1845.
20. J. B. Martin, *Call It North Country*, p. 96; for high grading in Colorado *vide* Marshall Sprague, *Money Mountain*, pp. 203–5.
21. The name is spelt Clema in Foxes' bill to the Association, and Climes in letters of the Association, on which the following account is based; these letters and other papers of the Association are in the Michigan Historical Collections at Ann Arbor.

thence to Liverpool on a vessel owned by the Rathbone brothers, and there embarked on the ship *Mary Ann*, reaching Boston on Midsummer Eve. About the same time William Aspinwall, the Treasurer of the Company, appointed one Daniel Webster to be general manager of the mining location. Webster tried to recruit men to go to Lake Superior from the Maine lumber camps, finding it especially hard to get a really good cook; he then travelled to Detroit where he stayed for some days making arrangements about supplies and shipping out copper ores before going to the actual mining location. Three days after they reached Boston, Clemo and his Cornish miners left for the Copper Country, under the charge of another agent of the Company, George W. Peck. Aspinwall, two days after meeting the Cousin Jacks off the *Mary Ann*, gave Clemo a bearer's note for Webster, in which he twice said that the miners had arrived in Boston in good health, that he trusted they would arrive safely and that Clemo had impressed him with his 'gentleman-like manners and intelligence'.

Aspinwall reckoned that the Cornish mining captain was 'a person able and disposed to promote the interest of this association', and that he would co-operate with Webster 'cordially in all plans necessary to this object', but George Peck, with the task of seeing the party safe from Boston to the remote mining camp, found that task no sinecure. A staid, puritanical clerk, Peck found the boisterous youthful Cousin Jacks, loose from the cramped restraints of a long ocean voyage, a most difficult party to manage. The evening after their departure by train from Boston, with his temper by no means improved by the hot, sultry weather and by getting grit twice in the same eye, he peevishly wrote to Aspinwall from Bennet's Temperance Tavern in Buffalo:

We reached here tonight, and took lodgings in this miserable hole in order to be with the men and keep them from evil communications . . . Had to delay the train at Albany to pick up the men, and narrowly escaped a row at Rochester on account of that Ivey.[22] By the way last night brought another of similar appearance and type. In general they behave well. Climes appears to understand them and control them very easily. He is a good fellow, but a regular John Bull full of national prejudice and rather too much disposed to let it show itself. I mean to beat it into him that W(ebster) won't like this.

22. The name is hardly decipherable in Peck's letter; an alternative name is 'Joe', but the tone of his correspondence hardly suggests that Peck was likely to get on Christian name terms with working miners. As much of this letter is concerned with travelling and freight costs, in terms indicating cheeseparing economy, it seems likely that he lodged at Bennet's 'miserable hole' simply to keep expenses down.

A second, hasty note from Peck to Aspinwall, sent from Sault Ste Marie on 2 July, five days later, contained the laconic but suggestive comment 'it is necessary to watch the men constantly'.

Peck's letters show that in the early copper-mining days on the Lakes there was some friction between the immigrant Cornish miners and 'Yankee' managers and bosses. Left to run the mines in their own way the Cousin Jacks were happy enough; they were prepared to 'make shift' in some respects and to accept frontier-life disadvantages and short supplies, but they often cavilled at the inferior, cheap and inadequate materials which many mining companies provided. The Boston and Lake Superior Mining Association kept a strict watch on the dimes and cents which contrasted sharply with the rather feckless financial management of many mines in Cornwall. Last, but far from least cause of friction where the Cousin Jacks were concerned, was the American tendency to draft the weakest or most incompetent worker into serving as the cook of the mining camp. Within a fortnight of the arrival of their Cornish hands, Webster was writing to the Trustees of the Boston and Lake Superior Mining Association:

I was happy that you were so fortunate as to get a person who seems to be well competent to take charge of your mining operations as Captain Climes. —I regret that I was not better prepared for them when they arrived, but all had been done that I could do in the time, and with the number of men that I had.[23] I find the miners rather hard to satisfy. They say that you agreed to have their washing done for them and furnish them here. They find a great deal of fault with their living and the cooking and threaten to quit work once or twice if they did not live better. In order to keep them quiet I shall send to Detroit by Mr. Peck for a small bill of supplies of such articles of which I do not get a large supply, and I shall be obliged to employ another cook as the one I brought (from Boston) cannot cook to suit them and do their washing; although he tries to do his best. I do not think he can get along with the work. For supplies I bought pork, beef, fish, lard, butter, flour, bread, corn-meal, molasses, sugar, tea, coffee, rice, potatoes, dried apples, raisins, peas and beans, pepper, seasonage, etc., all of which are good articles.

Such a list of provisions seems adequate enough; a good Cornish cook could have furnished reasonable enough meals from them, even down or, rather, up to a 'vitty' Cornish pasty. Considering that practically all supplies had to be brought in that first mining season, save fish and game, Webster had acquired a good variety of the necessities of life. Time

23. Presumably clearing brush and forest, putting up some sort of living quarters, and so forth, before actually starting mining operations.

would have to pass before fresh vegetables could be raised in the vicinity; till that time Cornish palates would have to forego turnip and onion in whatever their unfortunate cook managed to perpetrate in the semblance of a pasty. The beef and pork, too, was salt not fresh, and it is likely that a New Englander brought in an undue proportion—in Cornish eyes—of pork, beans, molasses, and coffee.

Webster then went on to tell the Company, hardly enthusiastically but with fair candour, that he was

> well pleased with Captain Climes; he seems to be very much of a gentleman and a good companion. He thinks their living is not good enough and that I ought to get a better cook. He also complains about the mining tools a great deal, and I suppose he will recommend to you to go to considerable expense in furnishing machinery and other things which he has been accustomed to have in England. If you decide to grant all these requests your expenses will be much greater than you expected, for he has been used to having everything he wants, let the expenses be what it will. I am aware of the importance it is to keep the miners and shall do everything reasonable in my power to please them, and for the sake of the company I hope we shall succeed in keeping them.

Before long, however, Clemo and Webster had some differences of opinion. Clemo, claiming to be in sole charge of actual mining operations, dismissed a man Webster had employed to do some prospecting work. Webster was acting as business manager on the location, but Clemo, in all likelihood, looked on him as a mine purser who, in Cornwall, would only deal with financial accounts, leaving complete powers of 'hiring and firing' men to the 'Captain'. In America, however, the captain's position was merely that of a foreman in the Old Country. Then, within a month, the men insisted on an eight-hour day although they had originally contracted to work twelve; as there were so many mines starting up in the district, the Company had to give way or be left stranded without a labouring force. A little later Clemo demanded that the miners be provided with steel instead of iron drills, and put a Frenchman Webster had hired to cook, to work in the mine. As winter came on, the Cornish mine captain demanded that the Company provide a house near the second shaft as he thought a mile and a half was too great a distance for them to travel to and from work in midwinter.

Doubtless other mining concerns on the Lakes were to have similar troubles with the Cousin Jacks they hired. Certainly the winters of Upper Michigan presented serious problems and hardships to mine

July 7   On the Westren Seas 1848

The Day was very fine Thousands of People was on the warps to witness our departure Some, that I had taken sweet Counsel with for many years others that I had given Employment to but among the multude was my Poor Aged Father & Mother them that had Promised themselves we should be a comfort to them in theer old age Bathed in tears to see no more

At 6 Oclock in the Evening Tuesday the 4 July 1848 we was tow'd out by the Brilliant Steamer to the Bay of St Ives we could see the sand Banks Covered with our Friends waving their anchiefs &c untill we graguly diseppared. tho I join to sing hyms with our Party on board. it would not drown that great affection that was so deeply stampt with in my hart. it was as the Deviding the sinews from the marrow, thinking on those that with Pain & sorrow brought us to manhood to see Props. no more now in their Decline of Life. the Evening rom on a Calm sea sickness soon come on some very ill. all of our Family was very well Except two of our little Girls they was very bad. the next morning at 4 Oclock little Breese move on the water at 12 next day Cape Cornwall graguly Disepared one thing greatly Comforted me to know I had left some behind that would not forget to Pray for us. at noon fine Breese from S East we crowded all sail about 5 Knots P hour. on our true Cource W. N. West 10 at Night fresh wind 6½ knots P hour

July 6. at 4 this morning heavy swell Down Main Royal sickness soon increased Some almost Dying Better Elementation at 10 Dull and foggy weather b ̊ W N West noon Clear & fine o that men would Praise the Lord for his goodnes & for his wonderful works unto the Children of men we saw two ships one bound to England the other we soposed to be her that left Penzance the same day we left Hayle but we soon lost sight of her at 8 Blow fresh Down top sails 9 knots W N West Cource

July 7 at 4 this morning wind increased more to the West chainge our Cource S W ½ W with aterable sea our utensels was flying in all derections at 11 oclock

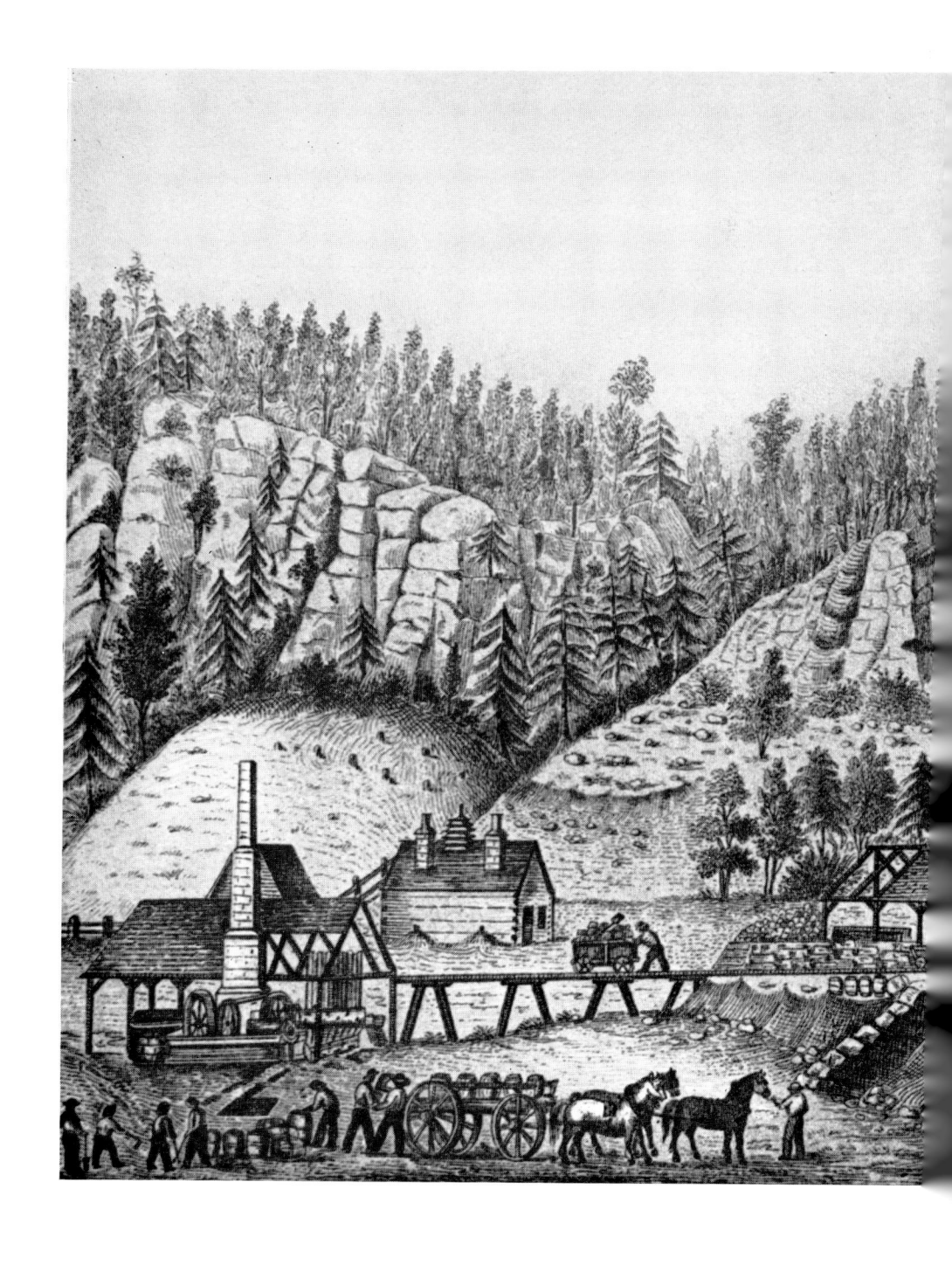

2 Cliff Mine, in the Keweenaw district of Upper Michigan

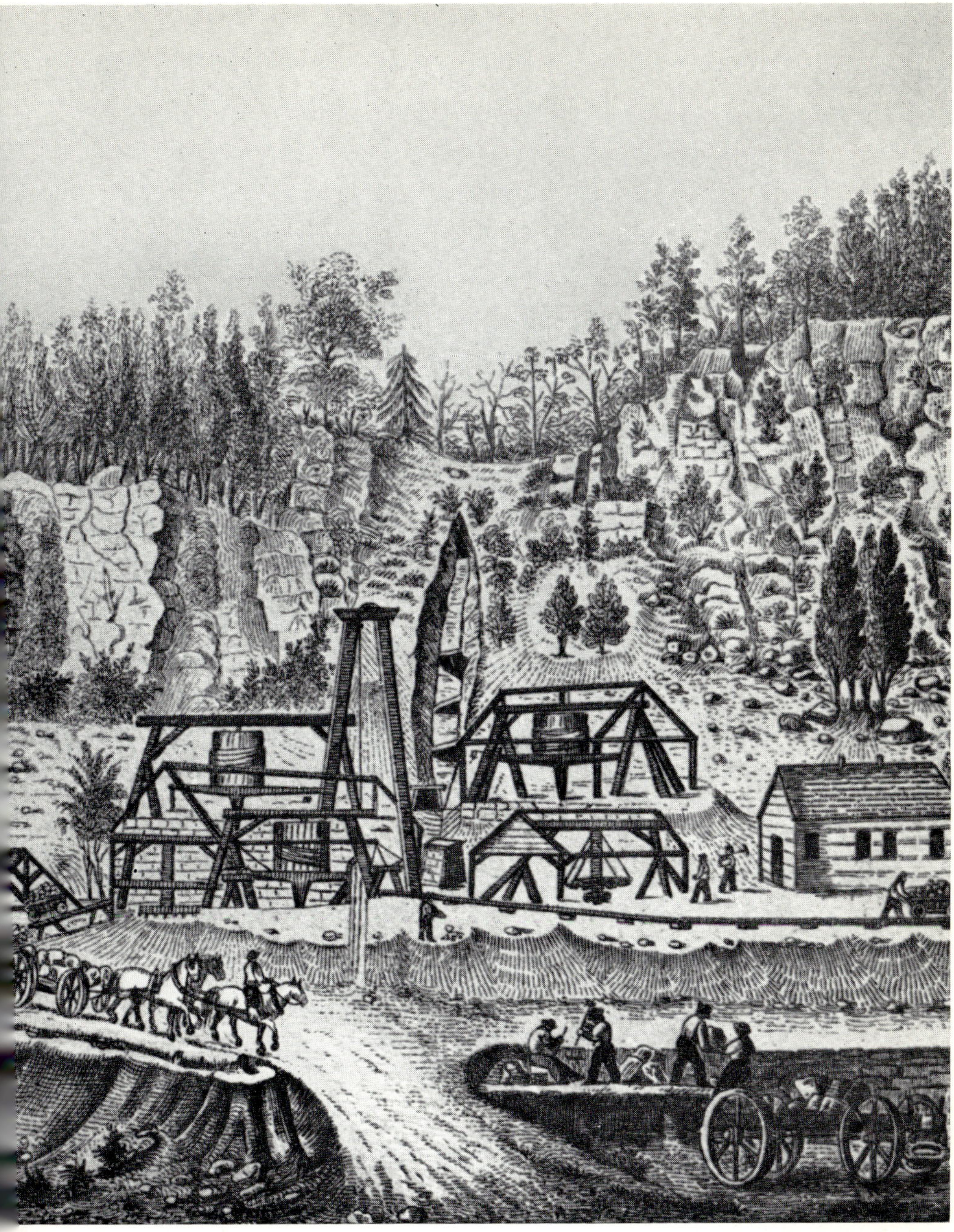

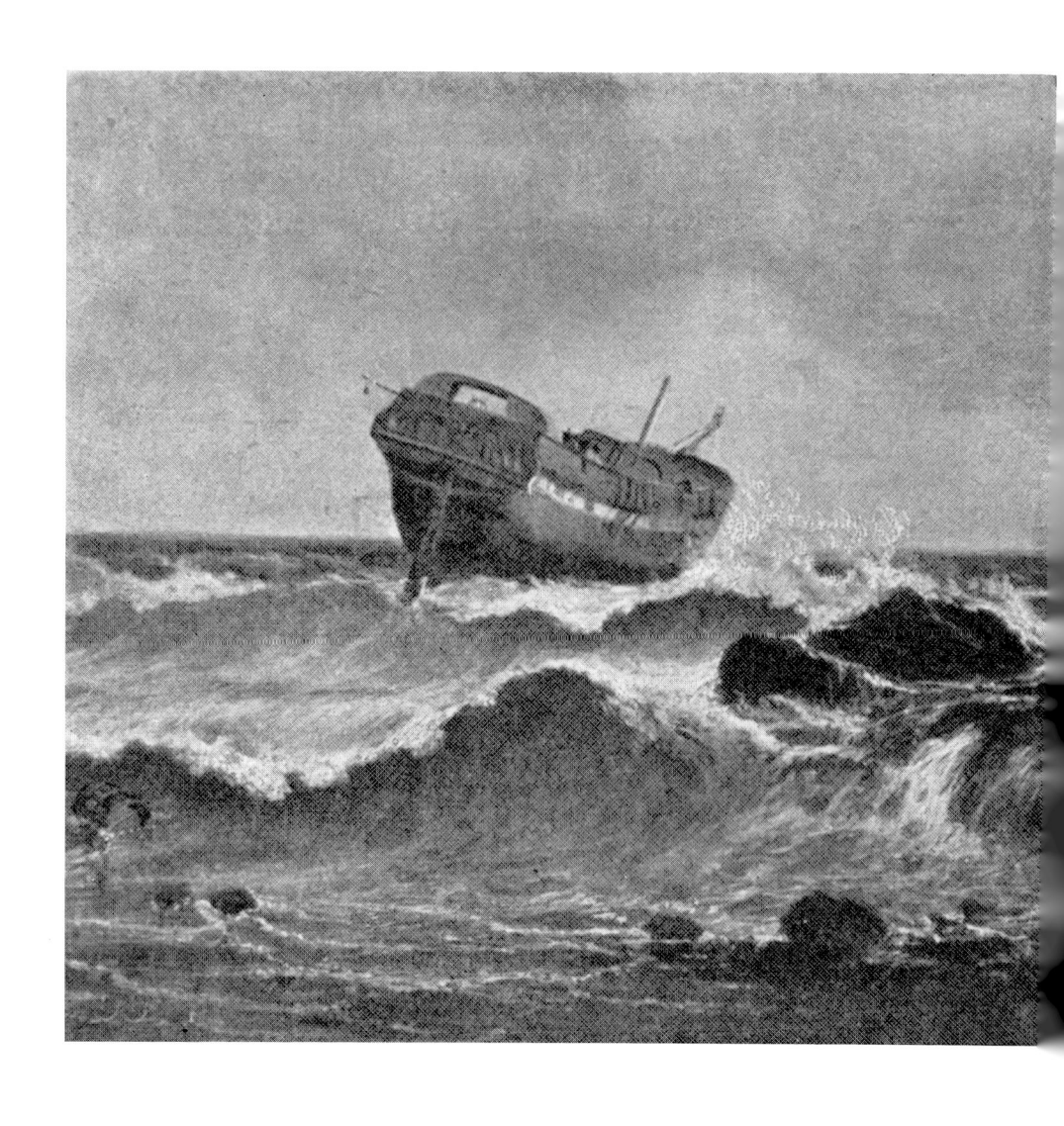

3 The wreck of the emigrant ship *John* on the Manacles, May 1855

managers and working miners alike. If hands were laid off and went away for the winter, there was no guarantee that they would come back next spring. If they were kept on, the companies had to arrange for adequate supplies of provisions to be brought in to last four or five months, and orders for these had to reach Detroit by early September to be sure of their being delivered to the mining camps before snow and ice cut off the remoter shores of Lake Superior from contact with the outside world. Webster had thought of getting a horse for incidental draught purposes on the Boston and Lake Superior's location, but decided that with little hay being to hand, the cost of keeping the animal would be far more than the value of the work it could do; in any case hand-sleds could be used to haul in all the lumber they needed for fuel. In time a far from regular winter mail service by dog-sleds and human carriers was organised to keep contact with the 'outside', but no bulky or heavy goods could be brought in or out from December through to April.

For a variety of reasons many early mining companies on Lake Superior failed. Some had inadequate capital or credit to meet the costs of bringing in supplies from Detroit and the East, and the very number of companies promoted in the first 'copper boom' days, of course, sent up the prices of such necessary provisions and equipment considerably. The physical hardships on the pioneer mining frontier were enough to convince some men that a single winter at Keweenaw or Ontonagon was more than enough for them. Over-sanguine early hopes were not realised, and many disappointed immigrants lost heart and departed. There was, too, not a little mismanagement by relations and friends put in to run mining ventures by influential individuals connected with the companies regardless of the fact of their utter lack of practical experience in mining.

Still, a start had been made. The first excitement subsided, but it had brought many immigrants into the Upper Peninsula, including several Cousin Jacks who 'knew copper' and who must have found new and promising locations in a district which stretched more than a hundred miles south-west from Keweenaw Point to the Ontonagon River. Prospects had been opened up on many points along the trappean Keweenaw Range. Only a ton or two of rough copper was shipped from Lake Superior in the mining season of 1845, and barely thirty tons the following year, which seemed meagre indeed compared to all the fury of speculation in Michigan copper that had raged in the East. How irrational that speculation had been may be judged from a report that one August day in 1845 the agent of a single company, by none too scrupulous methods, sold $12,000 'worth' of stock in a Michigan mine which was four times the value of copper shipped out by all the mines operating in the Upper Peninsula for the two seasons of 1845 and 1846.[24] Over the next decade copper production rose sharply at first, then fell away with miners taking themselves off to California in 1849 and with copper prices dropping, but again recovered with the completion of the Sault Ste Marie canal in 1855. The Civil War checked production again, and it was only in 1868 that Michigan surpassed the copper production of south-west England which had been falling since 1855 and slumped drastically in consequence of the financial panic of 1866.[25]

Through improved technological skills, too, employment in Cornish copper mines had been dropping in the years when Lake Superior ores began to exert influence on world copper markets. Furthermore, the discovery of the Marquette iron range in 1844 afforded miners in the Upper Peninsula far more opportunities of employment, if anything went amiss with copper, than either the long-established tin-mining or the infant china-clay industries of Cornwall. Early in 1850 too, government geologists had surveyed the riches of the more western Menominee iron range, but the extent and richness of the Marquette Range delayed its being developed before the early eighteen-eighties. In brief, the Old Country was an old country of declining resources, whereas Upper Michigan was a new country of seemingly infinite prospects in the way of mineral wealth.

As in Cornwall, so in Michigan, two or three highly productive and profitable mines tended to dominate the copper-mining industry, with half a dozen or so moderately successful 'second-class' mines lagging far behind though, perhaps, more often than not, returning reasonable dividends to their shareholders, and several smaller ventures, mostly of a very speculative nature, encouraged into activity but not into remunerative production by the very success of the outstandingly productive concerns. In the period before the Civil War the great Michigan copper mines were Keweenaw's Cliff and the Minesota Mine in the Ontonagon district; later Calumet and Hecla gained primacy. About two-thirds of the copper raised in Ontonagon in the three years from 1856 to 1858, came from the Minesota, and at the same time the Cliff was producing over seventy per cent of the copper mined in the Keweenaw Point district. The secondary Ontonagon producers were the Rock-

24. *Madison Express*, 25 September 1845.   25. See Appendices, pp. 297-9.

land, which immediately neighboured the Minesota, the National and the Adventure, accounting for roughly twenty per cent of the total output of the district, while about a dozen other small mines raised the remainder of the copper mined there. In Keweenaw, Copper Falls, Central and North West together aggregated some twenty per cent of the total output, the remaining eight or nine per cent being the contribution of ten or more small mines.[26] Around Portage Lake there was less disparity between the amounts raised by the three main mines, the Pewabic, Quincy, and Franklin, but this district was of considerably less importance before the Civil War, and only boomed with the discovery of the Calumet conglomerate lode some years later. In the Old Country at this time the great mine was Devon Great Consols, just east of the Tamar border, which was producing an eighth of the ores sold at the Cornish 'ticketings', a larger quantity than any two of its nearest rivals; long before the collapse of 1866 there were, in south-west England, only ten first-class copper-mines, about twenty second-class concerns and fifty or more speculative ventures with but feeble chances of survival in the event of a major economic recession. Both before and after the Civil War, Upper Michigan and Cornwall had each an inordinately large proportion of mining ventures that, hopefully and over-hopefully promoted, could only be regarded as gambling speculations; 'knacked bals' were to become as much a feature of the landscape of Keweenaw and Ontonagon as they were that of Cornwall.

Having the experience of the Old Country to draw upon was not altogether advantageous to the pioneering mining ventures of Michigan. Cornish mining captains, like Henry Clemo, tended to be 'set in their ways' and pigheadedly opinionated. Such men often induced the companies that employed them to install unnecessary costly machinery— all the more costly since it had to be bought from eastern manufacturing establishments like that which the Cornish Vivians had established in Pittsburgh. In many instances the resources of a particular location or mining claim were inadequate to meet the cost of such plant. Upper Peninsula mining capitalists learned through the hard way of bankruptcy that only by amalgamation and consolidation of claims could they install and economically operate up-to-date machinery. The Minesota mine raising over nineteen hundred tons of copper in 1858, and even the Rockland with an output of two hundred and thirty tons that season, could afford costly plant, but that could not be said of the Norwich,

26. Based on returns given in the *Mining Journal* (London), 9 January 1858 and 12 February 1859.

producing forty-one tons, and the Toltec, raising a meagre thirty-one tons, even if their prospects of much greater returns in the comparatively near future were good.[27]

The 'practical' miners, among whom the Cousin Jacks were prominent, often attributed the difficulties in which many of the companies became involved to the managers and agents appointed by eastern 'Yankee' boards of directors. A Cornish mining captain in 1859 claimed that the Adventure mine would have paid handsome dividends to its shareholders 'had it not been for the late agent, who through neglect of duty or incompetence, got the mine involved', while the directors of the Nebraska Mine Company were criticised for listening, too often, 'to outsiders, which has caused a change in their agency almost every year, and as a universal thing, every agent in his opinion has an improved plan or system or work'.[28] Even the Minesota management came under castigation for not adopting the Cornish tribute contract system of working instead of persisting in the false economy of paying low daily or monthly wages—'an objectionable system which carries out laziness to perfection', and which could certainly be blamed for a strike late in 1859.[29] It was the famous Cliff Mine, in the Keweenaw district, that afforded the most glaring instance of mismanagement in this Cousin Jack's opinion, for about 1853 or 1854

a new manager was appointed who knew little or nothing of mining. The mine, then about eighty or ninety fathoms deep, was considered by him to be deep enough, consequently he confined his attention to drifting and stopping, the latter more particularly, which resulted for a year or two in an increased product, finally the mined amount of product failed, and, indeed, they failed to produce the original quantity under the former management. Ere long they ceased to declare dividends, and it was thought by many capitalists that the mine had failed, and it was no longer the Cliff Mine. The mine being

27. The Rockland is quoted as a second-class, the Norwich and Toltec as third-class Ontonagon mines. The actual returns of the eight main producing mines of 1858 in Ontonagon and their 1861 output for comparative purposes were:

	1858	1861		1858	1861
Minesota	1912	1880	Ridge	69	31
Rockland	230	409	Norwich	41	–
National	190	934	Ogiabic	36	9
Adventure	87	3	Toltec	31	2

Two other Ontonagon mines, however, made fair progress between 1858 and 1861. The output of Evergreen Bluffs in the former year was 5 tons, and in the latter, 62; the output of the Superior Mine went up from 1720 pounds to 39 tons. *Mining Journal* (London), 12 February 1859 and 18 January 1862.

28. *Mining Journal* (London), 1 October 1859.

29. Ibid., 25 February and 8 September 1860.

inspected by competent men in behalf of the Company, the error was soon disclosed; they worked out all the reserved ground for stopping, and had not opened any new by sinking. Sinking being at once ordered and followed up, it was now again the Cliff Mine, and regular dividends are fully expected. Too often it is the case that the difference between a sound practical man and one of no practice is lightly admitted.[30]

Successful or not, badly or well managed, the early Upper Michigan copper mines brought a host of Cousin Jacks to the Lake Superior region. They contributed not only to the economic but the social development of this remote frontier. In the summer of 1848 the Rev. J. H. Pitezel came into the Upper Peninsula as a missionary, and met several Cornish miners, including Jennings the underground captain of the Cliff Mine; he found that such Cousin Jack mining terms were current as *wim* or *whim* for crude hauling machinery, *deads* for broken rock waste, *country* for the rock surrounding the ore veins, and *sumpen* to describe the gradual draining and seeping of water into the bottom of a shaft.[31] At the Trap Rock mine in Ontonagon the manager, Buzzo, was a Cornish local preacher who probably helped to bring nearly sixty men to hear the itinerant missionary preach when he revisited the Copper Country in 1852.[32] Several years later, in 1866, Barton S. Taylor, a Methodist minister, came to Houghton, and wrote:

Cornwall has emptied her army of miners in among these hills. The Cornish constitute very largely the religious element of this country. They are devout, honest, old-fashioned Methodists, much attached to their particular modes. They make much of the class meeting, and know how to use it as a means for the advancement of Christ's kingdom. I have known conversion to be occurring weekly for six months at a time through the class meeting. The members would during the weeks talk persistently with the men they worked with, and invite them to the class meeting. If an unconverted man ventured into the meeting, they had no thought of letting him go till he was converted—converted so that everyone knew it.[33]

30. *Mining Journal* (London), 14 December 1861. Signed 'A Cornish Captain' these reports on Lake Superior mines in the London *Mining Journal* were written by Captain Henry Buzzo, who became manager of the Ridge Mine in 1850, and later was agent for the Toltec and other mines. *History of the Upper Peninsula of Michigan* (Chicago, 1892), p. 539.
31. J. H. Pitezel, *Lights and Shades of Missionary Life*, pp. 168 ff.
32. Ibid., p. 338. The Buzzo referred to was probably a brother or cousin of the *Mining Journal* correspondent; Pitezel calls him the 'Rev. J. Buzzo'.
33. Taylor MSS, Michigan Historical Collections, Ann Arbor. The quotation is from a draft of an article Taylor wrote on the Upper Peninsula; he was stationed at Houghton from 1866 to 1869, and again in 1880–81.

Elizabeth Gurney Taylor, the minister's wife, recorded in her journal[34] that on one occasion, at least, thirty or forty Cornishmen remained after a Sunday evening service, early in November, for a prayer meeting, and that it was 'a regular Cornish meeting, singing and everything. Their prayers were very earnest.' Brought up in the Baptist denomination, Elizabeth, who was Taylor's second wife, several years his junior and not always able to control the four teenage children of his first marriage, did not always find the Cornish very congenial company. She was over-conscientious about the duties of a minister's wife; her religion had a morbid cast and, far advanced in her first pregnancy, she wrote in her journal that she was 'hungering for some neighbour or friend who can talk of Christ and Heaven as well as of meat and drink, neighbours and dress, and leaving God and Heaven entirely out or last on the programme'[35]—a laconic statement revealing that Cousin Jennies gossiped over 'dishes of tea' on much the same themes that they did in the Old Country.

Cornish clannishness led to friction and ill-feeling in Upper Michigan which had, even before the Scandinavian immigration set in during the Civil War, attracted many races—Irish, French-Canadians, and Germans in particular. The Irish and Cornish got on as badly together in Michigan as they did later in Butte, and years afterwards tales were told of an Irish mob in Ontonagon threatening to lynch the Cornish saloon-keeper, Hocking, who shot Pat Nolan, an Irish rival in that line of business in February 1856. A year later there was a riot round the Rockland mine; after one Cousin Jack had been killed, his compatriots drove the Irish out of the camp, but retreated to the security of a vessel on the lake when rumours came that the Irish from Portage Lake were coming west in full force to drive the Cornish out of Ontonagon County; the Irish march, however, degenerated into a saloon-crawl which petered out before they had advanced a league let alone fifty miles through woods in which saloons were too few and far between for the liking of the hard-drinking Hibernian labourers and miners.[36]

Cornish–Irish affrays, however, were mostly sporadic drunken fights which were liable to occur in any frontier mining camp. Men of all races and nationalities, not even excepting the sober New Englanders, at one time or another were involved in brawls and even in what a

34. Taylor MSS: Journal of Elizabeth Gurney Taylor, 11 November 1866.
35. Ibid., 29 September 1867.
36. Orrin W. Robinson, *Early Days of the Lake Superior Copper Country*, typescript in the Michigan Historical Collections, Ann Arbor. J. B. Martin, *Call It North Country*, pp. 87–88.

more western frontier called 'shooting scrapes' up in the backwoods of the Upper Peninsula. More serious criticism of the Cornish immigrants, however, came from a German writer, Johann Georg Kohl, who published an account of his travels in the 'North-western' United States in 1857. After briefly stating that the Cornish, being the majority of the 'English' in the Copper Country, had not only in the main set the pattern of mining operations but were superintending most of the mines, and that, furthermore, much of the machinery employed and mining terms used were Cornish in origin, Kohl quoted the opinion of a 'much travelled mine superintendent'. This suggested that some Michigan mining concerns would have welcomed a more tractable, less independently minded—that is, German—labour supply. Kohl wrote that this superintendent had told him that:

The Cornish had had more or less experience in mining, and also that they were in general very clever, handy workmen. Yet their mining experience and knowledge were always very limited, and they insisted on holding fast to their old methods brought from abroad. Moreover, they were ordinarily completely uneducated, for instance, they could for the most part neither read or write, which often proved a serious obstacle to their advancement.—In undertakings as new as these (mines), where absolutely everything had to be built up from the beginning, people had to take over all sorts of work and be adaptable. But if you gave such a Cornish miner a team of horses to drive, he did not know what to do, either with his hands or his feet. He did not want to learn anything more than he had been trained to do at home.[37]

Doubtless several Cornishmen were fast 'set in their ways', but by no means all of them, and the real trouble was that clannish Cornish mining captains favoured their compatriots and were prejudiced against Germans and other nationalities.[38]

The charges of illiteracy and conservatism levelled against the Cousin Jacks, however, demand consideration. Universal compulsory education was only introduced in England in 1870; for twenty years or more after that an iniquitously laxly administered system of school-leaving certificates meant that many children left school before they were twelve years old, while for those of even more 'tender years' compulsory attendance requirements were far from strictly enforced by local school boards. Such lack of schooling was to provide Upper Peninsula raconteurs with a rich store of dialect stories, dialect 'plods' in which standard English was rendered nigh as strange as Chippewa or Ojibway, yarns

37. J. G. Kohl, *Reisen Im Nordwesten Der Vereinigten Staaten*, p. 390. I am indebted to the late Mrs C. Paul for this translation.
38. Ibid., p. 397.

coloured with a rich streak of profanity more often than not. One such tale, preserved among the papers of Orrin W. Robinson relates that a young Cornish miner Tretheway was called to give evidence in an Ontonagon court about a shooting affray that had taken place outside the saloon kept by the old pioneer, Jim Paull. Tretheway got confused in the cross-examination, and first said that Paull, then that Mrs Paull, and finally that both the Paulls had fired shots. The judge, losing patience, told him to tell the jury 'in plain English' who did shoot. Tretheway, in a high state of excitement, turned towards the jury and exclaimed—'I! I seedin' all right. He shooten'—she shooten'—both shooten'—shooten' all a scatter, scatter all abroad—taurin' all to rags—shooten' sure enough, damme, Judge!' It was hardly surprising that with such a principal witness the jury disagreed, six for and six against the defendant, and that the judge dismissed the case.[39] Richard M. Dorson tells another Upper Peninsula 'plod' about a Cornishman consulting his friend about the meaning of the word 'category' which he had come across in a newspaper and could not understand. The friend, who had been attending a night school to 'improve himself', said that his teacher had laid it down that the right way to get at the meaning of a strange word was to 'analyse' it, and therewith proceeded to do so—'cat' was plainly 'cat', 'e' was 'e', so that meant ''e-cat', and 'gory' was 'bloody', so 'category' meant 'bloody tom-cat', which was good, plain, if somewhat forcible, Cousin Jack English.[40] It might be stressed, however, that the analytical Cousin Jack was making a determined effort in his spare time to make up his defective education, even though his efforts made him a butt for those who had enjoyed far better schooling opportunities than he had back in the Old Country.

Many Cornishmen who attained fair prominence in the Upper Peninsula had had very little formal education in Cornwall, and can only have reached the positions they did through personal grit, initiative, and shrewd native wit. Josiah Hall, who served as captain in Cliff, Pewabic, Central, and Calumet and Hecla mines, started his mining career in 1838 when only twelve years old; although born in Devon, he seems to have been of Cornish parentage.[41] Captain W. W.

39. Robinson Papers, Michigan Historical Collections. Robinson said this incident occurred in the early eighteen-sixties; the journal of W. W. Spalding (Mich. Hist. Colls.) indicates that the shooting affray took place on 26 May 1848.
40. R. M. Dorson, *Bloodstoppers and Bearwalkers*, p. 120.
41. *Memorial Record of the Northern Peninsula of Michigan* (Chicago, 1895), p. 342. He first came to the United States in 1851, working in Pennsylvania for a time before coming to Keweenaw.

Stephens of the New Port mine, Ironwood, was born in Cornwall in 1841, and started work in the tin-mines of his native county when only ten years of age, walking five and a half miles to and from work, but he managed to acquire some rudiments of education by trudging another three miles to a night school after working in the bal all day. He emigrated to Michigan from the Old Country in 1865, although he only came to the Ironwood mine after spells of working as a miner or as an engineer in Virginia City, New York, Baltimore, and California.[42] Another Ironwood mining captain towards the end of the century was John Tregumbo of the Pabst Mine, who started working in Cornish tin-mines in 1853, when only eight years old, and emigrated to Michigan in 1869.[43] Richard Hoar of St Austell started work as a merchant's clerk at the age of twelve in 1843, and came out to start his own business in Houghton in 1860; thirteen years later he was to be elected to the State Legislature.[44] Instances like these could be cited in considerable numbers, but although most of the Cousin Jack immigrants to the Upper Peninsula had had little in the way of schooling in Cornwall, there were some notable exceptions. There was John Vivian, in later life one of the chief clerical officials of the Osceola Mine, who came to the Copper Country in his late twenties, after spending some time copper mining in Connecticut and a period as ship's steward besides being employed for a fair time by Pennsylvanian and New York engineering firms; he had then worked underground in the Cliff Mine for three years before he met with an accident which led him to turn to school-teaching. Then he had become County Clerk for Keweenaw County in 1860, and a year later was elected Justice of the Peace, holding that office well over thirty years. In his boyhood Vivian, the son of a fairly prominent mining captain, had had the benefit of education in a private school.[45] Captain John U. Curnow, of the East Vulcan Iron Mine, had gone to school till he was sixteen, and had then worked for six years in Wheal Providence before coming to Michigan in 1865.[46]

Indeed the charge of illiteracy levelled against the immigrant Cornish folk by none too friendly critics and observers who had, perhaps, run foul of their prickly clannishness for some reason or other, was by no means as important as might at first appear. Many of them had qualities of perseverance and shrewdness that more than compensated

42. *Memorial Record of the Northern Peninsula of Michigan*, pp. 387–89.
43. Ibid., p. 456.
44. *Biographical Record of Houghton, Baraga, and Marquette Counties*, pp. 219–20.
45. *Memorial Record of the Northern Peninsula of Michigan*, pp. 429–30.
46. Ibid., p. 427.

for their inability to read or write at all or with difficulty; others had had a fair education which some of them had the wisdom to turn to good account. In any case the proportion of the totally illiterate seems to have been small, judging from such census returns that did make note of that fact. Even in the States, too, the children of Cornish immigrants—and of others as well—started work at a very early age; William Walls, born in 1846, and brought to America when only about four years old, was working in Pennsylvanian iron mines when he was only twelve years of age, and in later life became a prominent Upper Peninsula lumber merchant;[47] Edward T. Abrams started work in 1873 as a blacksmith, at the age of thirteen, but that did not stop him from qualifying in medicine at Detroit College some years later.[48]

In many ways conservatism, being too wedded to their accustomed working methods, might be regarded as a more serious defect of the Cornish than lack of school education—had the charge been justified. The copper miners from Cornwall had had no previous experience of masses of native copper which were the main source of the red metal in the Upper Peninsula in the early mining days of that region. Special rock saws and chisels of the toughest available metal had to be devised to deal with the intractable material,[49] and Cousin Jacks used to hand-drilling and blasting with black powder, or even to cracking off portions of metal-bearing ore by burning brushwood against the rock face, found such tools gallingly tedious to use. In the Minesota Mine, in the eighteen-fifties, the cautious German Kohl was struck, if not positively scared, by the reckless fashion in which the Cornish used explosives in their attempts to break up the copper masses, writing:

Their drill holes are very deep and wide and they put into one blast as much as two, three, or even four kegs of (black) powder, each keg being about twenty-five pounds. Indeed, sometimes they shoot off over a hundred pounds at once. The rock flies out like spray from below the walls of solid copper, yet these masses are affected very little, only now and then . . . they may be lifted somewhat from their original position.

Sheets of copper which are more than one and a half inches thick are not riven. Small thin pieces I have seen displaced (by blasting) but this clean unsmelted native copper has an unusual ductility . . . They showed me a big piece weighing about three tons that swung from a larger mass above it by a ribbon of metal no thicker than one's finger. They left it hanging there with

47. *Biographical Record of Houghton, Baraga, and Marquette Counties*, pp. 383-4.
48. Ibid., 283-4.
49. J. G. Kohl, *Reisen Im Nordwesten Der Vereinigten Staaten*, p. 394.

no fear that it would pull loose of its own weight, until they were ready to cut it with the use of tools.[50]

The underground captain of the Minesota at the time of Kohl's visit was William Harris; however much such tough masses of copper dismayed other men, they had no terrors for this Cousin Jack. When, in 1857, his men came up against a five-hundred-ton mass of native copper he put forty hands to work on it; it took nearly twenty months to cut up, and in the process the 'waste' chippings carefully saved made an aggregate weight of sixteen tons.[51] In later times pneumatic drills and chisels dealt far more speedily with the masses of native copper, but in these earlier times Cousin Jacks, 'set in their ways' showed that if they found copper in a mine they would do all in their power to 'bring it to grass'. Even if their methods smacked overmuch of 'brute force and ignorance' in the eyes of their critics—they won through, which was more than could be claimed for many so-called labour-saving inventions conceived by academic scientists and technologists whose devices had been tried, failed, and fortunately, forgotten, in Cornwall, Michigan, and other mining camps of both the Old World and the New.

Where the 'rule of thumb' Cousin Jacks erred, however, was in overhasty condemnation of rock formations and mineral specimens bearing superficial resemblances to those they knew at home. When Hulbert first showed a group of Cornish miners a few pieces of Calumet conglomerate, they told him that it was 'braave and 'andsum stuff' but that there was not enough of it in the whole wide world to keep a man in tobacco[52]—and a heavy smoker in those days would have paid only a few dollars a year on the weed. Within half a century of that remark being made, Calumet and Hecla had returned close on a hundred million dollars in dividends to its fortunate shareholders. 'Puddingstone' might have been very rare in the Old Country, but it had certainly proved plentiful in the Upper Michigan and had paid handsomely.

Ultra-protagonist of Old Country 'rule of thumb' methods was Captain John Daniell, twenty-six years of age when he turned up in the Copper Country in 1865 after two years in the Californian gold-fields. He soon gained a reputation as a man who 'knew copper' and the

50. J. G. Kohl, op. cit., p. 408
51. *History of the Upper Peninsula of Michigan* (Chicago, 1892) p. 314. Harris, who was born in Cornwall in 1818, came to Ontonagon in 1850, and was associated with the Minesota as captain for fourteen and as agent for a further eight years; he then ran the Allouez Mine for three years. He served in the Michigan Legislature for two terms, from 1871 to 1875.
52. Angus Murdoch, *Boom Copper*, p. 128.

vagaries of copper lodes, as well as of being profanely critical of 'damned geologists' and others of that ilk who were unlikely to run the risk of any more serious occupational disease than writer's cramp and who took every care to avoid galling their hands with pick and shovel. Working in Calumet and Hecla, Captain John saw with his own eyes the angle at which the famous conglomerate lode dipped, and reckoned that beyond the limits of the Company's claims the lode would be found at the depth of 2,260 feet. The geologists, however, calculated that the depth could not be less than 3,500 feet, and, since their calculated reserves were so great, hardly reckoned that that was worth talking about, especially since many of them thought the lodes would 'pinch out' or, at least, deteriorate greatly in quality before reaching such depths. Daniell, however, shared the conviction of the majority of Cornish mining men that the deeper a lode went the richer it became. He left Calumet and Hecla, got a group of Boston capitalists to back his hunch to the extent of a million-and-a-half dollars, and formed the Tamarack Mining Company. Daniell then hired miners to work three shifts night and day for three and a half years, sinking a shaft which struck the lode at 2,270 feet—ten feet deeper than he thought, or a percentage error of only ·44 in his calculations. Not only were contemporaries amazed at Daniell's success in finding the lode at that depth, but he had driven his men at such a speed that they averaged sixty feet a month; in the Old Country the most outstanding mining captain of the day, Josiah Thomas of Dolcoath, reckoned he did well to get shafts sunk at the rate of five or six fathoms, thirty to thirty-six feet, a month; when sinking a second shaft on the Tamarack claim in 1889, Daniell's men managed to attain the record rate of sinking a hundred-and-five feet in a month.[53]

Daniell himself was asked from the Old Country how the Americans managed to sink shafts so much faster. His answer was both a rebuttal and a confirmation of the charge of conservatism.[54] It was the Cousin Jacks—or, rather, their employers—in Cornwall, who clung to obsolete methods. In Upper Michigan he saw that his men used the most up-to-date machinery to drive a shaft of three compartments, twenty feet by seven feet inside timber. While Cornwall mining companies were still using hand-drills, practically every mine on Lake Superior was using drilling machines, and this at a time when it was admitted that practically every underground mining captain, or 'boss miner' in

53. *W.B.*, 5 December 1889.
54. Letter of John Daniell, dated 2 January 1890; *W.B.*, 30 January 1890.

Michigan's copper region was a Cousin Jack.[55] On the Tamarack, too, Daniell went on to project and sink three more shafts after first striking the lode, and confidently expected to mine copper a mile deep.[56] The Calumet men who had scoffed at the 'crazy Cornishman' a few years later were planning to sink their own shafts down to five thousand feet levels.[57] A few years later still it cost Calumet and Hecla three-and-a-half million dollars to buy up the Tamarack, after it had paid the men who had backed Daniell over twelve million dollars in dividends, or eight times their original investment.[58]

The first Tamarack shaft was completed at midsummer, 1885, nearly thirty years after the critical Kohl had levied his charges against the Cousin Jacks. Yet there is no reason to believe that the Cornish-American mining captains of John Daniell's day were any more progressive than the first Cornish immigrants who had come to the Upper Peninsula in the eighteen-forties and fifties. The first Cousin Jacks in Michigan, were men who ventured out to a remote mining frontier; no man 'set in his ways' was at all likely to have done that, no matter how alluring the prospect of comfort, wealth, or fortune had been; and they continued to come long after accounts reached the Old Country about conditions of life in the Upper Peninsula that were grim enough to deter all but the stoutest-hearted and most adventurous of men. Furthermore, a glance at Cornwall itself in the mid-fifties proved that it had a breed of mining men who had nearly doubled the output of its copper-mines during the past twenty or thirty years, and who had made ores yielding barely six-and-a-half per cent produce pay as against the eight per cent or richer ores of a generation earlier.

In brief, although some Cousin Jacks were reluctant to use new methods and tools, yet from the first in the Upper Peninsula they adapted themselves to new and strange circumstances and made the most of any advantages nature offered. Michigan mines were generally much drier and easier to drain than those of Cornwall, while an apparently limitless supply of lumber for fuel and timbering in the early days was right to hand on the mine locations. Even the rather querulous German admitted that the Michigan mines were as well ventilated and that hoisting machinery was as good as any to be found in the most progressive German mining districts.[59] Michigan, however, lagged behind in one respect; Michael Loam's 'man-engine' was working in

55. T. A. Rickard, *History of American Mining*, pp. 246–48.
56. *W.B.*, 30 April 1893 and 3 September 1894.   57. *W.B.*, 7 March 1895.
58. A. Murdoch, *Boom Copper*, p. 44.   59. Kohl, op. cit., p. 409.

Tresavean Mine as early as 1842,[60] but twenty years passed before a similar device, 'invented' by a man called Rawlings, who may well have been a Cousin Jack, was carrying men up and down the main shaft of Michigan's Cliff Mine.[61]

Cornish immigrants, whether educated or not and whether progressive or 'set in their ways', contributed much to the development of Upper Michigan in the decade and half that preceded the outbreak of the Civil War. The opening of the Sault Ste Marie Canal enabled them to speed the exploitation of the Copper Range so that, within a few years of the end of hostilities, it surpassed the Old Country's production of the red metal. Cornish miners, too, played an important part in opening up the iron ranges of Upper Michigan, first in the Marquette Range in the late eighteen-forties and through the fifties and sixties, then, from 1870, in the Menominee, and, later still, in the more western Gogebic Range and on the Mesabi Range over the Minnesota state line.

Local demands of miners for food and fuel, however, soon diverted many early settlers to agricultural and lumbering pursuits. Poorish soils and hard, long winters limited farming in Upper Michigan to raising potatoes for human food, breeding both horses and oxen for draught purposes, and producing some milk and dairy produce. Twenty years after mining developments began in the Keweenaw Peninsula, there were barely seven thousand acres of 'improved' land, representing less than a fortieth of the taxable lands of the three counties of Keweenaw, Houghton and Ontonagon. Nevertheless, every bushel of potatoes grown—and over sixty thousand bushels were harvested in the Copper Country in 1864—helped to ease the problems and costs of bringing supplies from outside and to reduce the risks of 'short commons', possibly of near famine, in the early months of the year when the lake shores were locked in the grip of ice. Farming in Upper Michigan, however, was a hard life, although it is possible that one or two Cornish agriculturalists, little versed in geographical knowledge, had been lured thither by the letter sent home by Richard Gurney from Grand Rapids in the Lower Peninsula in May 1849. At a time of agricultural depression, when poor crops, low prices, high rents, heavy taxes, and burdensome tithes had reduced many Cornish tenant farmers and small holders to despair, Gurney had written:

60. *W.B.*, 4 November 1842.
61. Newspaper clipping preserved in the Brockway Papers, Michigan Historical Collections, Ann Arbor, dated 2 January 1908 (probably from the *Daily Mining Gazette*), giving an abstract of a paper read by W. J. Uren on 'Early Life in the Copper Country'.

We have arrived in time to plant potatoes and every other kind of vegetable, which grow in profusion here, so that we shall be partly provided for. The country has not yet put on that richness of verdure in which I saw it clad last summer, but everything is beginning to bud. The ground is carpeted with violets, lupins, and strawberry blossom, which, contrasted with the delicate green of the oak, pine, beech, maple, and a variety of other trees, has a very beautiful appearance. Wild gooseberries and currants in profusion, groups of plum and cherry trees and hazels had such an extraordinary effect on our people that they can scarcely imagine that nature has done so much. There is not a doubt that I shall soon get a competency, and should thousands, nay tens of thousands of my countrymen come here, they would soon be in comparative affluence.[62]

Lumbering in the early days was limited to supplying local demands, and a dozen sawmills, which, in 1863, employed a hundred-and-seventeen hands, provided most of the timber the Copper Country required. The mines, besides employing carpenters, blacksmiths, engineers, masons, and other skilled artisans, gave a living, directly or indirectly, to clerks, storekeepers, teamsters, and mariners for the Lakes traffic. Lawyers, ministers, teachers, and doctors, some of whom were of Cornish birth or parentage, also followed the mining pioneers into Northern Michigan. Nor should the womenfolk of the miners and other immigrants be forgotten; at first, as on every mining frontier, men greatly outnumbered women, but by 1854 twenty-seven per cent of the population of the Copper Country was female and ten years later nearly forty per cent.[63]

Most of the women were housewives who had accompanied or followed their men to Upper Michigan. Many Cousin Jennies found that they had come to a hard country. They had to eke out scanty stores of bacon, salt, pork, potatoes and flour to last through the long winters, and often had miner boarders to look after besides their own husbands and children. In Ontonagon County in 1860, the twenty-eight-year-old wife of John Hoskins had five children, all under eight years old, and five miners boarding with her family; George Karkeek's family also had five children, rather older than the Hoskins' youngsters, and two miner boarders; besides five children ranging in age from seven to seventeen. James Opie's wife had as boarders four young miners, between nineteen

62. *W.B.*, 22 June 1849.
63. Figures and estimates quoted in the above paragraph are based on the 1854 census for Keweenaw and Ontonagon Counties and on the 1864 census for Keweenaw, Houghton, and Ontonagon Counties. The Civil War and some drift of miners further west helped to redress the disproportion of the sexes to some extent.

and twenty-six years of age, and, in addition, a young married couple with an infant child.[64] Boarders were often close relatives of the families with whom they lodged, but this did not always turn out at all happily, particularly in some of the smaller houses and in cases where the housewife was much younger than her husband's people boarding with them; indeed, family squabbles must be reckoned among the causes of further migration of miners both within and outside the boundaries of the copper range. Many Cousin Jennies grew old before their time harassed in the long winters by children and boarders huddling round the stove trying to keep warm in cramped, hastily built wooden structures which hardly deserved the name of house. Still, miners' cottages in Cornwall had been just as bad, if not worse, for many of the immigrants had spent their early days in houses that had, at most, two small bedrooms 'way up in the roof' and, downstairs, a kitchen-living room and a parlour that was rarely used from one year's end to the other.

A few women were engaged in other pursuits than housekeeping especially as dressmakers, milliners, and school-teachers; girls helped in the housework and looked after younger children when reckoned old enough to do so—which tended to be very young indeed. Richard Trevarrow's wife acted as matron and nurse of the hospital he 'kept' in Ontonagon County in 1860;[65] some widows kept boarding-houses. Most girls married before they were out of their teens. Although several men went from Upper Michigan to the mining camps of the Pacific Coast, it does not seem that so many wives and families were left behind as there were in south-western Wisconsin. In any event, the mining families of Northern Michigan mostly lived by mining only, whereas many of the lead miners of the Upper Mississippi owned small farms and kept dairies in a more fertile and climatically equable region; several of the Cousin Jacks who went off to California from Wisconsin in 'forty-nine' and later years were farmers who, having the draught animals on their homesteads, took them and trekked off to the West, leaving wives, daughters, and younger sons behind to look after the rest of the stock and their crops.

Among the Cousin Jennies who came to the Upper Peninsula was the orphaned Rebecca Jewell Francis who, at the age of seventeen, came in 1862 at a Methodist missionary teacher to the Ojibway Indian settlement at L'Anse. Apart from a missionary's wife across the bay, she was the only white woman for miles around in that wilderness country. Her fortitude there during a severe winter, trying to get the Indians and their

64. Federal Census, 1860.   65. Ibid.

children to understand strange white notions, earned her the Ojibway name *Swangideed Wayquay*—Lady Unafraid—which the Indians who brought her by canoe from Houghton gave her.[66] Still, all the men and women who came in the early days to Northern Michigan had qualities of courage and endurance which passed unchronicled. Often, as in the case of Elizabeth Gurney Taylor,[67] hearts must have grown faint, and feelings of homesickness, loneliness, fear, and despair must have made living conditions on the primitive mining frontier seem almost unendurable. Yet had many immigrants, whatever their country of origin, given way to despair, Upper Michigan would have remained a wilderness, and its mines and forests contributed nothing to the transformation of the United States from an agrarian homesteading economy to the paramount position it was to attain among the industrial and commercial powers of the world.

The contribution of the hard-rock miners from Cornwall and of the families that came with them to this development was great, although it is almost impossible to estimate how many Cousin Jacks had settled in the Upper Peninsula before the outbreak of the Civil War. Their clannishness meant that certain settlements were almost entirely Cornish and these led passing travellers to exaggerate the total Cornish numbers in the entire mining region. The early census figures afford only a rough indication. It may be safely assumed that, by 1865, there were nearly twenty thousand people living in the three copper-mining counties of Keweenaw, Houghton, and Ontonagon, and that at least three thousand of them were of Cornish birth or parentage. The Federal Census of 1860 attempted to enumerate countries of origin, and there were a few Cornish Jacks who filled in the appropriate column with the word 'Cornwall', but the majority answered 'England'. Apart from 'Tre', 'Pol', and 'Pen' prefixed names, the numbers of English born whose surnames were familiar in the mining districts of Cornwall in the five census districts of Houghton Village, Hancock Village, Portage Township, Houghton Township, and Ontonagon County exceeded fourteen hundred, including young families born to them since their arrival in America; there were in all probability between four and five hundred more in Keweenaw County. By far the greater number of them were 'single' men, mostly living with other families or in boarding houses, though it was not unusual for two, three, or four miners to live together without any female 'housekeep'. Among those with young families by far the greater proportion of the children had been born in either

66. J. R. Nelson, *Lady Unafraid*.   67. See above, p. 82.

Michigan or the Old Country, but some of the younger members of such families had been born in other states, more particularly in Wisconsin and the Atlantic iron-mining states of New Jersey and Pennsylvania; a few had been born in Illinois and in Canada. In many cases the older children were English, the younger American born. While the census returns indicate the wanderings of immigrants by recording the ages and natal countries of their children, they do not show whether gaps of several years between the ages of older and younger children reflected parental separations and husbands emigrating before their families from the Old Country, or whether, as was all too common in those times of heavy infant mortality, children had died.

Decennial census returns, too, missed transient miners who only spent a season or two in the Upper Peninsula; some 'birds of passage' were certainly returned, but there must have been many more of them in some years than in those when the census was taken. The Michigan State Census of 1864 recorded a total of 5,447 miners employed by sixty-three mining concerns in the three counties of Keweenaw, Houghton, and Ontonagon; in 1854 a similar State Census had returned 2,304 miners for the area which then only comprised the two counties of Keweenaw and Ontonagon. The French traveller Rivot, whose remarks were far more flattering to the Cornish than those of Kohl, stated in 1856 that

from 1845 to 1853 most of the explorations made in search of copper mines were conducted and superintended by energetic and practical Cornish miners; the largest part of our mining population are from the county of Cornwall (the total amount must be over five thousand). The mining captains of our principal dividend mines are all Cornish gentlemen, and are a very valuable body of men, possessed of a large amount of practical knowledge and much relied upon by the companies.[68]

That figure was an impressionistic guess, while the falling off of copper production in the Upper Peninsula from the high level of over ten thousand tons in 1861 should be associated with three currents of migration out of the district in the early days of the Civil War—to the Union armies, to Pacific Coast mining camps, and back to the Old Country. Many Cornish immigrants felt that the American fratricidal quarrel was no concern of theirs and were determined to have no part

---

68. Quoted by Francis A. Altwalt in the *Mining Journal* (London), 16 March 1861. As Altwalt was engaged in promoting the French Ontonagon Mining Company and seeking British financial support, he probably overstressed the theme that Michigan copper-mines had been opened up by reputable Cornish mining men.

in it. Nevertheless, whether they left or stayed in Upper Michigan, hundreds and even thousands of Cousin Jacks had already helped that region pass the frontier phase of its history.

Movement away from Michigan was but part of the saga of the American mining frontier and of the Cousin Jacks who played their part in that drama of human progress and endeavour. The Federal Census returns ten years later from Grass Valley, California, indicate that a number of Cornish families had moved on from Michigan to this gold-mining town on the Pacific Coast region.[69] Out of rather more than a hundred Grass Valley families of 'English' birth with names that can with fair certitude be associated with the mining districts of Cornwall, nine had had children born in Michigan, whereas only four had Wisconsin-born offspring—a lower number even than Pennsylvania whence six Cornish families had come. These numbers may seem insignificantly small, but Grass Valley was only one Californian gold-mining camp, although the one to which the Cornish went in greatest numbers. In any case, family men, having migrated once and settled were less ready to move a second time, and probably ten times as many single men moved on from the Upper Peninsula to the 'Coast'. Fugitive records of individual cases, and the marked recession in Michigan copper production in 1850 would justify the assumption that at least ten per cent of the Cornish pioneers on the copper frontier of the Northern Peninsula left for the golden frontier of California, although remote, winter-bound Ontonagon got the news of Marshall's strike at Coloma weeks after the 'California fever' gripped the rest of what now, including Wisconsin and Northern Michigan, had become the 'East'.

69. The designation 'Pacific Coast', or the 'Coast' in mining circles meant any district west of the Rockies.

5 California— here we come

The news of the Californian gold discoveries reached Cornwall more quickly than it reached Northern Michigan, and almost as quickly as it reached south-west Wisconsin. President Polk's Annual Message to Congress in December 1848, which announced that gold had been found in California roused feverish interest not only in the 'East' but in the Old World as well. Before that time there had been rumours of the existence of gold in the distant south-west lands that Polk had seized from Mexico, but American soldiers who served in California had little good to say of it, one writing home in June 1847, that it was 'a most miserable, God-forsaken country, the jumping-off place, fit only for Mormons, Millerites and monomaniacs'.[1] Not until December 1848 did really enthusiastic reports on California appear in Eastern papers the first being, in the main, based on the official report Colonel Mason had sent to Washington and which Polk had submitted, along with his Annual Message to Congress on 5 December 1848. The *Wisconsin Tribune*, published in Madison, first gave prominence to the gold discoveries in its issue of 21 December, but its reports were mostly copied from Washington and New York papers that had appeared nearly three weeks before. Even then some editors were far from enthusiastic cautioning their readers that

the reports from California are no larger in proportion to the distance which they have to travel, than were those respecting the mining region of Wisconsin a few years ago, and we doubt very much whether persons going there for the purpose of obtaining gold will find it as profitable even now as digging lead in Wisconsin.[2]

Within a matter of days rather than weeks similar reports and warnings were appearing in the English press; the cautionary note in English editorials, as well as in those of the Eastern states, is a significant indica-

1. Letter dated San Francisco, 25 June 1847, and copied by the *Madison Express* of 14 December 1847 from the *Detroit Advertiser*, n.d.
2. *Mineral Point Tribune*, 29 December 1848.

tion that the gold 'fever' was raging and that many were determined to go to California as quickly as they could.

The excitement over the discovery of gold seems almost inexplicable until it is remembered that the finds in California were the first extensive gold deposits to be struck in modern times. Over three centuries earlier the Spanish conquest of Mexico and Peru had started the circulation of all the stories of El Dorado, but apart from the Spanish discoveries of rich silver-mines in various parts of the vast domains which had once comprised their empire, remarkably little in the way of extensive deposits of precious metals had been unearthed by the hundreds of adventurers that had gone forth from the Old World in high hopes. The first permanent Virginian settlers, in 1606 entertained high hopes of finding gold, but were no more fortunate than the Frobisher expeditions of miners to Baffin Land a generation earlier. A few minor discoveries kept hope alive, but it was only in the earlier decades of the nineteenth century that a little gold was found in Virginia, North Carolina, and Georgia in deposits just rich enough to pay modest returns provided labour and a certain amount of capital was diligently and carefully applied. At least in the nineteenth century men knew gold when they saw it, which was an improvement on the 'miners' who had accompanied Frobisher on his expeditions to the far northern regions of America in 1576 and 1577; not all the skill of the alchemists—still a moderately respectable profession in Elizabethan London—could turn the mica (amphibolite and pyroxemite) they brought back into gold.

When news of Marshall's discovery on the American Fork of the Sacramento River leaked out there was a rush to the region. Men already in California—traders, seamen, soldiers, and a few ranchers—were quickly on the scene, to be followed by Mexicans, Chileans, and a few Indians. The first-comers had the opportunity to select and work the most promising-looking alluvial 'diggings', and, naturally, it was the fortunes made by the luckiest men that were reported, usually first by word of mouth alone and losing nothing in the telling. Mason's report and the President's message had been sober enough, but people probably took more account of reputed first-hand accounts from California, and such reports seemed to grow in proportion to the distance they travelled.

Both in America and in the Old Country there were two types of California 'fever'—the adventurous and the speculative. The first drew men away from their homes and their employments to go gold-hunting in the West. The second was more complex; it can be described as a desire to profit personally by more or by less legitimate means which did not, necessarily, involve undertaking personally a hazardous journey to California. Trading and mining companies, with grandiose titles which generally included the words 'California' and 'Gold', were promoted not only in New York and Boston but also in London. Such concerns issued prospectuses which depicted the riches of California in exaggerated terms and their own prospects in all the colours of the rainbow. Some of them were genuine, but there were others that were simply schemes to batten upon the credulous investing public. Shipping companies eagerly offered passenger and freight transportation to the gold regions. In the advertisement columns of the London *Times* a self-styled 'well-educated gentleman' asked for two others 'to share expenses with him' in a trip to California; another asked for 'respectable persons' in possession of not less than a hundred pounds sterling to apply if they wished and were willing to make up a 'party' for the gold diggings; yet a third announced that 'a seafaring man is ready to go equal shares in purchasing a schooner to sail (to San Francisco) on speculation'.[3] Whether such advertisements evoked responses is impossible to say. Some of them may even have been genuine, but people who ventured on those that were not, in the hope of easy gains and were beguiled out of their money, were naturally reluctant to admit their folly and their greed. But this type of speculative mania was nothing new; all that was new was that the bait was now the gold diggings of California instead of mines in Cornwall, South America, and even, within the past five years, on the shores of Lake Superior.

The tales that attracted and often deceived were fantastic yet they originated from genuine discoveries of gold in California where, in 1848, many persons had certainly 'struck it rich'. Those who reported such stories of fortune-making, however, rarely went to the mathematic trouble of correlating the actual amounts of gold discovered and the numbers actively engaged in searching for the precious metal. It must, however, be admitted that in the first hectic months no one had much idea how much gold was being produced from the diggings nor how many men were there working for it. From the outset, too, the men in the diggings fell into two classes—the secretive and the boasters. James Marshall and his employer, Sutter, kept the first discovery secret as long as they could, perhaps in the hope of averting the catastrophic effects of a swarm of locust-like gold-seekers upon Sutter's ranch, per-

3. *The Times*, 12 January 1849, quoted in *Royal Cornwall Gazette* 19 January 1849.

haps merely to get out as much gold as they could for themselves before others came to attempt to claim a share of the mineral wealth.[4] Many other adventurers kept or tried to keep lucky strikes secret for the same motive, but usually such secrecy defeated its own ends by rousing suspicion. On the other hand there were those who bragged and boasted of their gains and exaggerated the magnitude of their finds. The secrecy of the one class and the volubility of the other inevitably led to the circulation of contradictory rumours about the richness of the diggings. Generally tales of rich finds were repeated and repeated again, growing with the repetition; when they crossed the continent or crossed the Atlantic, it was the more sensational discoveries that had the greatest news value and were published in the papers.

The printed word in those times, when so many could not read or only read with difficulty, was regarded with far less suspicion than in a later age hag-ridden with propaganda. Some who read it were probably astonished but did not question the veracity of a letter from Monterey, in all likelihood copied second- or third-hand from other newspapers, which appeared in the *Royal Cornwall Gazette*, in January 1849. The writer, perhaps protesting a little too frequently that he personally knew the fortunate individuals concerned, quoted a number of vast fortunes being found in the Californian diggings. One man employing sixty Indians was making a dollar a minute; a group of seven men, employing fifty Indians, got out two hundred and seventy-five pounds weight of gold in forty-four working days;[5] ten men in ten days made fifteen hundred dollars each; one man got two and a half pounds weight of gold from a shallow rock basin in a quarter of an hour; the generality of miners were averaging sixteen dollars a day, and many were making from five times to ten times that amount daily.[6]

Such stories, whether true, exaggerated, or false, were only a snare and a delusion to the credulous. If ten men were doing well and a hundred making nothing at all, average gains were meaningless. It is doubtful if one man in twenty who read or heard these accounts thought of questioning whether 'average earnings' were the lot of the generality

4. Such secrecy by Marshall and Sutter was justifiable, if selfish, when it is remembered that they had no indications that the finds Marshall made were more extensive than rich pockets which had been earlier found in Virginia, North Carolina and Georgia.

5. Worth about $60,000 or approximately $8,500 per individual white at the rate of $18 per ounce. The writer seemed to imply that the diggers did not work on Sunday, a day devoted not to religious observance but to trading, washing clothes, cooking, and the other necessary odd jobs.

6. *Royal Cornwall Gazette*, 19 January 1849.

of diggers. A 'bonanza' strike, too, might often be no excessive reward for long fruitless months spent prospecting it. The 'lucky strike' stories, however, were enough to inspire all sorts and conditions of men in many places with an avid and irresistible desire to go to California as quickly as they could.

Gold-seeking adventurers did not believe the adage that it was better to travel hopefully than it was to arrive. Nevertheless the most heroic, romantic and perilous part of the Californian gold saga was, for the majority of the latter-day Argonauts, the journey thither. The Cornish emigrants to California fell into two groups: those who came out directly from Cornwall and those who were already in the States, mainly in Wisconsin and Upper Michigan, when the news of the discoveries became generally known.

To both groups California was remote and could only be reached by long, costly and hazardous travel. The cost, indeed, might seem wellnigh prohibitive to home-country miners. The cheapest passage offered came to about twenty-five pounds sterling steeerage direct from London and Liverpool to California, advertised to include provisions; the luggage that a labouring miner emigrant might take added a pound or two to this, depending how much heed he paid to the advice of earlier emigrants that, ship supplies being far from good, it was advisable to take a fair store of personal provisions along with him. The direct passage by sailing vessel round Cape Horn took eight or nine months, and for a working man to deliberately forego loss of remunerative employment for such a long period was no small sacrifice. Right from the first, however, shorter and quicker passages were offered, but there was no guarantee that a passage from an English port to ports in the Southern states would save much time or money in the long run; feverishly anxious to reach California as quickly as possible, however, few stopped to think how long it would be before they would manage to get further on than Galveston and other ports on the Gulf of Mexico or on the Atlantic coast. All they thought about was getting on the way as quickly as they could; starting off, however, was one thing, reaching California another matter entirely.

Actually it cost about the same to go direct by sea from Cornwall as it did from Boston and New York whence, in the early rush days, the cheapest passage to California was about a hundred dollars. A fair proportion of the Cousin Jacks in Wisconsin and Michigan chose to go by sea rather than by land, and came back to the Atlantic ports for transport to the diggings. Whether in the Old Country or the New, they all had the initial problem of raising the money to pay for the journey and to live on until they reached the gold-fields. A few had savings, but it is unlikely that many miners in Cornwall were very affluent in the closing years of the 'Hungry Forties'; conditions had improved slightly from the starving times of 1846 and 1847, and, of course, there were many who now felt it wise to get out of the Old Country before another period of bad times began. Some were able to borrow money from relatives and friends, while another expedient was resorted to in the western mining parish, Breage, where, in the summer of 1850, a company was organised to send out parties to the Californian gold diggings. Before late August, that year, enough capital had been raised to send out a party of eight men equipped with mining tools and tents, and they were given a farewell supper in the village tavern, when

about thirty sat down to the roast beef of old England. Several parties addressed the meeting on the state and prospects of the California territory, and after a number of complimentary toasts were drunk, the meeting separated, the evening having been spent to the enjoyment of all present.[7]

This party or others from Breage did quite well in California if one is to judge from a substantial terrace of houses in the parish 'church-town' built from the proceeds of Californian mines. In another Cornish parish, Roche, it was to come about that moderately successful persons instead of being described as being 'in clover' were said to be 'in California'.

A less flattering account of California was given at the December monthly meeting of the Truro Institute later that year. The speaker, a Mr Rosser of Rugby, admitted that he had never been in the gold-mining country himself, but that he had assembled his information from 'government official documents' and other sources. Having said that the climate was, for a newly explored region, not extremely unhealthy, he went on, perhaps because of the 'self-improvement' aims of the Institute, to state that

to preserve health in California, a man must be temperate, which is the great secret of health in all climates. A person going there must make up his mind to work hard, to be persevering and energetic; and he must also make up his mind that there is equally as great a chance of losing his life as of getting back again to England.[8]

Rosser's further remarks about the dangers of hostile Indians and of white desperadoes lent weight to his declaration that he had no intention of going to California himself, and that, though he did not wish to

7. *W.B.*, 30 August 1850.   8. Ibid., 13 December 1850.

deter anyone from going there he would regret if anyone decided to go to the Pacific Coast from hearing the description he had given. At the end of his talk, however, he had to answer questions about the cheapest and best routes to California, and about the extremely confused system of land titles and mining legal rights then existing in the Pacific Southwest. The chairman, however, came to his rescue by calling a halt to proceedings, moving a vote of thanks, and finally announcing that the next lecture at the Institute would be given by a local clergyman on the safer and less controversial subject of 'the literary attractions of the Bible'.

Naturally it was the Wisconsin Cornish who left most in the way of records of Cousin Jacks going overland to California, crossing the Great Plains, the Rockies, the arid and rough semi-desert country west of the Rockies, and the Sierra Nevada. The long sea voyages from British ports to the Pacific coast were slow, monotonous, and long even by the fastest vessels; landfalls at Rio de Janiero, Valparaiso, and other places were welcome even to those most anxious to arrive at the diggings, but to spend more than half a year in the cramped quarters of sailing ships tried the tempers and the physical endurance even of men whose working lives had been spent 'deep down' in the narrow, dark levels of Cornish mines. Partnerships entered into eagerly in England were dissolved long before the parties to them reached California. Many suffered agonies of seasickness and the terrors of tempests at sea. Provisions ran short; stale water was an additional misery that had to be endured. Those going by the shorter, but dearer, route across the Isthmus of Panama ran added risks of yellow fever and cholera. But many lived through and endured all in the quest of gold.

One voyage deserves passing mention. The Cornish colony at Mineral Point may have participated in it. A local capitalist, Oscar Paddock, in the winter of 1848–49, put up the money for a boat which was built by Henry Butler, a carpenter. The latter was helped and advised by a seaman called Vance who subsequently captained the vessel on her adventurous voyage for California. This ship was much the same size as the Newlyn fishing boat, *Mystery*, which, five or six years later, with a crew of seven, successfully reached the gold regions of Australia direct from West Cornwall, being only about thirty feet long and seven in breadth. In one respect the Wisconsin venture was the more amazing, even though the projected ocean voyage was shorter. Newlyn was on the sea; Mineral Point was not. In fact, it was not even on a navigable river, and two ox teams were required to get the vessel to the Mississippi at Galena—a distance of over fifty miles. There masts and rigging were provided and fixed, and the river launching took place. The boat got down to New Orleans without incident, but whether it was simply impatience to get to California or whether the adventurous crew had some special reason to evade making the acquaintance of Federal officials, Vance put off to sea without getting the necessary clearance papers from the Customs House. Off Cuba the vessel was seized by a Spanish cruiser; a prize crew was put on board to take her into a Cuban port; there the Spanish, deceived by the apparent docility of the Wisconsin adventurers, failed to keep a close watch: under cover of darkness the Americans seized, gagged, and bound the guards left on board, sailed off and, next morning, put the Spaniards on board the ship's boat and set them adrift off a deserted district of the Cuban coast. The same thing happened a second time off the Mexican coast, but on this occasion the Wisconsin men gave the Mexican naval men the slip, and then reached and sailed up the Nicaragua River, where they sold the ship to a local merchant for a thousand dollars.[9] They then made their way overland to Aspinwall[10] where they secured a passage to San Francisco.

The last lap of the journey for those who came by sea to San Francisco was by no means easy, and the account given by another man from the lead mining country, Colonel Collins of White Oak Springs, may have deterred many. Writing from the North Fork of the American River in August 1849, Collins told his family that he had had a 'long and wearisome trip', but that he was enjoying good health, going on to relate that

We had a hard time getting our baggage and provisions to this place distant from San Francisco about two hundred miles. We came within some forty-five miles of this place by water, and then purchased a wagon and yoke of oxen between seven of us, with which by packing them up the mountains, we brought about a hundred and twenty pounds to this place, and then sent the team back and brought up nine hundred (pounds) more. The price of carrying this forty or forty-five miles is from twenty to twenty-five cents per pound, and can scarcely be had for that. We were lucky in getting a yoke of cattle cheap that had just been driven from Oregon. They were small size, about such as you could get about twenty dollars for in Wisconsin, but we were glad

9. Based on the account given in *History of Iowa County*, pp. 676 ff. No record survived of the name given to the vessel.

10. Colon; this town was originally named after the New York transportation magnate by the Americans who built the port to serve the traffic to California; the Colombian government refused to accept this name, and finally their 'official' name was adopted. O. Lewis, *Sea Routes to the Gold Fields*, pp. 185–86.

to get them at a hundred and fifty dollars.—Our wagon—oh such a wagon! —why you would have to give a boy two bits to burn the tire off, and yet it was worth a hundred and fifteen dollars.

Collins, when he wrote, admitted that he had not been at all successful in the diggings; he had hopes, however, of doing better by employing Indians who could be hired for their food and clothing—the latter item being shirts costing a dollar-and-a-half every fortnight. Indians were, unfortunately, not very reliable, since 'they get tired and quit'. Provisions, too, were dear, and he meant to return to Wisconsin once he got enough gold which he was quite confident of doing; yet he would not advise any man, particularly any one with a family, who was making a fair living in Wisconsin to migrate to California, although

to those, however, who are like myself out of business, and poor, I would say if you are a good worker and can stand a sight of the Elephant, in all his fury, you may come here, and in one or two years lay the foundation upon which to make a fortune.[11]

By the time this letter reached the 'East' many accounts had already arrived there from those who had gone to California in search of fortune. These varied in tone: hopeful individuals were still hopeful whatever short of death befell them; those of volatile temperaments, who had been most feverishly excited at first by prospects of speedy fortune, out of luck, tended to be despondent and utterly disillusioned: still, some people were finding gold in California and more were preparing to follow them out there. One of the first Cousin Jacks to reach the diggings at Weber Creek was Edward Dale, who had emigrated to the States from St Agnes some time previously. Within three weeks of his arrival in the mining camp, he wrote reassuringly to his wife back in the States on 25 September 1850:

After a long and tedious journey I have at last got the pleasure of writing you from the land of gold. I know that very melancholy accounts have reached the States, relative to the Californian emigrants, which I am afraid have caused you a great deal of uneasiness, therefore I only write you a brief and hasty letter to let you know that all is well.[12]

The long delays in getting letters from and to the Californian gold camps certainly caused much anxiety both to those left behind and those who left. Dale intended to come back East after giving the gold diggings a trial or, alternatively that his wife join him in California when he had

11. *Wisconsin Express*, 6 November 1849, reprinted from the *Galena Gazette*.
12. *W.B.*, 29 August 1851.

found a suitable place to settle. Others who had left families behind either in the Old Country or the States had similar ideas. The partings were agonising when men left wives, parents or children behind, for none knew when if ever they would meet again. Richard Wearne less than four years after his arrival in Wisconsin, briefly recorded in his journal on 3 April 1852:

> I was last sunday taking my farwell of Brother Zach. Richd my son left Wensday 4 in company 7 another Company for California. May God Bless them. When shall we meet again: if not on Earth, may us in Heaven.[13]

Not until November did he get a letter from the younger Richard to say that he had arrived in California; then another of his sons, William, and one of his daughters, Biddy, left for California on 24 November; another letter came from Richard and other news came to the Cornish settlers in Wisconsin from the gold-seekers. On 29 January 1853, Wearne who had been reluctant to see his children go, recorded in his journal:

> I hear of very distressing times at California want of Provisions snow to the depth of 8 and 9 feet many perishing through want and cold.

Then, on 23 March news came that William had died on 19 December, less than a month after he had left Linden,

> on the Isthmus at the 7 Mile House with the Pyanama feaver; 75 others died on their Passage.

Edward Dale, however, had reached California safely by the overland route, though not without irritating accidents and troubles on the way. Three or four days beyond Fort Laramie he and his companions found the ferries of the Platte River so inadequate to cope with the numbers of travellers that, rather than be held up a week, they

> had better build a boat and put ourselves across; a few of our men swam across, the rest of us set to work and in little more than a day we had a ferry boat of our own, ready to cross the swift and once-dreaded Platte, which at this place is a quarter of a mile wide. By this time there were a number of emigrants applying to us for a passage, who were so anxious to get over that they would pay almost anything for it, so we agreed to stay a day or two and work our ferry boat; we need not do anything but take the passage-money, as they would row their own waggons over and pay besides. We got from three to four dollars per Waggon, the first day we made a hundred dollars. I would not wish for a better California than that, but notwithstanding, we sold our boat at a high price, and pursued our journey.[14]

13. Richard Wearne MS.   14. *W.B.*, 29 August 1851.

It was California or nothing for these Cousin Jack miners. The ability with which they improvised a ferry-boat for themselves showed their ability to deal with unforeseen difficulties and even to profit thereby. Still, it was not the sort of gold-seeking and fortune-making they wanted, so Dale and his partners pushed on, although they learnt that, a little further up the Platte River, another ferryman was making up to four or five hundred dollars a day ferrying California-bound emigrants across the river.

Like others on the overland route, Dale's party soon came into country where it was hard to find grass and water enough for their oxen, but they reached Salt Lake City where they stayed fifteen days to rest their hard-driven animals, though Dale thought it too long a stop. Unlike other overlanders, Dale said nothing about the Latter-Day Saints, the city they had established in the wilderness, or even their marital idiosyncrasies. His party set off again, and Dale's impression of the country they then traversed differed from hundreds of others who passed over that vast region of desolation only in the graphic, forthright description:

We reached the sink of the Humboldt River; we travelled down this disagreeable river for three hundred miles, to where it spreads very wide and sinks into the earth. Here, one of my oxen that I would not have sold for a hundred dollars died after two hours' illness. From the sink to Carson River, a distance of forty-five miles, it is a complete desert, there being neither wood, water, nor grass, and some part of the road very sandy; it is a trying place for teams; we started on it, taking with us about fifty gallons of water and a considerable quantity of grass, and our teams went through very fine. The destruction of property in this desert is beyond my description.... I do not think a quarter of a million dollars will cover the loss of property on this forty-five miles; dead horses, mules, oxen, waggons, harness, and all kinds of outfit were strewed all over the place; the stench from so many dead cattle was almost insupportable. The next day we struck Carson River—we found the inhabitants here taking every advantage of the poor emigrants—they charged two dollars per pound for flour, a dollar and seventy-five cents for a little pie, and everything else in proportion. I think they were the greatest land sharks I ever saw.[15]

With no further mishaps, taking a rather easier southern route than most overlanders, the Dale party crossed the Sierra Nevada and reached the diggings six weeks after they left Salt Lake City. Dale's letter concluded with complaints about the high prices charged for the common

15. *W.B.*, 29 August 1851.

necessities of life in the mining camps, so high that for several days until his party found a little pocket of four ounces 'of the prettiest gold I ever saw', they had hardly enough to live on; he also made a far from cheerful reference to the numbers of emigrants who had died on the way or soon after their arrival in California.

The impressions of Edward Dale were shared by many who 'crossed the Plains' to California. John T. Grenfell, another Cornishman, left Wisconsin in March 1850 and reached the gold diggings late in July. Grenfell's party got through without mishap save jettisoning possessions along the way to get along the quicker; in fact they were in such haste to get to the gold-fields, that they crossed the Sierra Nevada in a place where the snow was lying from fifteen to twenty feet deep. Fifteen months after his arrival, Grenfell wrote to a relative in Sancreed. He briefly outlined the searchings that late arrivals in the gold regions had to make in order to find only modestly remunerative claims during the third mining season on the Pacific Coast. At first Grenfell had only made three dollars a day, which hardly met his expenses; he had then moved south to Wood's Creek and reckoned that, at the end of more than a year, he had averaged out six dollars a day,

but the diggings here are not so good as they have been and, from what I can learn, the generality of men are not making more than three or four dollars per day now. There are a great many men doing well while thousands are doing nothing. There have been a great many fortunes made in this country, and it is likely a great many will be made yet, but not so many nor yet so fast as they have been. I think the chance of making money is better here than in England.[16]

Grenfell went on to give a rough estimate of the extent of the gold country, but he wrote in very different tones from the writers of the exaggerated descriptions of 'a land all gold' which had reached the East and the Old Country barely three years before. Men had to look for gold, and it might be weeks or months—if ever—before they might find it in payable amounts; it was not a matter of simply scooping up a panful of sand and gravel any place one chose and washing out an ounce or two of the precious metal in a few minutes. The first over-bright colours of the Californian dawn had faded, and John Grenfell was even contemplating the agricultural possibilities of the Pacific Coast. In the more rugged mountainous mining regions he thought that only a little gardening would be possible in a few localities, but

16. *Royal Cornwall Gazette*, 16 January 1852.

that 'there is some of California first-rate for farming and stock-raising'. He hoped that, with more families arriving, society would improve, and there was room for improvement, for

> we have all kinds of men here, gamblers, robbers, murderers, and men capable of doing everything that is bad. There are bull fights, gambling, shooting, horse-racing, and all kinds of exercise carried on here, and Sunday is the greatest day for business. There have been a great many murders committed in these diggings since I have been here, but not so many lately.

Of one thing Grenfell was convinced—he was not, if he could help it, going to travel overland again; if he undertook such a journey a second time he was going by what he called the 'timber-ing horse'.

The Camborner, John Roberts, also went from Wisconsin to California, going by the Isthmian route in 1851. He first located at Shone Flat, near Sonora, and shortly after his arrival wrote, in August 1851, a lengthy letter to his brother, Thomas, in the Old Country. He described in detail the success he and his partners had had, though never working so hard in their lives before, with a 'long Tom', and discussed the possibilities of quartz lode as distinct from alluvial gold mining in California. Prices were comparatively high, though not invariably so, and taking into account the chances of fortune and the high prevalent wages, Roberts found it impossible to advise his brother

> either to come here or to stay at home, but weight the matter before you leave; of course you will think of returning again some time. It will cost sixty pounds[17] to come here, besides the time in coming and returning, four months at least,[18] so it will take a man a good while to be as well off as he was before he left; there is also the sacrifice of home and family, for any person coming to California should stop at least three years.[19]

Roberts also described mining frontier life at some length; Sonora town was more than adequately supplied with gambling houses; Sundays were hectic, wholly unsabbatarian days on which most business transactions were done, bull-fights and circuses held, the greater number of the miners coming in from outlying camps to get the supplies they needed and the services and amusements lacking in the diggings. Hardly a Sunday went by in Sonora without a fatal shooting or stabbing affray. In the short time Roberts had been in the district he reckoned that at least thirty men had been killed or hanged, mostly, however,

---

17. Approximately $300; Roberts estimated the dollar as worth 4s. 2d.
18. Via the Isthmus route, which was comparatively well organised by this time.
19. W.B., 9 January 1852.

characters of whom the community was well rid. What Roberts thought of 'due process of law' in western mining camps was summed up briefly:

The civil laws cannot govern the people as yet; they are too impatient to wait for the decision of civil government, so they set Judge Lynch on the throne, and produce all the testimony possible on both sides of the question, after which the impartial judge gives a hasty but generally correct decision; therefore sometimes in one hour after the committal of a murder, the culprit receives his punishment by being hanged to a tree.[20]

Edward Dale, John Grenfell, and John Roberts all went to California with the single object of finding gold. A Truro man, W. E. Gill, went out for the same end. His letters home contained similar descriptions of mining activities and lynch law, although noteworthy in instancing a case of the latter which took place on a ship bound for San Francisco.[21] Gill, however, wrote at some length, though in general terms, of the agricultural resources and prospects of the Far West. In discussing the quartz deposits he made a not altogether friendly allusion to the abilities of American miners and mining engineers compared to those of the Old Country, when he wrote:

Some of our Yankee cousins persuaded John Bull to buy 'gold quartz veins' where no veins existed, and 'veins' without consulting their rightful owners. Johnny himself has not been without his errors in these matters; gold was the bait which excited his cupidity, and the bland representations of interested parties, written or oral, betrayed him out of much of all his business like caution. Of course, he will benefit by the experience now reaped, and the imposter must stand aside, to make way for the honorable, that the recognized practical and scientific may identify and select that which is good.[22]

It may be remarked that the British companies which acquired Californian mining properties, whatever their value and however legal or the reverse their title, sent many skilled miners and engineers from the Old Country to the Pacific Coast. Gill went on:

I have seen abundance of gold-bearing quartz veins from three to six feet thick, ranging over mountains some two and even three thousand feet high. How far below the valley, of course, I do not pretend to determine. The richer veins have good walls, and are regular, with very little dip, lying in a bed of

20. *W.B.*, 9 January 1852.
21. This incident probably occurred on one of the river steamers, but Gill's account does not specify, and it may have happened on the high seas.
22. *W.B.*, 29 October 1852.

soft schist, killas, interstratified with an abundance of quartzose nodules—'sugary spar'. Gold is also found plentiful in a kind of decomposed or rotten slate. Every miner will recognize the value of these details. As to working such veins, the economy is at once evident, as no shafts have to be sunk, no steam engines erected for either pumping water or raising the ore to the surface, as all can be done by a simple adit level driven at once into settled ground. The mountains are covered in the North with splendid fine timber. Many valleys, too, afford streams of water sufficient to drive the machinery necessary for stamps. Here some ability must be displayed in separating the gold without a loss, and here John respectfully takes leave of Jonathan. The latter is evidently a better huxter than a mining engineer.[23]

Gill's references to quartz gold indicated the direction in which Cornish miners were to be most active in Californian gold mining. The first discoveries had been alluvial or placer deposits, similar to stream tin workings in Cornwall. Like the shallow-lying lead deposits of the Upper Mississippi they had been 'poor men's mines', affording mineral that could be worked by individual laborers or, more generally, by small partnerships of two, three, or more labouring miners. Even such workings, however, could be more profitably exploited by some measure of capital expenditure which was far beyond the means of the ordinary working miner. On the richest placers considerable amounts of gold could be 'saved' by washing the pay-dirt in a tin pan or an Indian basket; then the wooden rocker or 'cradle' which dealt with larger amounts of auriferous sand and gravel by the same technique, though rather less laboriously, came to be widely used. The 'long Tom' could be made by anyone with a saw, hammer, and nails from a few pieces of board and a piece of wire mesh,[24] but if it speeded up the processes of saving the gold particles from the pay dirt, it needed far

23. *W.B.*, 29 October 1852. A further letter from Gill, published in the *West Briton* on 1 April 1853, criticised the lack of social tone and distinction in Western society, the lack of dignity and decorum in the San Francisco courts, theatrical performances and the improvement of communications in the mining regions. The *West Briton* in all published three letters Gill wrote, all dated from San Francisco, in its issues of 29 October 1852, and 4 February and 1 April 1853. Such letters served to keep people in the Old Country well abreast of the progress of the new El Dorado.

24. John Roberts, in his letter home to his brother in Camborne, described the long Tom thus: 'it is nothing but a washing strake such as we use in the lead mines of Wisconsin, with a sieve at the end of it, under which is fixed a box called the rifle (riffle) box, set in a diagonal position. It is about four feet long and two feet wide, and four inches deep; on the inside of the rifle box there are placed two bars, about two inches high, across the bottom called the rifle bars, which catch all the gold, and the poor stuff washes over them. This is all the machinery necessary for washing gold; the cost of it is from twenty-five to fifty dollars, depending on the quality and size.' (*W.B.*, 9 January 1852). The cost seems high, but boards, nails and wire were all scarce articles in many of the early diggings.

more water than the pan and rocker, and opened a wide door for capitalist investment in alluvial mining. Water in many of the Californian diggings was scarce; 'fluming' companies were organised to bring water from a distance, and the cost was high. It became higher still, when 'hydraulicking' was resorted to in order to wash down the gravel bluffs, once the beds of archaean rivers, to get at auriferous beds lying beneath sand, earth, rock, and gravel.

There was no difficulty or cost where streams flowed by or over placer deposits. In other places the water had to be brought to the gold, long ditches being dug and wooden flumes constructed to bring the water where it was needed. When it had to be brought a considerable distance only men with capital at their command could do it. Artificial watercourses had been used by several mining enterprises in the Old Country, and their construction did not demand any very advanced engineering skill. In California, however, they came to be used on an unprecedented scale, especially after hydraulic mining began in the late eighteen-fifties, and some were costly enterprises involving tunnelling through mountains and taking years to complete. As early as 1855 the Blue Gravel Company in the northern mining region of Nevada and Yuba Counties began work on a tunnel twelve hundred feet in length, which took over seven years to complete and cost over a hundred thousand dollars.[25] Later, in 1874, the Blue Tent Consolidated Mining Company of Nevada County, to bring water a distance of twenty miles, started digging a 'ditch' six feet wide and four feet deep, and hoped to complete the work in a year.[26] The same year the 'V Fluming Company' was organised in the same district to construct an artificial watercourse 'of sufficient capacity to run through it timber of all sizes, lumber, wood, logging and other material required in quartz mining'.[27] Such enterprises not only entailed the damming of rivers, but sometimes the enlargement of lakes to act as reservoirs. In 1872 the North Bloomfield Gravel Mining Company employed a labour force of forty-five white men and a hundred and fifty Chinese enlarging a dam to increase the capacity of the 'reservoir' Bowman Lake.[28] Water was essential for many types of gold-mining operations, and the men who managed to get control of water supplies were in a fair way to making their fortunes; miners, however, sometimes complained of excessively high charges demanded by the fluming companies for water, and in April 1874, the miners of Moore's Flat resorted to strike action in an

25. *Grass Valley Union*, 30 January, 1873.   26. Ibid., 8 May 1874.
27. Ibid., 4 June 1874.   28. Ibid., 27 June 1872.

attempt to force the price of water down to what they deemed a fair price.[29] The variable rainfall and snowfall on the eastern Sierra Nevada, whence the streams of the gold regions had their source, facilitated and gave added advantages to such capitalistic ventures, and curtailed the days of the 'free' miner labourer in that country.

What brought most capital, and with it Cornish miners, into California was the discovery of quartz gold in lodes that ran far into the depths of the earth. The first extensive quartz discovery was made in 1850 at Gold Hill, near Grass Valley, and, more finds being quickly made and confirmed, the rush of capital into the gold diggings began. From speculating in grandiose trading schemes to California, the investing public, began investing in quartz-mines, undeterred by the bruises and shocks it had sustained by over-sanguine ventures in previous schemes bearing the glittering name—California. Eager and gullible speculators were easily found by 'Californian' company promoters in London as well as in Boston and New York, although many of the concerns were genuine enough. While there were many Cornishmen in California before the London capitalists advertised for skilled Cousin Jacks to come and develop their mining properties on the Pacific Coast, they induced many miners and engineers to come out directly from Cornwall in the early days of Californian mining industry.

The first quartz-mining ventures in California met with very mixed success. Many early companies had far more capital than financial management or technical skill As placer deposits began to show signs of deterioration and exhaustion, however, California became increasingly dependent on the development of underground hard-rock mining to keep gold production from falling away too drastically. Hydraulic working of the gravel bluffs did not develop until after many promising quartz locations had been prospected, and, in any event, this type of working also demanded heavy capitalisation. Still, in the declining years of alluvial mining, new placer deposits were being discovered in the vast auriferous region; these, and the hope of still many more, meant the maintenance of wages at a high level; to attract labourers, companies had to have capital to meet heavy payrolls. Managing some of the quartz companies, too, were dogged individuals who were convinced that the gold was there and that, cost what it might, they were going to get it.

Just how many Cousin Jacks came to California cannot be estimated with certitude. Probably at least a thousand of those who came on from

29. *Grass Valley Union*, 28 April 1874.

Wisconsin did not return to the Badger State but settled in the Mother Lode country. Possibly as many more came direct from Cornwall during the first years of California's mining history. Yet, just as so many left Wisconsin and Michigan for California, so within a few years many Cousin Jacks chased the golden mirage elsewhere—to Australia in the early eighteen-fifties, to Fraser River in 1858, over the Sierras to 'Washoe' and the Comstock a year later, and a few, doubtless, to strikes reported from the Apache-ridden country of Arizona, New Mexico, and Sonora.

British records of emigration did not differentiate the Cornish from those who emigrated from other shires. In the years from 1852 to 1860 a total of 871 passengers sailed from British ports to San Francisco,[30] but this was only a small proportion of those who went by the Isthmus or overland 'across the Plains'. Many passengers to New York and other Atlantic ports of the United States and Canada almost certainly did not stay there long but moved on, by land or water, to the Pacific Coast. Federal Census returns, too, did not distinguish the Cornish from other English-born immigrants, but there were some mining districts in which the 'English' settlers were nearly all Cornish.

In California the greatest concentration of Cornish people by 1870, possibly even by 1860, was at Grass Valley in Nevada County, the centre of the quartz-mining district. In 1860 Grass Valley had a population of 3,940, of whom 530 were 'English'; no less than 470 of the latter were miners. Ten years later the population had risen to nearly seven thousand of whom 1,245 had been born in England, and of these most of the 855 men were miners; women numbered 200, the rest were children. The Cornish almost certainly contributed two-thirds of the 'English' miners, and the total numbers of Cornish-born in Grass Valley must have been at least eight hundred, and probably was well over a thousand. Although several of them had spent some time in Wisconsin and other parts of the East, a close analysis of the ages of the miners—half of them were under thirty-one years of age—and of the birthplaces of the children of those who had their families with them, indicates that three-quarters of the Cornish in Grass Valley came out direct from the Old Country. More Cousin Jacks were located elsewhere in Nevada County and throughout the Mother Lode district, there being quite large communities particularly at Soulsbyville, in Tuolumne County and also at the New Almaden quicksilver-mines. A low estimate would be that by

---

30. N. H. Harries, *Cornish and Welsh Mining Settlements in California*, unpublished thesis, Berkeley, California.

1870, at least two thousand Cornish immigrants had come to California, and that probably another thousand came thither in the declining days of Old Country mining after that date before, in the mid and late eighteen-eighties, the Witwatersrand became the golden magnet for Cousin Jacks. Such numbers, in California, might seem very small, until it is remembered that they came out from a single English county, that they were almost notoriously prolific, and that their descendants were numerous. Nevertheless no Californian locality could ever claim, like Butte City and Johannesburg, to be the most populous 'Cornish City'.

Many of the English-born miners in Grass Valley in 1870 can be identified by their surnames. The old local jingle

> By Tre, Lan, Pos, Car, Pol and Pen
> You may know the most of Cornishmen

can be dismissed as leaving out of account eight or nine out of every ten Cousin Jacks. There are, however, numerous surnames that in Cornwall are associated with the mining regions, and the Grass Valley census returns might easily be confused with lists of residents of Camborne or Redruth, Truro, or St Just. Rogers, Bice, Richards, Andrew, Mitchell, Bennett, Arthur, Tangye, Jenkins, Spargo, Oppy, Scobel, Sparnon, Northey, Kinsman, Hocking, Coomb, Crase, Glasson, Job, Paull, Retallick, Stevens, Goldsworthy, Angove, Sampson, Pascoe, Deeble, Bennallick, Menhennick, Rowe, Bray, Robbins, Vivian, Odgers, Hosking, Rule, Coad, Pearce, Grenfell, Bowden, Prisk, Gluyas, Kitto, Uren, Eudey, Eddy, and Truscott are almost as certainly Cornish names as those with the Cornish prefixes like Tresize, Trevillion, Trebilcock, Trezona, Rosevear, Carlyon, Polmear, Penglase, and Penrose. All these and a host of others of probable Cornish associations were found in Grass Valley in 1870, specified as of 'English' birth.[31] Out of 517 returns, which may fairly be regarded as a sampling of these immigrant miners, 28 were under twenty-one years of age, the youngest quartz

---

31. While some of these residents of Grass Valley may have come from other parts of England, the list deliberately excludes a great number with the surnames Thomas and Williams, which, in Grass Valley, was a Cornish rather than a Welsh name; Richards, too might be Welsh; similarly Pearce is possibly Irish. In addition some Cornish names have been lost by mis-spelling by those making the returns, and, in Cornwall itself there was a tendency to drop the 'Tre' prefix, and names like Tremar, Trekieve, Treworrick, and Tredinnick have been cut down to Mar, Key, Worrick, and Dinnick. There has also been drastic elision of certain names; visitors to Cornwall have got into difficulties seeking Launceston until they discover that locally it is Lanson, although the superb example of this is in West Cornwall where Boscaswell has come to be locally known as Scale.

miner listed as such being sixteen; 246 were between twenty-one and thirty-one; 169 from thirty-one to forty-one; 42 between forty-one and fifty-one; and only a dozen over fifty-one years old. Nearly four hundred of them were 'single' men, though many had left wives and families in the Old Country or the East; only a hundred and eleven had wives and families with them. A number of the returns listed the possession of varying amounts of real and personal estate, usually only two or three hundred dollars in value, but whereas sixty-four of the family men made such returns, only eight of the single men did. If the possession of such property indicated that the owners had either become naturalised American citizens or that they intended to settle permanently in Grass Valley, it shows that, as in 1849 and 1850, so in 1870 Cousin Jacks had come and were still coming to California with the hopeful notion of making a fortune and then returning to the Old Country. The social significance of this hardly needs emphasis.

In the previous census, in 1860, not a single 'English'-born woman had been listed in any of the Grass Valley returns; in 1870 there were at least a hundred married women, mostly falling in the age-groups of the twenties and thirties, and the majority of them with young families; there were only a few unmarried girls in their late teens. The dangers of the hard-rock miner's calling was not yet reflected in Grass Valley by the high proportion of widows found in British census returns from the Cornish mining districts. Offsetting this, however, was the fact that four out of every five quartz miners in Grass Valley was either single or a 'grass-widower'; widows, too, probably quickly remarried on the mining as they did on the American farming frontier. By 1870, however, the female element had definitely appeared in Grass Valley, which was becoming transformed from a temporary mining camp to a permanent settlement; but men still outnumbered women two to one and hardly any spinsters were to be found among women of marriageable age.

As time passed, an increasing proportion of the 'single' men decided to settle in California. Those who had left wives behind sent home for them, and bachelors for their old sweethearts to come out and marry them. The frontier was changing into a settled community, but as long as they remained single, however, miners were less reluctant to 'up and quit for diggings new'. There may even have been fewer Cousin Jacks in Grass Valley in 1860 than there had been in 1857, for news of the Fraser River discoveries reached the Sierra foothills in the spring of 1858, and the 'Washoe fever' broke out there a year later.[32] Every second Cousin

32. See below, p. 179.

Jack in the Californian diggings was probably a 'bird of passage', either a man who had come with the set purpose of remaining only a season or two or a man whom California could not hold if he saw a mere glimmering possibility of quicker and greater success elsewhere. Not that the Cornishmen were unique in this—the same could be said of adventurers of all races who came to the gold-fields in the early days whether from Cornwall, New England, Wisconsin, Chile, or China.

The predominantly male element in gold-mining districts had certain social consequences. Only as women and families came did there develop a communal life in part centring round chapel and school instead of round saloon and gambling hall. Many of the first Cousin Jacks in California, as in Wisconsin and on the Lakes, were devoutedly religious, and set about organising Methodist churches and meeting places almost as soon as they arrived. With others, over-strictly 'brought up', religion was to dwindle to decorous and respectful attendance at the funerals of fellow workers, and a tendency to strike up not always melodious singing of Wesleyan hymns liquored-up no matter whether they were in Gunnislake or Red Dog, in Redruth or Rough and Ready, in Chacewater or Caribou. And in Grass Valley round Midsummer Eve

or perhaps yesterday morning early, our reporter saw five of the old time men of Grass Valley sitting on the high planks of the sidewalk, in front of the bridge, singing in five different keys and in five different tunes, the good old Presbyterian ditty commencing

> 'Rock of Ages, cleft for me
> Let me hide myself in Thee.'

There was not a light visible in the town, save at Protection Hose Company's house, or a feeble flicker now and then from the dark lantern of the watchman who stood across the street to see that the old owls did nothing worse than catch the tune of that song. The feelings of that old crowd were harmonious whatever their voices may have been, and they kept the song running three hours. The five singers are all splendid men, if not splendid singers, but we wish they would, next time, get a tune which has not quite so much of melancholy in it.[33]

Others, no less fond of singing, in Grass Valley, had organised choirs, the first Cornish Carol Choir having its beginnings in 1854 and continuing to this day. The musical leanings of others led to the formation of brass bands of the type so popular in the Old Country. For several years Soulsbyville was famous for a silver cornet band, and there were others

33. *Grass Valley Union*, 24 June 1870.

up and down the Mother Lode country. It was not uncommon for three or more members of one of these bands in Cornwall to emigrate together, taking their instruments with them, and then, as soon as they settled, start up a band in their new location.

Murdering a hymn tune can hardly be called criminal, even though the Grass Valleyans whose slumbers were disturbed may have regarded it as anti-social. Cousin Jacks in California were, however, guilty of far worse offences. Their antipathy to the Irish led to several brawls and, on occasion, to bloodshed. Daw's Union Saloon in Grass Valley, on the night of Sunday, 11 August 1867, saw a free-for-all fight that ended in a Michigan Cousin Jack, Edward Scobel, being fatally stabbed by Joseph Lawrence, trouble having started when some Cornishmen taunted Lawrence with being Irish. Scobel, apparently, had done nothing more than tell the short-tempered Lawrence, who may have had more than enough liquor, that 'no-one wants to fall out with thee, but if you want to kick up a rig, I'm your man!' When Lawrence was tried for murder, it was stated in evidence that a group of Cousin Jacks 'had it in for him' since he was half-Irish, his father being Cornish and his mother Irish. He was found guilty of second-degree murder, sentenced to the penitentiary for ten years, but released when he had served about a third of that term. At the trial one of the jurors from Moore's Flat was challenged; he replied that he knew neither Scobel nor Lawrence, and had read of the case in the papers, but had formed no opinion about it since, 'the killing occurred in Grass Valley, and it was decidedly rough down there, and so many men being killed, I thought very little of it'. This juror was accepted, though Grass Valley, in fact, was a comparatively peaceful town which, however, in the five or six weeks before the affray in Daw's saloon had had two fatal shooting scrapes and one less lethal stabbing affair.[34] On the whole, the Cornish rarely got involved in anything more serious than fist-fights; the comparatively few more serious affrays in which they became embroiled, apart from the racial feud with the Irish, usually developed from quarrels over women which was hardly surprising in predominantly male communities.

The popularity of Cornish wrestling matches on the Fourth of July and other festive occasions in Grass Valley might also be instanced as a trait of masculine society, but more significant sociologically were the

---

34. *Grass Valley Union;* 13 August 1867, 9, 15 and 16 January 1868 and 5 May 1871, for the Scobel murder case, and the issues of 6, 11 and 27 July and 27 September 1867, for brief reports of the other 'scrapes'.

numbers who lived in boarding-houses; the 1870 census returns show, however, that in Grass Valley comparatively few miners lived in such big boarding-houses and hotels as those often found in the Michigan copper country. With less capitalistic company organisation in the early days and a larger percentage of free labourers uncontracted to employers, men tended to take time off from the diggings to build small cottages; they did not bother much about the niceties of substantial and comfortable construction since so many of them hoped to strike it rich soon and go home. In the Sierras there was, as in Upper Michigan, plenty of timber for building, and after the first seasons there was a decided tendency for the single miners to board in family cottages with relatives or friends rather than in hotels. The 1870 census recorded several cases of two, three or four miners sharing a house and a number of families with one or two miner boarders living with them. Most of the miners who were living together in a house seem to have looked after themselves, doing their own cooking, house-cleaning and laundry work; a neighbouring housewife may have helped them occasionally, and Chinese done the laundry for them, but hired female servants were hardly to be found anywhere. The substantial tradesman, John Bennett, had no servants living in his house in 1870, nor had the mine superintendent Henry Scadden, or the quartz-mill superintendents, William H. Crase and William Clift, or the rich merchant, Henry Coleman, the brewer, Charles Mitchell, and even the physician, Richard George. Men of their profession and social standing in the Old Country would almost certainly have had two or three servants, at least, 'living in' their homes.

The appearance of capital and highly skilled technical management in quartz mining did not bring much social distinction or class division and stratification to the Pacific Coast. The more staid and decorous trades-people and professional men, particularly when married, may have looked upon the boisterous sports and hard-drinking habits of many of the miners with some disapproval, but the 'Jack's as good as his master' creed was widely prevalent in the mining regions. The Cornish and Irish had their antipathies, but they could unite in anti-Chinese agitation against cheap coolie labour on the mere hint of an economic recession throughout the decades following the end of the Civil War. The Cornish tended to be clannish, partly because their dialect was, at times, almost incomprehensible to other English-speaking immigrants and Americans, who laughed at them because of it.

As miners, the Cornish were generally held in high esteem in California. When they first came to the quartz-mining districts it was usually conceded that neither Americans, Germans, nor Irish could equal them in anything connected with hard-rock mining. They were adept at sinking shafts and driving underground levels. Tunnelling under the Atlantic in St Just in Penwith and through sands at Perranzabuloe had taught them the most practicable methods of timbering shafts and levels to reduce risks of rock falls or 'caves' to the minimum. In Grass Valley it is related that an Australian syndicate laboured in vain to sink a shaft down through a quicksand to reach a rich 'ledge' of goldbearing rock; a Cousin Jack then came along and showed how it could be done by using an iron cylinder with a valve-like device at its bottom; this was steadily sunk down through the shifting sands, and the shaft securely collared or timbered as it was sunk, until the hard-rock and gold ledge was struck. In the Empire Mine, at Grass Valley, waste rock was used as it was used in Cornwall as wall props, while at the nearby Idaho-Maryland a Cornish mine carpenter showed the way to install a pump which went down vertically at first, took a right-angle bend to run along a level, and then took a forty-degree dip to get to the level it was to drain.[35] In drifting or lead-driving it even came to be said that a Cousin Jack could put a drift through muddy water, a statement on a par with the Pacific Coast saying that, to get an American mine prospector to Heaven it was necessary to start a gold excitement up there.[36]

In the district around Grass Valley and Nevada City mining conditions were very similar to those in the Old Country. The metallic lodes were even narrower than those in Cornwall; they dipped and faulted in equally unpredictable ways; they ran through hard metamorphosed 'country' rock; there were similar difficulties in mine drainage. What worked in the Old Country worked in the new, and as early as 1855 the first Cornish pump in California was installed at the Gold Hill Mine.

35. N. Harries, op. cit.
36. *Grass Valley Union*, 28 May 1871, quoting the *Owyhee Avalanche* (n.d.). The writer, after describing the American, Irish and German miners on the Pacific Coast, concluded: 'The Cornishman is probably the most skillful foreign miner that comes to our shores. For this he deserves no particular credit, because it is a calling to which he has been accustomed from his childhood. The Cornish miner is of a very quiet disposition, although very headstrong, partly on account of his being a Johnny Bull. Generally speaking, he is satisfied to work for others, but insists on being promptly paid for his labor, and doesn't care about engaging in mining on his own account. They are mostly stalwart, goodlooking fellows, dress better than any other class of miners, and are very fond of women. They also appear more clanish than other foreigners and a majority of them are good singers. When they visit a saloon, they generally range themselves around a table, call for a pot of ale, or porter, or beer, and pass the time away with anecdote and song.'

Cornish 'stamps' were constructed to crush rock-stuff, but in the course of a generation these were so improved that 'Californian' stamps were put up by some mining companies in the Old Country.

One development in Californian mining was not welcomed at first by the Cousin Jacks. They had gained a high reputation for their prowess with gunpowder as a blasting agent, and also for rock-drilling with a 'double-handed' drill in which one man held and turned the drill while his working partner 'scat to 'en' with a heavy sledge hammer. In the late eighteen-sixties, for motives of economy and more effective working, dynamite or 'giant powder' and a single-handed drill was introduced in some Californian quartz-mines. Early in 1869 some mining agents and superintendents wanted to adopt these new practices in Grass Valley. Experiments with dynamite were made at the Empire Mine in February and March, and also some single-handed drilling that bored a three-quarter inch hole through rock as quickly as the double-handed drill favoured by the Cousin Jacks put down a one-and-a-quarter-inch-wide hole.[37] The Cornish prejudice against working alone underground was reasonable, for if an accident occurred there might be fatal delay before help came to the injured solitary worker. No man, too, could be blamed for wishing for company in the deep, dark, underground levels. Furthermore, when there was a question of possible danger, 'two heads were better than one', since they could consult each other and, if need be, call others to their aid; a single individual might easily miss a natural danger warning, although some accidents occurred when a reckless miner overbore the more cautious counsel of his working partner or partners. The other objection the Cousin Jacks raised was about the fumes caused by dynamite explosions; these were certainly nauseating to some men, until they got used to them, and miners naturally asserted that 'giant powder' fumes were dangerous to health.

The tests convinced mine managers, however, that dynamite was a far more effective blasting agent than the old black powder, and that single-handed drilling saved labour. This last naturally aroused suspicions that the capitalist interests meant to reduce the numbers of men employed. Times had not been good in recent months owing to the collapse of the speculative mining boom that had followed the end of the Civil War, and ordinary working miners feared unemployment. The mining managers and superintendents, however, thought that by increasing production they would regain prosperity, and that dynamite and single-handed drills used in conjunction would facilitate the explora-

37. *Grass Valley Union*, 27 March 1869.

tion and exploitation of more auriferous ground and thus increase, not reduce, employment, although, of course, fewer men would be engaged in drilling.[38]

'Giant powder' was the main topic of conversation in the quartz-mining region of Nevada County in the spring of 1869. Trials of the new explosive at the Empire Mine on 30 April, with single-hand drills, were claimed to have been most satisfactory by those who made the tests and many who witnessed them agreed.[39] 'Outside' drill men who had put down the holes for the dynamite in the Empire Mine on this occasion were reported as asserting that, although they had been using 'the giant' for some time, they had never felt any ill effects from its fumes. A day or two later, however, a Miners' League was formed to resist the introduction of dynamite in the Grass Valley mines. With this, all the bitter and complex issues of rights to labour, rights to property, individual freedom, and social restraint that have been associated with trade-union history burst in upon the gold region of the Sierra foothills. Striving to be fair to all parties, but with a leaning towards individualistic ideals and notions, the editor of the *Grass Valley Union* commented:

The (miners') league is all right enough, and those who are opposed to working Giant Powder are at liberty to join the League. But does organisation mean that those who wish to work with Giant Powder will be prevented from doing so? Is the league going on a mission of mercy to prevent men from risking their healths? Why do not leagues interfere with working in quicksilver mines, where men are salivated and mercuralised more or less all the time? We imagine that leagues to prevent sickness amount to nothing and will accomplish nothing. A thinking man should hesitate a long time before he enters into any combination to interfere with labour, on the ground of protecting the health of the labourers. Each man can attend to his own health, but the league has for its object keeping Giant Powder out of the mines. If it is profitable to work with this powder, no league can prevent it.[40]

On Monday, 10 May, the men of the Empire Mine, probably the most predominantly Cornish-staffed one in the quartz region, went on strike, and the same morning the managers of the North Star mine discharged all their employees who had joined the leagues 'against giant powder'.[41] Both sides had pushed matters to extremes and forthwith the league coupled with its demand for banning dynamite demands for higher wages for underground carmen and shovellers. Within a matter of days, almost of hours, the crisis had reached the point

38. *Grass Valley Union*, 9 April 1869 (Editorial).   39. Ibid., 2 May, 1869.
40. Ibid., 8 May 1869.   41. Ibid., 12 May 1869.

where league meetings were passing resolutions declaring that they would have the use of dynamite stopped in Grass Valley mines 'peaceably if they can, and forcibly if they must'. On the other side, opponents of the league were declaring that if the strikers prevented any man who wished to work from doing so, the sheriff's posse comitatus would be called in to arrest them; if that failed recourse would be had to the State Militia, and, in the last resort of all, to the Federal military forces.[42]

A week went by in which there were several public and secret meetings of strikers. At the end of the week the mills associated with the Empire and North Star mines had to close down for lack of ore. Both sides continued to utter threats and counter-threats. Rumours that workers were to be brought in from other districts irritated the strikers, but mines where dynamite had not been introduced continued to work without trouble. More time passed. The employment of French-Canadian and Australian single-hand drillmen only served to stiffen the determination of the strikers, as did the arrival of some 'broke' miners ready to do anything for wages after their fruitless hegira up to White Pine on a fool's rush.[43] There were rumours that the strike was spreading to other parts of the gold-mining region and to the New Almaden quicksilver mines. The owners of the North Star and Empire met in San Francisco, and published reports about the superior efficiency of dynamite along with testimony from working miners declaring that its use had not been deleterious to their health.[44] Towards the end of May a few men went back to work in the Empire, but the 'peaceful suasion' of league agents dissuaded them from working more than a single shift.[45] Agents and agitators, some without authorisation from the league, tried to extend the strike to other mines. A 'Law and Order Association' of Nevada County met on May 25th, and passed resolutions avowing that it would use all its influence, and 'if necessary all the power of the law of the land', to maintain the 'sacred and inviolable' rights of mine-owners to work, manage and control their own property as they thought best, and the 'inalienable right' of every labourer to 'make contracts for the disposal of his own labour'.[46]

All through June the strike went on. At the beginning of that month the Empire had managed to get about a score of hands to work, a mere fraction of the force it normally employed, but on the night of the tenth

42. *Grass Valley Union*, 14 May 1869.
43. Ibid., 19 May 1869.
44. Ibid., 21 May 1869.
45. Ibid., 25 May 1869.
46. 27 May 1869.

of July, two of these men, returning to their cabin after work, were attacked and beaten up with bludgeons, while there was another incident of an attack on 'scab' labourers at Gold Hill the same night. Supporters of 'law and order' talked of 'assassinations' and the law of the 'slung shot and the bludgeon', and a group calling itself the 'Grass Valley Law and Order Association' met to pass resolutions that it was determined to protect 'the rights of the labouring men and the lives and property of the inhabitants of Grass Valley'.[47] The fact that only two men were injured, neither apparently very seriously, in a strike which had been going on for nine weeks would suggest that the alarm of the self-styled defenders of 'law and order' was excessive, but by then even a very trivial incident was as dangerous as 'giant powder' itself in that tense atmosphere, and there was no reason to predict that other and worse incidents would not follow. The leaders of the strike made, probably in all sincerity, professions of abhorrence of violence. So did their opponents. Yet a few firebrands on either side could easily have precipitated an appallingly ugly situation. That they did not could be regarded as proof of the law abidingness of the quartz-mining community and of the Cousin Jacks who seem to have comprised the majority of the strikers.

The strike ended towards the end of July, undramatically. Men had drifted back to work, perhaps fearful of cheap Chinese coolie labour being brought in to operate the mines; early on in the strike there had been a disposition by the Miners' League to stand only on the dynamite issue and the mine-owners made some sort of promise that, for the time being, dynamite would not be used in their mines.[48] Single-hand drilling, however, went on, and the experimenting captains of the Empire started using rifle powder as an explosive with fair success. The mine-owners, too, made a public statement that they had the right to use what materials they thought best in their mines. Nevertheless, for the time being it could be regarded as a victory for the men and for the League since dynamite had been excluded from the Grass Valley mines.

In other places, however, the 'giant' was used, and apparently without injurious effects on the health of the men using it. The 'law and order' people, however, may have had misgivings when some miners, who had probably graduated in the salmon and trout poaching schools of the Old Country, out fishing near Omega on the Yuba and losing patience with sportsmanlike hook and line methods, exploded some dynamite in the water and killed an incredible amount of fish. Envisag-

47. *Grass Valley Union*, 13 July 1869.   48. Ibid., 25 July 1869.

ing that such methods on the mountain streams of the Sierra 'would soon depopulate our inland waters of the finny tribe', there was a demand that the law intervene to prohibit such practices.[49]

It was impossible to prove that the use of dynamite was injurious to health, and so, when in February 1872,[50] it was re-introduced in Grass Valley at the Eureka Mine the strike against it soon collapsed. At that time the managers had the advantage of labour looking for work locally, while the strike movement was discredited by early incidents of violence and intimidation which alienated the law-abiding element in the ranks of the miners themselves. Furthermore, there was a general realisation that in the former prolonged dispute miners had lost wages, investors dividends, the mining industry generally had suffered and the trades-people dependent upon it, too, had felt the pinch. In this later dispute some of the mine-owners, notably those of the Eureka where the trouble started, evaded difficulties by resorting to the Cornish system of tribute working, allowing the contracting miners to work for themselves and to use whatever materials they thought best. The Miners' Union, or League, was active enough, but apparently it had not kept together since 1869, while the continual movement of miners from one mining district to another on the Pacific Coast was unfavourable to the development of strong local trade-union organisation and effectiveness. The threat of the owners of the Empire Mine that, this time, they would close down rather than submit to dictation by the men was decisive, since the industry was depressed and employment none too easy to find in February 1872. If the earlier strike against the introduction of dynamite had been successful it could be attributed to its breaking out in a period of economic recovery from depression; the later strike coincided with the onset of one of America's major slumps. Times, in fact, were becoming so bad, that the Eureka mine, which had first re-introduced dynamite to Grass Valley in an attempt to retrieve its falling fortunes, and then had resorted to tribute labour, was, by May 1872, likely to become, in Cornish phrase, a 'knacked bal': only a few months earlier it had been paying dividends amounting to two thousand dollars a day.[51]

There were other disputes and strikes in the Californian gold-fields in which some Cornish miners were involved. The 'giant powder' troubles in 1869 affected the New Almaden quicksilver-mines where working conditions were even normally unhealthy but high wages and

49. *Grass Valley Union*, 20 August 1869.   50. Ibid., 18 February 1872.
51. Ibid., 5 May 1872.

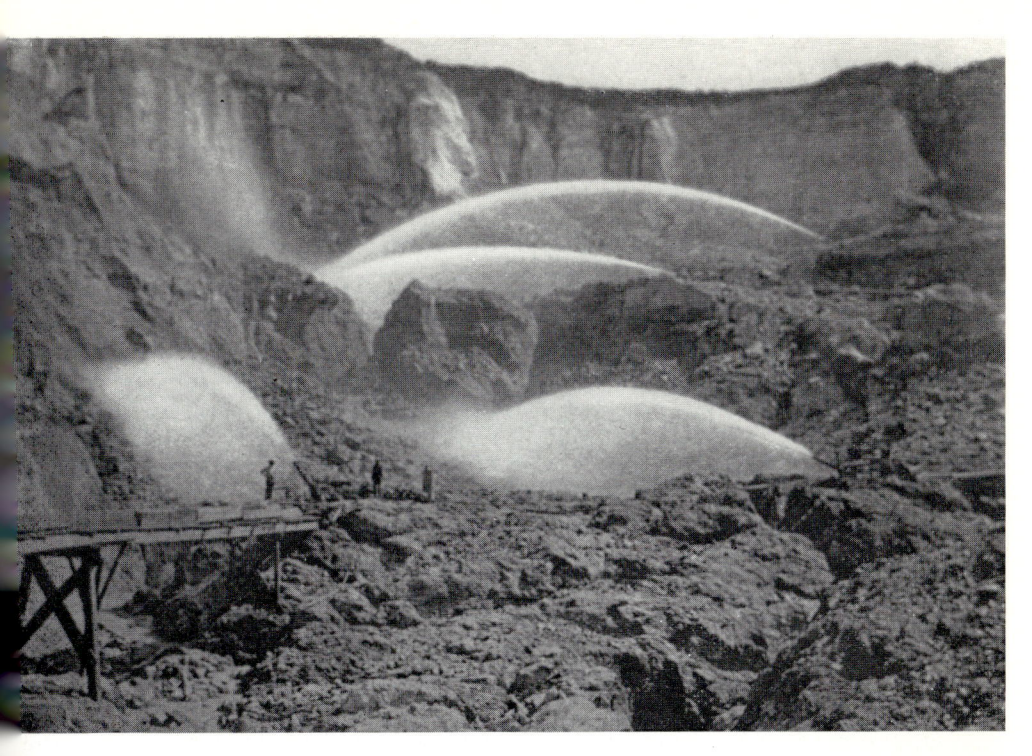

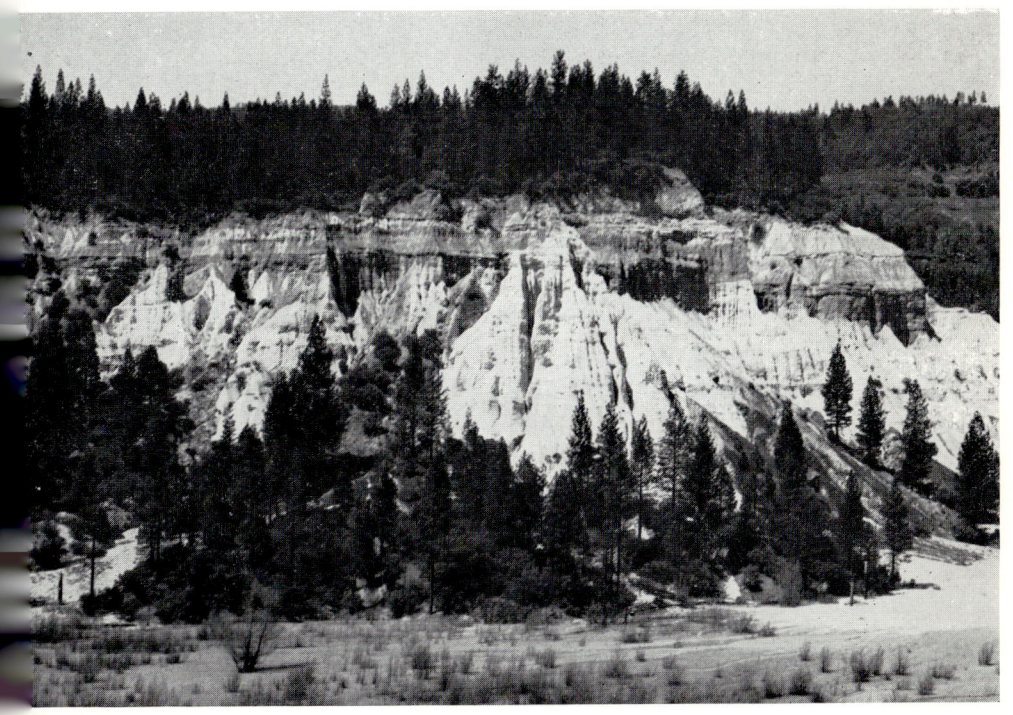

4 Hydraulic mining at the Malakoff Diggins in 1870, and the same site today

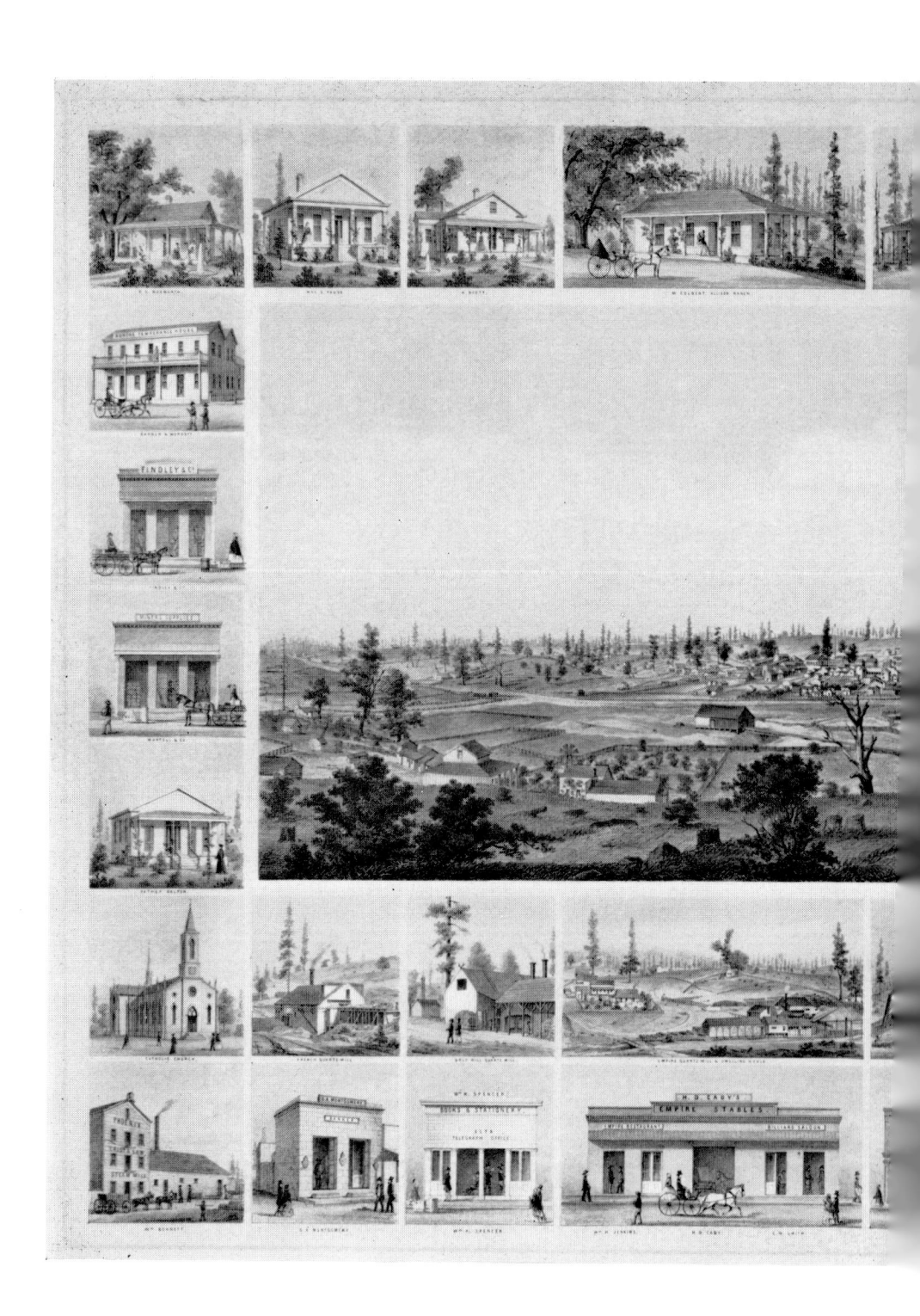

5 Grass Valley, Nevada County, California, in 1858

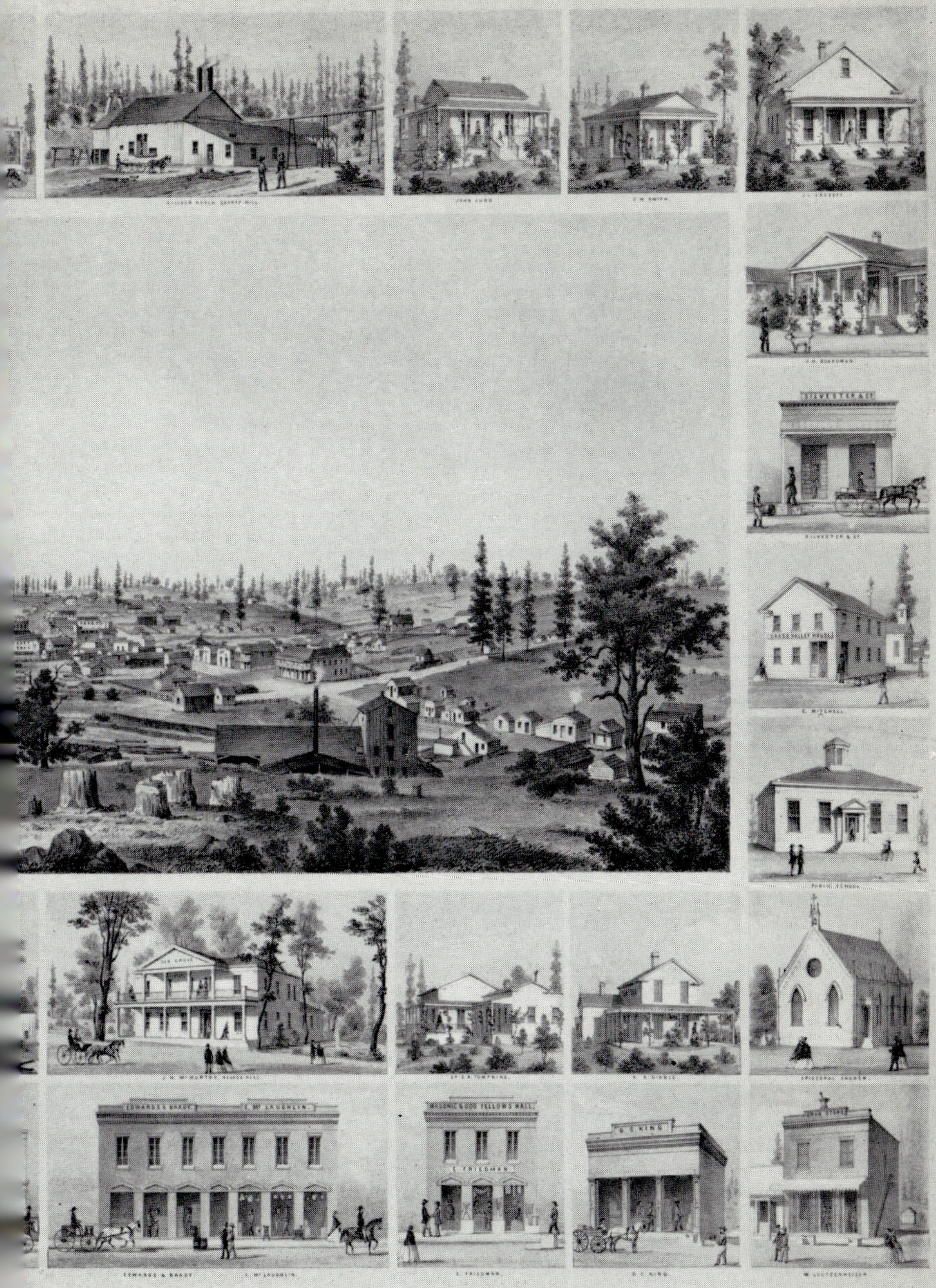
JOHN JUDD
C. W. SMITH
SILVESTER & Co
GRASS VALLEY HOUSE
PUBLIC SCHOOL
EPISCOPAL CHURCH
EDWARDS & BRADY
C. McLAUGHLIN
MASONIC & ODD FELLOWS HALL
EDWARDS & BRADY
C. McLAUGHLIN

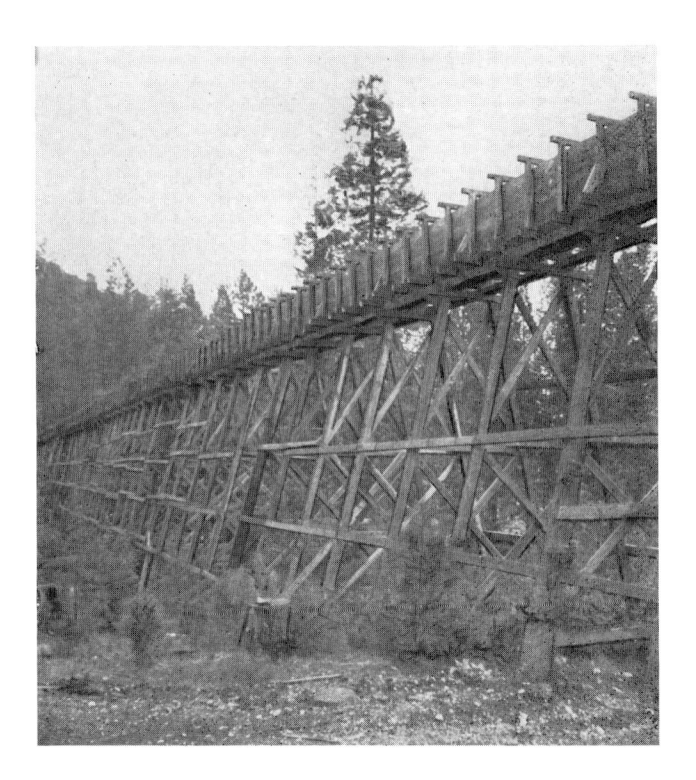

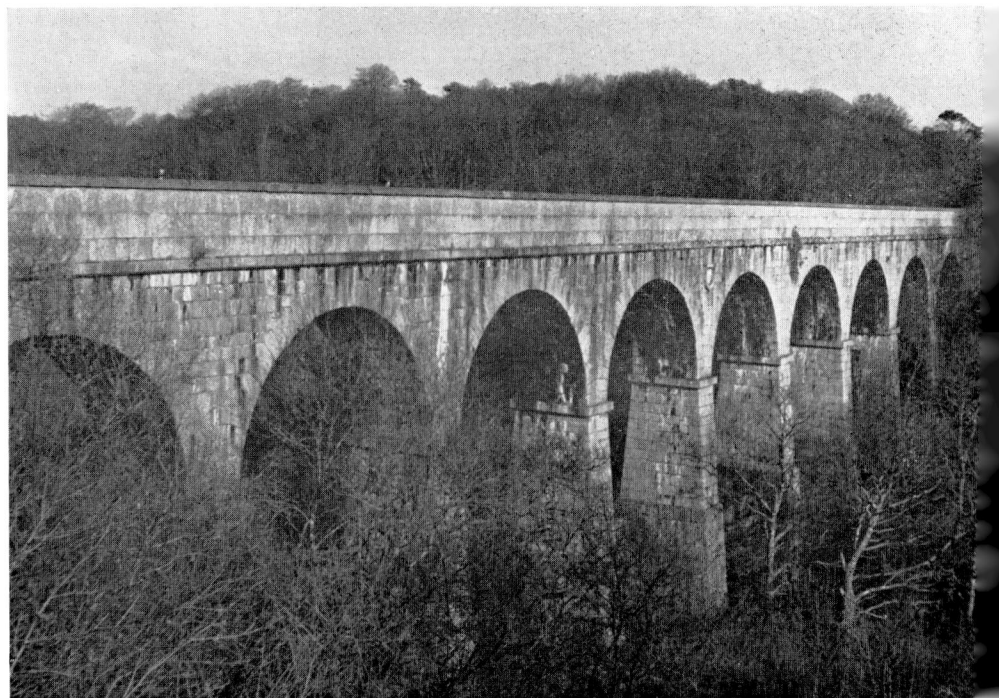

6 Bringing water to the mines. A flume near Quaker Hill, Nevada County, California, and Treffry Viaduct and Aqueduct, Luxulyan, Cornwall

the deliberate policy of excluding Mexican labour had attracted a considerable Cornish contingent thither. Disputes over wages caused trouble in the Amador County mines in 1871, matters becoming so desperate that troops had to be brought in to preserve even a vestige of what passed for law in the gold-mining camps of that country. In 1877 there was a strike at Smartsville in Yuba County, largely caused by the exorbitant prices charged by the company-owned truck shops, and the attempt of the company to exact six dollars a year from single and twelve dollars a year from married men to maintain a 'miners' physician', besides endeavouring to make the men pay for board in company boarding-houses irrespective of the fact whether they lodged in them or not.[52] In California as in the Old Country, however, the Cousin Jack, as a rule, was not a good 'Union man', and the 'giant powder' dispute in Grass Valley was one of the very few successful strikes in which they played a prominent role. There were individuals among the Cornish who came forward and gained some prominence as labour leaders, but most of them took no more kindly to organisation, dictation and advice by fellow workers than they did from their 'bosses'. If told to 'work or quit' many, as lief as not, would 'up and quit' and in fact when the dynamite dispute broke out the second time a fair number promptly left Grass Valley.[53]

However, Cousin Jacks had been coming to and leaving Grass Valley ever since the discovery of quartz gold on Gold Hill in 1850. In the eighteen-fifties they had come, worked awhile and then gone off on the news of gold strikes elsewhere. Not half of them, perhaps hardly a quarter, had come to the Pacific Coast with the idea of settling there permanently. Cornish Grass Valleyans were to be found in many places in the eighteen-sixties and seventies. They had swarmed over the Sierras to Nevada, rushed to such boosted locations as Meadow Lake and White Pine, headed for Montana and the Black Hills of Dakota on rumours of gold discoveries there, found their way to the mining camps of Arizona and Colorado, even drifted across the Pacific to Australia and New Zealand, and when gold was found on the Witwatersrand decided to try their luck there. Many returned to the Old Country—several of them to come back to California again. Whether they left California or stayed, they made their contribution to the social and economic growth of the Golden State. That growth was not so

---

52. *Grass Valley Union*, 14 February 1877. The immediate cause of the strike was an attempt by the mine-owners to reduce wages.
53. Ibid., 20 March 1872.

steady or as uninterrupted as statistics of population increase and of economic outputs suggest. All up and down the Mother Lode are ghost towns; some old-time settlements have vanished, others have the mere shadow of their former bustling life and activity; Rough and Ready and Red Dog are one with Nineveh and Tyre. Other callings than mining have saved Grass Valley and Nevada City from a similar fate. But there was a time, in the early summer of 1858, when some alarmists began to express fears that the entire Mother Lode country would be rapidly deserted. News had come up from San Francisco that riches exceeding those of the Feather River in the days of its greatest glory had been found far to the north on the Fraser River.

6 Toil, trouble, and tribulation

Less than a decade after Wisconsin editors were lamenting that the 'California fever' was sweeping away the best and most enterprising manhood of the Badger State, California's own turn came with the outbreak of 'Fraser Fever'. Reports that rich deposits of gold had been found on Fraser River reached California early in 1858 and caused an 'excitement' that was unprecedented, even in the Mother Lode country, where, ever since the fall of 1848, fortune hunters had been scurrying in droves from one diggings to another on news—or rumours—of rich strikes. Now it seemed to be 1849 all over again and it was placer gold again—alluvial gold which a man could pick up for himself with the crudest of hand-tools, without the costly gear now used to maintain gold-production in California, which only rich capitalists could afford. A new northern El Dorado raised the hopes of hundreds of men who thought they had come to California too late to get the best harvest of gold or whose experiences, in that state, had been utterly unfortunate. Shipping and outfitting firms in San Francisco quickly saw opportunities of profiting from the new furore, and did nothing to check upon or contradict the most exaggerated accounts of gold finds and fortune making in the northern British possessions.

Apart from fur-traders and a few enterprising merchants of both Old and New England, hardly anyone had given much consideration to the region until towards the end of the fourth decade of the nineteenth century. Since 1821, so far as the British were concerned, the region was under the control of the Hudson's Bay Company, which avoided all unnecessary publicity and kept a veil—even a shroud—of secrecy over the potentialities of the sub-continent over which it had gained and extended exclusive commercial rights. Concerned with trading in furs as its main source of wealth, it wanted no interloping agriculturists or miners in the Pacific Northwest, and was reluctant to publish any accounts of its 'territories' that might attract such, to it, unwelcome settlers. It was able to restrict agricultural 'colonisation' within the vast region in which it only, by law, possessed exclusive rights of trading with the Indians, to its servants and veteran employees; most of these were Indians or the half-breed offspring of British agents and clerks, who in that lonesome country had turned 'squawmen', and they only cultivated a little land and raised a few cattle round the trading posts. Stuart in its original foundation, the Hudson's Bay Company continued to be operated in the main by Scots who, in the vast wilderness north and west of the Great Lakes, developed a society not unlike the clan system of ancient Caledonia, hardly, in fact, less primitive than that of the aboriginal nomadic Indian hunters and fishermen who still comprised the greatest part of the population of an extremely sparsely peopled region.

It was these Indians that aroused active American, particularly New England, interest in the region in the third and fourth decades of the nineteenth century. Seamen and fur-traders came back to New England, and in their religious meeting houses told of the heathen they had encountered, thereby stirring the proselytising missionary impulse that at times, and particularly at that time, was a prominent trait of 'Yankee' character. The rivalries, fortunes, and misadventures of the American fur-trading companies were also given much publicity, particularly those misfortunes and losses which could, justly or not, be attributed to the activities of the privileged monopolistic Hudson's Bay Company which became to Anglophobe, egalitarian Americans as abhorrent as the East India Company had been to their grandparents sixty years before.

Such was the beginning of the 'Oregon fever', but it was only the mere symptom of an irritating rash until the depression of 1837 bankrupted many struggling New England farmers who had obeyed the scriptural injunction to increase and multiply to such an extent that their land could no longer support them and their families. In any event hard times came just when fur-traders' and missionaries' tales about the resources of the Oregon country were gaining considerable publicity. Such reports did not, as a rule, boast of a land flowing with milk and honey, but at least they told of a better watered and a better timbered country than the prairie regions of the Great Valley of the Mississippi to which many New Englanders had already migrated, and of a country enjoying a less extreme climate than Illinois and Wisconsin. Overland to Oregon they made their way, few at first, then, in the early eighteen-forties, in scores and hundreds. The migration brought to the Pacific Northwest not only New Englanders but some who had attempted to settle in the Middle West but who had found it not to their liking. Even when California gold fever broke out, Oregon still attracted a fair number of those who made their way to the Pacific Coast, including a few Cornish from Wisconsin, who probably had not done as well as they had hoped to do by coming out from the Old Country to the lead-mining region of the Upper Mississippi. Furthermore, the Californian diggings offered ready markets for the forest and farm produce of the early settlers in Oregon.

Among the Cornish immigrants who came to Wisconsin in the early eighteen-forties had been Samuel James of Trelan in St Keverne.[1] In the Old Country he had probably incurred the dislike and distrust of the local Tory landowning gentry of the Lizard district by the active part he had taken in the movement for Parliamentary Reform. Largely self-educated, and with an inquiring philosophical bent of mind, he had written a few pamphlets and newspaper articles before coming out to farm in Racine County, Wisconsin. When he arrived most of the best lands on the shore of Lake Michigan had already been taken up by settlers who had swarmed in during the past decade,[2] and after a few years he decided to move to Oregon, a region which had been attracting considerable notice in the Wisconsin press for some time. Possibly the climate more than anything else decided him to move on, for, in 1860, he wrote in a letter to another intending Cornish emigrant:

The climate of a new country is the chief thing to be regarded (by an intending emigrant). If the climate be suitable, we need not despair of being able by time and industry to make other things so. All through the great country of the United States, from the Rocky Mountains to the Atlantic Ocean, the climate is extremely severe, such as I am sure no-one can form a just idea of who has not been out of England: in winter, it is almost as cold as Kamchatka, and in summer as hot as the West Indies; and the changes in spring and even in summer sometimes from heat to cold and the contrary, are most sudden and excessive and the water even of running springs is seldom very good ... It is also for the most part very sickly there.[3]

But, James went on, west of the Rocky Mountains, the climate was better, healthier, and free from extremes of heat and cold; the air was pure, and the water so soft and clear that he often drank from a spring just for the pleasure of doing so and not from thirst. In Washington Territory, where in Thurston County he finally settled,[4] the climate

1. *W.B.*, 20 April 1866.
2. *Wisconsin: A Guide to the Badger State*, W.P.A. (New York, 1961), p. 551.
3. *W.B.*, 3 August 1860.
4. The letter was dated from Grand Mound, Thurston County, 7 February 1860.

is nearer like the English than is any other that I am aware of, but there is some difference. We have no such heavy storms, and very rarely a beating rain. There is, so far as I have been able to judge, generally about two weeks of pretty sharp cold at some period during the winter, that is, in some part of the time from late in November to the end of January; the rest of the winter is mostly warmer and far pleasanter than it is in England. In the beginning of February, spring begins to come a little; the lent-lilies, tulips, hyacinths, paenies (*sic*) and rhubarb are now beginning to come up in the garden, and the grass starts a little in the fields; we are now (in early February) very busy in getting ready to sow oats and spring wheat, the weather being mostly as warm and pleasant as April in England, though sometimes a little frost or a flurry of snow, which quickly goes off, takes place now and then to the end of March. From the latter end of March to the end of June, is a most delightful time, gentle showers and warm sunshine; from the end of June to the middle of September very little rain falls, a scant shower now and then, but haying and harvest are not very often hindered by the weather. About the middle of September, gentle showers become more frequent, and this is the best time to sow wheat, although it is sown from this time till May following. The full spring of grass takes place, and the weather throughout the fall is very pleasant. The months of July and August are considerably warmer than they are in England. Wheat is ripe from the middle or end of July to the end of August, according to the time of sowing. In consequence of not being so much rain in summer, the grass does not grow so much at that season as it usually does in England, excepting quite near the ocean, where there are almost always night and morning mists which keep everything green and fresh.

James had settled in one of the heavily-timbered coastal districts of Washington Territory, which was very unlike the prairie or natural grasslands of Wisconsin. In many places, however, forest fires had left second-growth brushland which could be cleared without much trouble; the earth mould was very fertile while the subsoils being a mixture of sand and clay could be drained easily besides facilitating communications in what was, then, a roadless wilderness. Cattle had plenty of forage on which to live in the woods which provided abundant timber for building and fencing. Any settler coming into the region could buy a quarter section of a hundred and sixty acres for forty guineas sterling, and could almost select any location he chose since few had, at that time, arrived in the district, and a 'squatter' might not be called upon to pay for the land he settled upon immediately, perhaps not even for years to come. Many Americans, too, liked to reclaim 'wild land' and then sell it to a later arrival while they moved on to do the same again. On such holdings, James wrote:

every man is his own (land) lord, and, therefore, can do as he pleases with it, and it is astonishing to see how much work is done in a short time towards bringing the land under cultivation. No time is spent in hauling manure, or any extra cultivation; one ploughing and one harrowing, without any rolling is generally all that is done to raise a crop. The grain is all harvested by the scythe, and one man harvests (that is, cuts, binds, and shocks) one English statute acre for a day's work, and in some places this will yield sixty-five imperial bushels of wheat. It is bound in large sheaves (generally called bundles) about three hundred to the acre; after standing in shocks for a few days it is hauled into the barn, which is made large enough to hold all the crop. It is threshed and winnowed generally at once, by machines which travel through the country, or else it is trampled out by horses or oxen.[5] Grain thus threshed and cleaned costs about one-fifth of the crop to pay for it. Our fanning mills (winnowing machines) are of the very best description, so that three hands will clean up and put into the bin from one hundred to a hundred and fifty bushels of wheat in a day. In consequence of these and other things a great deal of time is left for clearing and fencing the land. It requires about five thousand rails to make a good fence a mile long. A good man easily makes a hundred fir or cypress rails in a day, some persons make from a hundred and fifty to two hundred and it generally costs about as much time to haul them and put them up into a fence as it does to make them. To lay up five hundred rails in a fence is reckoned a day's work, and from sixty to eighty are generally hauled on a waggon at a load, and commonly from two to four loads are hauled in a day with an ox team.

5. This description shows that Pacific Coast harvesting methods in these early pioneering days were no more modernised than those then prevalent in Cornwall. Reaping machines had been introduced in Cornwall during the decade at the end of which James was writing from Washington Territory, and there were a number of 'portable' threshing machines in use; these were hauled and operated by horses, although in 1812 Richard Trevithick had used steam power to operate a threshing machine he devised. An acre a day harvesting was deemed a good performance in the Old Country, too, at this time, and there are a few occasional newspaper reports of sentuagenarian and octogenarian labourers accomplishing such or even slightly greater days' work. Modern harvesting methods almost make some of James's terms unintelligible; a *shock* was a number of sheaves stacked together for support on their stubble ends, and although such a term was used outside Cornwall there were other local or regional names for it in Britain. The technique of shocking corn, too, varied. In St Keverne and most Cornish parishes the usual practice was first to stand a single sheaf upright, then to place single sheaves against it to make a row of three sheaves sloping upwards to a blunt point of grain-ears then placing two more sheaves on each side of these central three again leaning towards the centre, and then sometimes though not invariably, inverting a low-bound sheaf over the top of the seven as a 'hat' to keep off and drain off rain from the heads of grain. 'Hatting', however, demanded long-strawed wheat which was less common in the drier clime to which James had emigrated than it was in the country whence he had come. In drier regions of Britain six or eight sheaves might be shocked merely in double rows of three or four sheaves on each side; this formed a less compact shock which aired and dried more quickly, but it was a practice that 'asked for trouble' in areas where there was much risk of even moderately high winds in harvest time.

Most pioneer farmers in Washington Territory pursued a mixed type of husbandry, raising cattle and sheep, making butter and using the skim and butter milk for fattening pigs along with wheat grain. It was only the best lands that produced crops of sixty-five bushels of wheat to the acre, for later in his letter James called ten to fifteen bushels an average crop while richer loam soils, on the average, produced up to thirty or thirty-five bushels. Oat crops varied from twenty-five bushels per acre on poor to three or four times that yield on the most fertile soils. In the concluding part of his letter, James told his friend that the best way to Washington Territory was to come via New York, Panama, San Francisco, and Olympia, which by steam vessels and allowing certain extra expenses would, he thought, cost about forty pounds sterling; coming by sailing ship round Cape Horn would only cost twenty or twenty-five pounds, but would take three times as long.

This advice to the prospective emigrant to come by sea suggests that Samuel James had found the overland route such as he would not willingly undertake to travel over a second time or to advise any of his friends to take. He warned his correspondent to be wary of the lodging-houses in the ports and of strangers who would impose upon and rob 'greenhorn' travellers. A staunch Methodist, James recommended the immigrant if he stayed in any place for a few days to

> enquire out the nearest Methodist minister, and he will probably help you much with his advice. Bring certificates of church membership along with you—they will be of much service at such times. Make no agreement for passage, without finding out some respectable storekeeper or minister to see that all is right.

He ended his long letter, after remarking that what he had written applied to Vancouver Island and British Columbia as well as to Washington Territory, with the suggestion that it be communicated to the Cornish local newspapers so as to help others who had thoughts of emigration.

Probably Samuel James only desired to be helpful to others who, in 1860, were debating in their minds whether or not they might 'better themselves' by leaving the Old Country and going to America. There is no indication that his letter, unlike so many 'private' letters which found their way into newspapers were published by parties with selfish interests in promoting emigration. Nevertheless, his letter left an impression that the country wherein he had settled had as much claim to the epithet 'delectable' as had the Duchy of Cornwall, and that the only real difficulties and troubles that might befall the emigrant were those encountered on the way thither. Washington was a country where every prospect pleased.

Not unaptly, however, the over-often repeated quotation could be completed in this case—'and only man was vile'. Samuel James, too, may never have known Nature's ravaging terror and the blackened desolation left by forest fires. He had, however, when he wrote, been on the Pacific Coast long enough to have realised that where man was concerned it was a 'dark and bloody ground.' Before the miners came north from California, fur-traders and settlers had been embroiled in Indian wars. In the late eighteen-fifties, hardly a summer had passed without Indian unrest and outbreaks of greater or less magnitude. Racial antipathy between white and red man had not been virulent in the Pacific north-west in the old fur-trading days, but after the Treaty of Washington in 1846 and the consequent gradual but complete withdrawal of the Hudson's Bay men from the region south of the British line, the era of conflict began. Events on the far side of their now defined border more than justified the fears of British authorities that, if mineral or agricultural prospects attracted immigrants from the South into their preserves, tragedy and bloodshed would certainly follow. The finding of alluvial gold on other northern rivers, south of the Canadian boundary, at the same time as the Fraser discoveries, led to numerous, smaller, unchronicled re-enactments of the Black Hawk tragedy of the Upper Mississippi lead region. Although these racial conflicts were smaller in scale, in their totality they led to far more bloodshed among both red and white men.

Fearing what inevitably came about, the Hudson's Bay men tried to keep the news of the northern gold finds secret, but with so many wandering adventurers on the Pacific Coast after 1849 this was impossible. As early as 1852 or 1853 a specimen of gold from 'Vancouver's Island' had been sent home to England by a travelling engineer, but although displayed in the British Museum it had attracted very little attention;[6] it is possible that Hudson's Bay men knew that gold existed in their 'territory' even before that time. It would be an exaggeration to say that the finds in California started everyone seeking for gold, but they certainly inspired many to search for it in all sorts of likely and unlikely locations. James Hargreaves had hastened back to Ophir Creek in New South Wales because its topographical features were so similar to those of some rich Californian diggings and had struck a 'bonanza'.

6. *Mining Journal* (London), 29 May 1858.

A few in Cornwall recalling tales of little 'quillets' of gold found in the Carnon tin stream workings generations back, kept a keener eye for anything that glittered in the mines of the Old Country, while in Mineral Point a boy idling away his time and wearing out his boots the way boys are ever prone to do, kicked up a stone which was reported to be auriferous.[7] In California there developed a notion that the great Mother Lode was merely a fragment in a broken range of gold deposits which increased in richness the nearer to the North Pole, and this drew some prospectors northwards towards 'Oregon' when they heard descriptions of that country's topography given by the men who had brought down supplies and draught animals to the Californian mining camps. The paleface mania for gold had become known to northern Indians, who, in their hunting and fishing expeditions, had noticed the yellow dust and passed it by on the banks and sandbars of the rivers of their country. Indians in the North had certainly traded gold to the Hudson's Bay Company men, while the secrecy which the Company sought to maintain about it seems to indicate that this amount of Indian gold had increased substantially in the years immediately following the Californian 'excitement'.[8]

By the beginning of 1858 it was a secret no longer, and in April the exodus from California began. During the last few months accounts had been coming southward about rich discoveries of gold along the Fraser River, but a discordant note was struck by the self-styled 'Old Californian', Thomas J. Peters who wrote from Portland in Oregon on 18 April,[9] flatly stating that all the reports were 'humbug', a humbug worse even than the Kern River fiasco in California some years before, and that his trip to Fraser River was the crowning folly of one whose 'lot' it had been 'to run after humbugs' all his life. He had got back as far south as Oregon, which he grudgingly admitted to be a good country—if it did not 'rain six months of the year when they don't need it. If California only had such rains it would be the best country in the world.' Nor had Peters a good word to say for the northerners who looked on all Californians as rogues. Although some support was soon forthcoming to back up Peters's jaundiced views of the north country, his criticisms evoked a stinging reply which, if it was not deserved by Peters himself, certainly touched to the quick a certain class of 'miners'

7. *Mineral Point Tribune*, 6 May 1852.
8. Reports of rich gold finds as far north in the Oregon country as Rogue River were appearing in Eastern newspapers in the early summer of 1850. (*Wisconsin Express*, 11 June 1850).
9. *Daily Alta California*, 23 April 1858.

who were to be found in practically every mining camp on the Pacific Coast. The writer, concealing his identity under the initials 'A.S.', but claiming to be vouched for by prominent San Franciscans, declared that:

Peters never was within hundreds of miles of the so-called Fraser mines. He is one of a class of men to be found in almost every country in Christendom, who possess neither energy, industry, or perseverance, and who, if he knew of the existence of a deposit of gold within forty rods of his locality, sufficient to supply his wants for life, and the condition imposed for his procuring it was that he should dig it with his own hands, I do not believe that he would do so, at least not more than sufficient to keep body and soul together.[10]

Whether Tom Peters had been to Fraser River or not made little difference to gold-hungry Californians. In late April they started swarming north out of San Francisco in hundreds; by midsummer close on ten thousand had sailed out through the Golden Gate bound for the new northern El Dorado,[11] and in the same period less than a hundred passengers from the Puget Sound ports arrived in San Fancisco. Many more went overland up through Oregon and Washington Territory, risking conflict with the Indians who were causing considerable trouble to the federal troops in that region all through the summer of 1858.[12] River boats coming down from Stockton and Sacramento were crowded with miners from the Mother Lode country, and some inland journals were expressing fears that the Californian gold diggings would be left to the Chinese. The *Grass Valley Telegraph* reported round about the middle of May that:

The effect of the Fraser River excitement is beginning to be felt here severely. Not less than a hundred and fifty men have already left Grass Valley, and others are leaving daily, mostly men employed about the quartz mills. Scarcely a mill but has lost some of its best workmen by the prevailing Fraser River epidemic, and many of those who remain do so from want of funds to reach the new El Dorado. The consequence of so many drifters leaving us is, that those who remain are striking for higher wages, and some of our mill men are seriously thinking of stopping their works until men can be obtained who are

---

10. *Daily Alta California*, 29 April 1858.

11. The San Francisco Customs House recorded 6,251 passengers sailing north between 30 April and 23 June; the *Alta*'s city columns reported another 3,200; the numbers of stowaways and those sailing without tickets were notoriously high. The steamer *Cortez*, tonnage only 1,117, on one voyage took 894 booked passengers to Victoria from San Francisco. (*Daily Alta California*, 24 June 1858.)

12. News reached Fort Vancouver on May 24 that two companies of dragoons under Colonel Steptoe, had lost fifty men and three officers in a clash with Indians on the Snake River. (*Daily Alta California*, 1 June 1858).

willing to work for prices which the mines will justify. It must not be supposed that all mines can afford to pay extravagant wages. Indeed there are limits beyond which they cannot go without loss, and prices have been paid hitherto as high as the mines will warrant. Some of our mill men are talking about employing Chinamen—an experiment which might succeed, but of which we have many doubts. Others are taking green hands, Americans, who readily adapt themselves to the business, and soon become good drifters. A fine opportunity is now offered to those out of employment in the valleys to make from three to four dollars per day, or from fifty to a hundred dollars per month, and one of the benefits of the Fraser River excitement will be to give good and steady employment to those who are willing to work for fair wages at home.[13]

But the prospect of sure employment at three or four dollars a day was not enough to hold men in Grass Valley when there were rumours that men were making twenty, thirty, even fifty dollars a day up on Fraser River. Even the report of a quartz bonanza averaging two thousand dollars to the ton of quartz and paying out from sixty to two hundred dollars to the man was not enough to stay the exodus.[14] It can be assumed that the hands the 'green' Americans replaced included several Cousin Jacks. Furthermore, though gold was the main lure, many British-born miners welcomed the chance to leave the tougher mining camps, in which the only law was that handed down by Judge Lynch and executed by vigilante 'mobs', and to return to the protection of life and property guaranteed and enforced by British laws and institutions once more.[15]

With so many men leaving California in the early summer of 1858, it

13. Quoted in the *Daily Alta California* on 21 May 1858. On June 14 the *Alta* published a letter from the Grass Valleyan quartz miner Charles C. Roberts which stated that the Fraser fever made it almost impossible for the depressed quartz-mining interests to continue working, and that the only quartz mill then doing moderately well in the vicinity was that at Allison Ranch.

14. *Daily Alta California*, 21 June 1858 quoting *Sacramento Statesman*, n.d.

15. *Mining Journal* (London), 3 July 1858. The *Royal Cornwall Gazette* on 9 September 1858 copied an article from the London *Times* which praised the firm, just policy of Governor Douglas, particularly since 'amid so many conflicting duties he found time to travel to the distant and inaccessible gold-diggings, settled disputes between the Indians and the whites, and explained to the American miners the law under which they were to live. Whether it was from this novel and sensible step on the part of one who might have stood upon his authority and dignity, or whether Californians have had enough of the sweets of Democratic government, acting under the inspection of a Vigilance Committee, certain it is that they have conformed in an exemplary manner to the laws of the Colony, and even seem to have imbibed a taste for a strong Executive, which has established a vigilant and efficient police, and protects with a firm hand both life and property.' The Ned McGowan troubles, however, were yet to come.

was not surprising that many envisaged the depopulation of that state and its consequent decline. Victoria and other ports of Puget Sound had visions of their surpassing and eclipsing San Francisco. Canadian talk of a British transcontinental railroad linking the Fraser diggings to the ports on the St Lawrence seemed more likely to be put into effect than any railroad between California and 'the States'; Californians had long been urging such a project as vital to their interests, only to see it bedevilled and frustrated by Congressional feuds in Washington over the same issues which had delayed the admission of California to statehood in 1850 and which, then, had almost precipitated the disruption of the American Union.[16] Californian impatience with the controversy between the North and the South was intensified by the new gold discoveries, and to the flamboyant oratory of Southern slavocrats, who had not scrupled to call Western free labourers 'mud-sills', was hurled back the challenge that gold, not cotton, was king.[17] Although the Californians had banned slavery, their motives had been far from purely altruistic, and the racial prejudice against persons of colour in the Golden State led to the migration of many Negroes from California to Fraser River in the summer of 1858. Along with the Negroes went several Chinese whose departure was viewed with relief rather than with regret by white Californians. American racial prejudices, however, were to sustain further rude shocks when the British authorities on Vancouver Island enrolled immigrant Negroes into their police force and when certain law-disrespecting and peace-infracting Americans suffered the indignity of arrest at their hands.

For different reasons from those entertained by Americans, many British welcomed the rush of population into the northern regions as a means of terminating the exclusive trading and political privileges of the Hudson's Bay Company. The prospect of a firmly established British colony, however, was galling to many Americans particularly to such scapegrace adventurers as Ned McGowan, who spent eight or nine months in the Fraser diggings, creating trouble for the officials and police Governor Douglas had appointed by continuing his feud with the San Francisco 'stranglers' whose clutches he had narrowly escaped in the days of the 1856 Committee of Vigilance. McGowan possibly had designs of filibustering the annexation of the 'Oregon' country which

16. *Daily Alta California*, 24 February, 19 and 24 June 1858.
17. Ibid., 26 May 1858. Since James Hammond's notorious speech had only been delivered to the Senate on March 4, this Californian retort was speedily forthcoming, and it might be suggested that it was a pity it did not get more attention in the East.

Polk had apparently sacrificed by the 1846 Treaty of Washington, but there was to be no second 'Bear Flag Revolt' on the Fraser similar to that which had helped win California for the Americans in 1846. In fact, McGowan was lucky not to meet a similar fate to that of the unfortunate Captain William Walker in Nicaragua in September 1860. The majority of those who went north from California, however, were far more interested in getting gold than in trouble-making politics, and not much richer, somewhat sadder, but hardly a whit the wiser, McGowan left the Fraser country for the south early in 1859; a little later during the Civil War he did the Confederacy more harm than good in Arizona by attempting to win Apache support for the Southern cause. His sojourn on Fraser River, however, revealed that, in the mining camps located within their boundaries, the British intended to maintain law and order. The British, too, had learnt from the Eureka Stockade affair in Australia that the business of government in gold diggings was to protect and not to exploit immigrant miners, and Douglas and his subordinates did all they could to ensure the provision of the necessities of life at reasonable prices to isolated camps by road-making, trail-cutting, and bringing in steamboats to ply on the rivers wherever navigation was possible.

Even then, things were very difficult. Fraser River, like most of the gold excitements, showed that exaggerated, reckless early hopes invariably ended in disillusionment and bitter disappointment for the greater number of those who, without much reason or forethought, rushed off to 'new El Dorados' at the first whispering rumour of rich gold strikes. A few fortunate men had found a fair amount of gold late in 1858, and it had been reports of their luck, often wildly exaggerated, which caused the great exodus from California in April, May, and June. Most of these migrants had not taken any trouble to find out anything about the conditions they would meet on Fraser River, or even how to get there. Shipping companies can be blamed for advertising passages to the new gold regions which only took men to Victoria or to other harbours of Puget Sound. The diggings were many miles up the river, and men who thought that the Fraser offered such advantages of facile navigation as the Sacramento and the San Joaquin found out their mistake when they got to the northern ports. Hundreds arriving in Victoria could not wait for river boats capable of navigating the lower Fraser to be provided, but bought boats and canoes from the Indians and recklessly set off across the wide straits to the mainland and the outfall of the river. Many foundered on the way and were drowned;

others reached the river mouth and found the Fraser a treacherous torrent, swollen by the melting of the snows on the inland mountains, and worst of all, the spring floods had submerged many of the bars on which the gold had been found during the winter when the river had been at its lowest. The great question was no longer how to get to the gold or how to work it when they got there, but when would the river go down. It was only slowly realized that flood levels would only subside when the snows on the distant mountains ceased to melt. In June, men were confidently speaking of the river 'going down' by mid or late July; in July, less confidently, they reckoned that the gold-bearing sand bars would be dry sometime in August; in August they were somewhat less than hopeful that it would be possible to start looking for gold in September. By that time many had gone back to California again; some, however, were staying on in a grim determination to 'see it through'; others were staying simply because they were stranded, having used up all their supplies and money, utterly unable to go back home unless some charitable ship's captain took pity on their plight. When the waters finally subsided, there were so many adventurers in some places that there would not have been room to shovel out a spadeful of dirt had the authorities not strictly regulated the size of placer claims.

Gold did exist, but not enough for the throngs of adventurers that had come to Fraser River. There were about eight bars on the twelve-mile stretch between Hope and Yale,[18] the locations and names of most of which have been completely forgotten, but in October 1858, a considerable amount of gold was recovered from Cornish Bar four miles below Hope.[19] The name suggests that it was worked by Cousin Jacks, and among those who toiled on it may have been the Wisconsin Cornishmen Richard Wearne[20] of Linden and Christopher Clemens. By 1860 Clemens had made sufficient to return to his family in Grant County and buy 'four hundred acres of land all in one piece'.[21] It is unlikely that Cousin Jacks on Fraser River fared any better than the majority of gold-seekers who flocked into that region in the late eighteen-fifties; one or two did well, some cleared their expenses and assuaged their appetite for adventuring. Others had no luck at all and cursed the day they left Cornwall, Wisconsin, or California. Many, indeed, lost their

---

18. E. W. Howay, *Early History of the Fraser River Mines*, p. xvi.
19. Ibid., p. 7.   20. *History of Iowa County*, p. 879.
21. *Commemorative Biographical Record of the Counties of Rock, Green, Grant, Iowa, and Lafayette, Wisconsin*, p. 394.

lives on Fraser River, among them being the Wisconsin Cousin Jack, Nicholas Jenkins of Plattville,[22] while the only record of another Cornishman, John Carne Williams, is that he in March 1859, left about seven ounces of gold for safe keeping with the Police Superintendent, Chartres Brew, at Fort Yale, when he and his American partner, George White, went further up river in search of richer bars or of the mother lode which many miners were convinced must exist somewhere to the north.[23]

Mining experience in California suggested that the fine flakes and spots of gold found on the bars of the Fraser must have been washed down from auriferous lodes further upstream, and attempts to locate these lodes were dogged and determined but, for the generality of adventurers, in vain. The reckless folly of those who, in the early summer of 1858, set out in rowing boats and canoes to cross the stretch of water between Victoria and the mainland of British Columbia, was sanity compared to the daring disregard for personal safety of those who, disappointed in finding good claims on the bars of the lower Fraser, attempted to make their way up river to the country where they reckoned the mother lode was to be found. Northward towards regions of cold and snow they went; canoes capsized and men were drowned in the swift-flowing, icy-cold flood waters. Overland travel was every whit as perilous. For part of its course the Fraser flowed between the cliffs of a canyon whose crags were inaccessible even to mountain goats. Travellers had to pack all their provisions along with them, for the forests were poor in game. The Indians at best were suspicious, at worst hostile, while many of them seemed to have conceived the idea, possibly originating in the times when the Hudson's Bay Company had been the bitter rival of American fur-traders, that the British would not condemn them for any attacks they made on the hated 'Boston men' who were fighting their brethren down to the south. Some Indians and half-breed trappers were a little more helpful, and informed them at times of locations where the yellow dust and gravel they sought might be found. Occasionally they found signs of gold, usually just 'colour' enough to revive flagging hopes and to lure them still further on into the bleak, hostile northern wilderness.

Finally, gold in abundance was found some five hundred miles from Victoria at Cariboo. Among those who 'struck it rich'—at a second

---

22. *Commemorative Biographical Record of the Counties of Rock, Green, Grant, Iowa, and Lafayette, Wisconsin*, p. 460.
23. F. W. Howey, *Early History of the Fraser River Mines*, p. 91.

attempt—was the Cornish miner Martin Raby. At Lightning Creek he sank a fifty-five-foot hole without finding a particle of gold, and he began wishing himself back in the diggings of California and Australia where he had previously worked. His partner gave up but Raby decided to try again over at Williams Creek, some twenty miles further away. There he found a rich lode, staked ten one-hundred-foot claims, organized a company to work them, hired labour at the rate of ten and twelve dollars a day, and by the end of the short working season from early June to late September he and his new partners had got out a hundred thousand dollars worth of gold. He wrote home to his brother-in-law, who was then managing the Parys Mine in Anglesey, that

> the gold in this country is not disseminated like that of California and Australia; it is principally confined to the beds of creeks, or in such channels as make it difficult to find. If a man should be fortunate enough to strike a lead, he is almost sure to make a fortune, and that is what we are all after in Cariboo—but it must be remembered that there are many blanks to a prize.[24]

Towards the end of his letter, Raby added that no-one should come to Cariboo without at least five hundred dollars in his pocket when he arrived, and that the winters were long and cold, the thermometer standing at thirty below when he wrote on 17 February 1864.

It was, however, another Cornishman who discovered the greatest bonanza at Williams Creek. Billy Barker, a seaman, had been among the first adventurers who sought gold on the Lower Fraser in 1858. Not particularly lucky but still hopeful, he ventured further and further up the river. With seven others, including the Hankin brothers who may have been fellow Cousin Jacks, Barker found a rich gold lode in August 1862, and after spending a week celebrating their discovery, the party went to work; the claim is reputed to have been worth at least half a million dollars, but Billy's share looked pretty wan after a winter spent in Victoria honeymooning with a shipwrecked London widow. He brought her back to the diggings with him in the spring, but there she found the company of younger men more congenial than that of her stocky, grizzle-bearded, bow-legged husband who, now most of the money was spent, liked to pass his evenings when the day's work was done sitting like other Cousin Jack miners before a blazing fire in his stockinged feet, perhaps even reading the *Pickwick Papers* and regretting that he had not heeded Tony Weller's admonitory advice about

24. *Mining Journal* (London), 21 May 1864: letter of Martin Raby to Captain Mitchell, 17 February 1864.

'vidders'. When he had the gold, Billy Barker 'whooped it up' in the saloons with the best of them, but with his not uncommon weaknesses for wine, women and song, Billy and gold were apt quickly to part company. Up in the Cariboo country it almost became a legend how he would come into a saloon, put down a few drinks, and then start hopping and jigging about proclaiming

> I'm English Bill,
> Never worked and never will
> Get away girls
> Or I'll tousle your curls!

He lived to be seventy-four, old indeed for a Cornish miner who lived rough most of his days, dying in a Victoria poorhouse for aged men. The town named after him, Barkerville, through a disastrous fire in September 1868 and the exhaustion of rich 'pay dirt', can hardly be said to have survived him, though it did not become completely a ghost town. Billy Barker was probably the only Cousin Jack miner to have a town named after him, a town whose life was as turbulent, chequered, and marked by such varying fortunes as his own. But all along the Pacific Coast region, from Bering Straits down to the Sonora Desert, in those days were adventurous and luckless men like Billy who could truly carol

> Let their chorus loudly ring,
> The Broken Miner's lot they sing,
> Most bitter is the lot indeed
> Of him who cannot find the lead.[25]

In 1858, however, there was another American gold 'excitement', one that had drawn men to a location far less remote than Fraser River. Gold had been found some years previously in Rowan County, North Carolina, but in 1858 some mining company promoters took a hand, and organised operating companies to labour for certain profit to themselves and a few *douceurs* in the way of intermittent dividends for the investors whom they persuaded to 'put up the money'. Among those who came to work in these mines were a handful of Cousin Jacks, including James Eudie, Nicholas Snell, Thomas Penaluny, Matthew Matthews, Thomas Thomas, and a young engineer, James Gill, who, though born

---

25. Adapted from verse by John A. Fraser, published in the *Cariboo Sentinel*, 29 October 1866, reprinted in *Rhymes of the Miner*, edited by E. L. Chicanot (1937), p. 17. The Cornish *lode* in American mining terminology became *lead*. The account of Billy Barker is based upon Louis le Bourdais, 'Billy Barker of Barkerville', *British Columbia Historical Quarterly*, July 1937, pp. 165–70.

in Ireland, was probably of Cornish stock.[26] They came to mine gold and probably took no notice of the frantic excitement that swept the locality when news came of John Brown's raid on Harper's Ferry, or even of the feverish presidential electioneering of 1860.[27] When, however, the guns roared at Fort Sumter and, a few days later, North Carolina seceded from the Union and joined the Confederate States at war with Abraham Lincoln, the first thoughts of four of them were to return immediately to the Old Country, but

being unable to turn our paper money into cash and learning that some very large consignments of arms and ammunition had arrived in the Southern States from England, and reading in the newspapers of Carolina that the sympathies of England were also with the Confederates, we resolved to stay and watch the progress of events. Gold Hill township was inundated with recruiting parties, and such was the spirit created that in less than two months nineteen out of every twenty of the male adults were either in the service of the local or the general army. The English settlers generally kept aloof, which aroused such a deadly animosity against us that it soon became unsafe for us to venture out, unless in parties of four or more, and then the not very welcome cries of 'You damned Britishers' mingled with oaths and curses of no flattering description, were the escort with which we were invariably accompanied. So great was the excitement in April last that, after a muster of the Gold Hill volunteer force for practice, a party of thirty or forty surrounded our lodgings, and demanded that 'the damned Britishers' should be given up to them, which being refused, they demolished the windows, and forced an entrance into the house, and I decamped in my shirt and trousers for the bush... A party of eight or ten gave chase. However, after running about ten miles, I found that I had outstripped my blood-thirsty pursuers and I at once climbed into a tree to wait for the darkness of night before I would return to my lodgings.[28]

26. Rowan County Census, 1860. The returns also include Gribbles, Martins, and Moyles, who were possibly Cornish. The uncommon spelling 'Penaluny' and 'Moyall' can be attributed to the inability of their bearers to spell their own names, or to their being recorded by a local census official who took little pains to assure their correctness. Similarly the 1860 Census for Jo Daviess County, Illinois, returned Cousin Jack names Bolitho and Chenoweth in the strange forms 'Bliotho' and 'Chenywth', while the same Census for Hancock in upper Michigan found Francis Holman giving his name the way he pronounced it 'Ulman', thus possibly bewildering later genealogists into thinking him German, had he not given Cornwall as his country of birth.

27. It cannot be said for certain that any of these Cousin Jacks had at the time of the Census of 1860 taken out naturalisation papers and been qualified to vote. Many immigrants, it is true, voted without being so qualified. It is possible that the listing of 'personal property' in the return of one of these men, James Eudie, indicates that he was naturalised; even then his property was only estimated at $25.

28. *Royal Cornwall Gazette*, 4 October 1861.

The Cousin Jacks then thought it best to avoid further trouble by joining the militia, but they had no desire to get involved in the war. With the news of the first battle of Bull Run, and the subsequent return home of wounded men and the bodies of some who had fallen in action, Carolinian war-passions became even more intense, and the Cornish miners became anxious to return to Cornwall. When they heard that the Gold Hill militia company was to be sent to the front on active service late in August they decided to leave, despite the fact that the only money they possessed was paper currency which, they knew, would be worthless outside the Southern states. The authorities at Richmond did not cavil at giving them their passports, but when they reached the Potomac River at Weetman Court House, some three hundred miles from Rowan County, they

> found it was no easy matter to get to the opposite side of the river, which is at this place fifteen or sixteen miles broad, and no-one would take us across at any price; we were therefore obliged to purchase a small boat, for . . . twenty dollars, and off we started to cross the Potomac for Maryland. Our boat was so leaky that one of us had to keep bailing her by turns, another for steersman, and two to row. We passed several ships-of-war, but they would render us no assistance, and on reaching the shores of Maryland we made signals of distress, in hope that the garrisons would send us relief or help, but all to no purpose. After many fruitless attempts to effect a landing, we at length put into a small creek, and after hollowing for some time, we found that we were about forty miles too far down the river, and were directed to row up the river again. We accordingly re-embarked, but finding it impracticable to reach our destination the same night, we put ashore on a little island in the river, and after hauling our little boat above the reach of the tide, we lay down on the hard ground . . . for a few hours. . . . At daybreak, we resumed our journey up the Great Potomac, which was now very rough, and often nearly swamped our little boat. In the afternoon we landed in a place about forty miles from Washington. After rowing about on the Potomac in the course of two days nearly a hundred miles, we at once proceeded to New York, via Washington and Philadelphia, and . . . sailed for Old England.[29]

These gold-miners had certainly endured a hazardous journey to avoid being forced to fight in a war for a cause with which they had no sympathy. The dangers they faced acquits them of any suspicion of cowardice; all they wanted was freedom to carry on their profession without interference from anyone. Although they had no use for the Confederacy, on their return to England the man who narrated their

29. *Royal Cornwall Gazette*, 4 October 1861.

adventures said that in North Carolina, enthusiasm for the war was so great that he had no doubt that

in case of necessity three-fourths of the female population would shoulder rifles for active service. Amongst the foremost in the ranks of the North Carolina regiments are the free coloured niggers. The war-spirit is universal. I believe that, with a very few exceptions, they would die to a man rather than yield a point to the North. If the South is to be subjugated it must be by crippling their cotton trade. Specie is very scarce. Fifteen to twenty per cent is freely given to exchange paper money for cash, and even at that rate it is a hard matter to get cash. The British residents are all of opinion that it will be a long and bloody war.[30]

While at least four of the Cornish miners and settlers in the North Carolinian diggings made their way home to the Old Country in the early days of the Civil War, a few remained.[31] Still living in Rowan County in 1870 were the grocer Benjamin Martin, whose estate had dwindled to a third of what it had been ten years before,[32] an elderly housewife, Mary Gribble, who was living with Martin's family in 1870, and John Williams who had turned from mining to farming during the decade and had thereby trebled his possessions.[33] Other members of the Martin, Jenkins, and Gribble families were still in the districts, but Joseph Eudy, who may have been the elder brother of the James Eudie recorded in the 1860 census, had left North Carolina for California in 1856 or 1857.[34]

Another Cornish miner who ran into trouble with the Southerners, was C. P. Holton, of Probus. In December 1862, he wrote from Colorado

30. *Royal Cornwall Gazette*, 4 October 1861.
31. *The Royal Cornwall Gazette*, from which paper the above passages were quoted, did not give any names: the whole account was published as a news-feature article, merely prefaced with the words 'A young man, just returned to Cornwall from North Carolina . . . says—'. It had taken the four men rather more than two months to reach Cornwall since Bull Run had been fought on 21 July though the engagement mentioned may have been the Blackburn's Ford clash on 18 July which the Confederates called Bull Run, the second and far greater battle being referred to as Manassas in Confederate history.
32. Viz. $3,000 personal estate in 1860 and $500 real and $500 personal estate in 1870. Martin married early in the war and by 1870 had a family of three children.
33. He had $300 real and $125 personal estate in 1860 and $1,000 real and $500 personal estate in 1870; during the decade he had lost his first wife, who had borne him ten children, married a girl twenty-six years his junior who, by 1870, had borne him two more children.
34. The 1870 census for Grass Valley records Joseph Eudy as having one child aged fifteen born in North Carolina, and children aged thirteen, eleven, and eight born in California; his eldest child had been born in England in 1851; for some time he was captain of the Eclipse Mine at Independence, Owen's Valley, California (*Cornish Telegraph*, 1 June 1870). James Eudy may have been one of the four men who returned to Cornwall in 1861.

to his parents saying that he had just received a letter from them that had been sent from Cornwall on 1 March 1861; he had himself last written them when he had been in Mississippi State in June 1860, but he had had no reply to his letter; he thought, however, that letters had been 'appropriated' by the Southern Confederacy. Some emigrants probably wrote their parents more frequently than young Holton, but there were others who must have been no more regular in their correspondence than he had been, or apparently, for that matter, his parents at home. Since his previous letter the disruption of the Federal Union had affected his fortunes severely. The Secessionists had taken nineteen hundred dollars from him which he had brought from the North; then Holton

> managed to get out of the slave-polluted territory with my life, which is far better than thousands of others; even Jews in large numbers I saw impressed. Well, I came to Colorado Territory Gold Mines, expecting it would be quiet during the war, but after being there two months, the rebels became so numerous and over-bearing, I concluded to enlist on the right side, although a certain party in England thinks the North is wrong.[35]

With a companion, Alfred Hill, probably another Probus emigrant, Holton joined the Union Army in November 1861, and after about six weeks' training was moved to Fort Lyon, which was

> on the border of the Arrapahoe, Cheyenne, Appache, Kiowy, Indian Country. While at this post we drilled both night and day, to perfect us in the art of war. On March 2, we received orders to march to the relief of Fort Union, New Mexico, three hundred miles... After travelling from forty-five to seventy-five miles per day, on foot principally, but in a wagon or ambulance sometimes, we managed to get over the desolate plains and into Fort Union on the eleventh of March, although we laid over on the road two days and nights, but it killed a few men, and a large number of horses, mules, and burro. Governor Connolly of New Mexico, gave us a warm reception, and well he might, for the rebels under General Sibley had possession of the capital (Santa Fe) and the rest of the territory except Fort Union. March 23rd—Started south without orders from the Department Commander, and on the 26th engaged a part of the enemy, 1,700 strong, and after fighting all day with only a third of our forces, whipped them, took a hundred and seven prisoners, although they had the pick of the ground in the Appachi Pass. When our vanguard, under Major Chivington, (an elder of the Methodist Episcopal Church), had exchanged shots with the enemy, he sent an aid back stating that a fight was on hand; many of our poor boys, who had been limping along and had been

---

35. *W.B.*, 20 February 1863.

lame for days, straightened up like gun-barrels, and started on a double-quick, loading and fixing bayonets on the run, and I made up my mind then that we would be victorious. Well, that night we kept a strong guard out over our prisoners, etc., with pickets on the alert. The next day we expected to meet with the whole rebel force, but they did not come, and we rested until the next day. March 28th, at daybreak, our little band of 1,100 men and eight field pieces, marched five miles and separated, four hundred under Major Chivington, and seven hundred with the artillery under Colonel Slough. Colonel Slough went up the pass and engaged the enemy, while my company with three others under Major Chivington, marched over the mountains to engage them in the rear. At two o'clock we saw down the mountain in the pass the supply and ammunition train of the Texan rebels, a distance of two thousand feet at an angle of forty-five degrees. At a given signal of the Major's, we went rapidly down the mountain, the enemy, in ten minutes after we commenced, pouring shell after shell on us, which made us go quicker for descent; we had three cannons spiked, and all their train in our hands; of the two hundred mules we could not run off, we had to bayonet them to death, then we burnt their ammunition, any supply train of which is worth over one million of dollars. As soon as this was through, an aide arrived, stating that Slough was retreating, and ordered us to fall back twenty-five miles over mountains, without any road, to protect our train, instead of charging down on their rear, as was our cherished plan.

They got back to their camp at ten o'clock that night and two hours later the rebel commander asked for a twenty-four-hour armistice to bury the dead, to which Slough agreed, and

the next morning before daybreak we got up, eat our breakfast, and ten of us in each company started to examine for our wounded and dead in and near the enemy's camp. The enemy, poor fellows, had nothing but rawhide and a little parched corn for their breakfast. Their losses in the two pitched battles were five hundred killed and three hundred wounded: our loss was forty-five killed and seventy-two wounded. Our little band had engaged three thousand five hundred men with far more artillery than we had. So . . . we whipped them over three to our one.

Holton went on to describe the bravery of an Englishman, John Muselow, who though wounded early in the battle with a 'minie ball', fought on, killed nineteen men, and then had to be ordered back to get his wounds dressed.

The Northern forces then fell back to Fort Union but

a few weeks after we marched the whole length of New Mexico and into Arizona, driving the rebels before us, through mountains and plains, had one more battle with the Texan rebels with the loss of three on our side and, as far as we could see, sixteen on theirs. They, the poor deluded rebels, got back into their country with only nine hundred men, and they half starved and almost naked; since then our marches from one fort to another have been to this time (we have travelled chiefly on foot) three thousand miles. We are resting here and expect to start to Dinver headquarters the commencement of next week. Horses are coming from the States to mount us.

The force in which Holton had served, and which had fought and marched over such vast distances had done much to secure the West for the Union, but it is not surprising that he closed his letter by expressing the hope that 'as soon as I get through, I am coming home to locate permanently with you'.

Several other Cornishmen served in the Union armies, notably the St Just 'Major' John Williams who, after the War, was to be county superintendent of education in Madison County, Mississippi,[36] while the Truronian Thomas V. Keam after obtaining the rank of captain during the war had a most successful career in the Federal Indian Service in Arizona and New Mexico.[37] Just before his disastrous defeat at Chancellorsville blighted his reputation, there was some talk in St Just in Roseland that 'Fighting Joe' Hooker had been born in that parish of humble parentage, but after he lost thirty thousand men on that blood-drenched field there does not seem to have been any Cornish inclination to dispute his New England origins. Only slight Cornish interest was aroused by the success of the Confederate agent George N. Sanders in getting through Union territory to Canada disguised as a Cousin Jack miner, apparently carrying some mining tools with him.[38]

It is hard to say where Cornish sympathies lay in the fratricidal American war. While there was a general disposition to condemn slavery and the 'Slave Power' before the attack on Fort Sumter, there later came to be considerable sympathy for the Confederate cause. It was to be expected that the 'Tory' *Royal Cornwall Gazette* would be rather pro-Southern, but this became even more pronounced after the death of its editor Edward Osler, who had Canadian connections and distrusted 'Yankee' imperialist designs on the British possessions.[39] The failure of the north at the outset to take a stand on anything save the preservation

---

36. *Cornish Telegraph*, 30 November 1870.
37. *W.B.*, 26 February and 10 December 1883 and see below pp., 206–8.
38. *Royal Cornwall Gazette*, 26 September 1862.
39. Osler died on 7 March 1863; he was related to Sir William Osler.

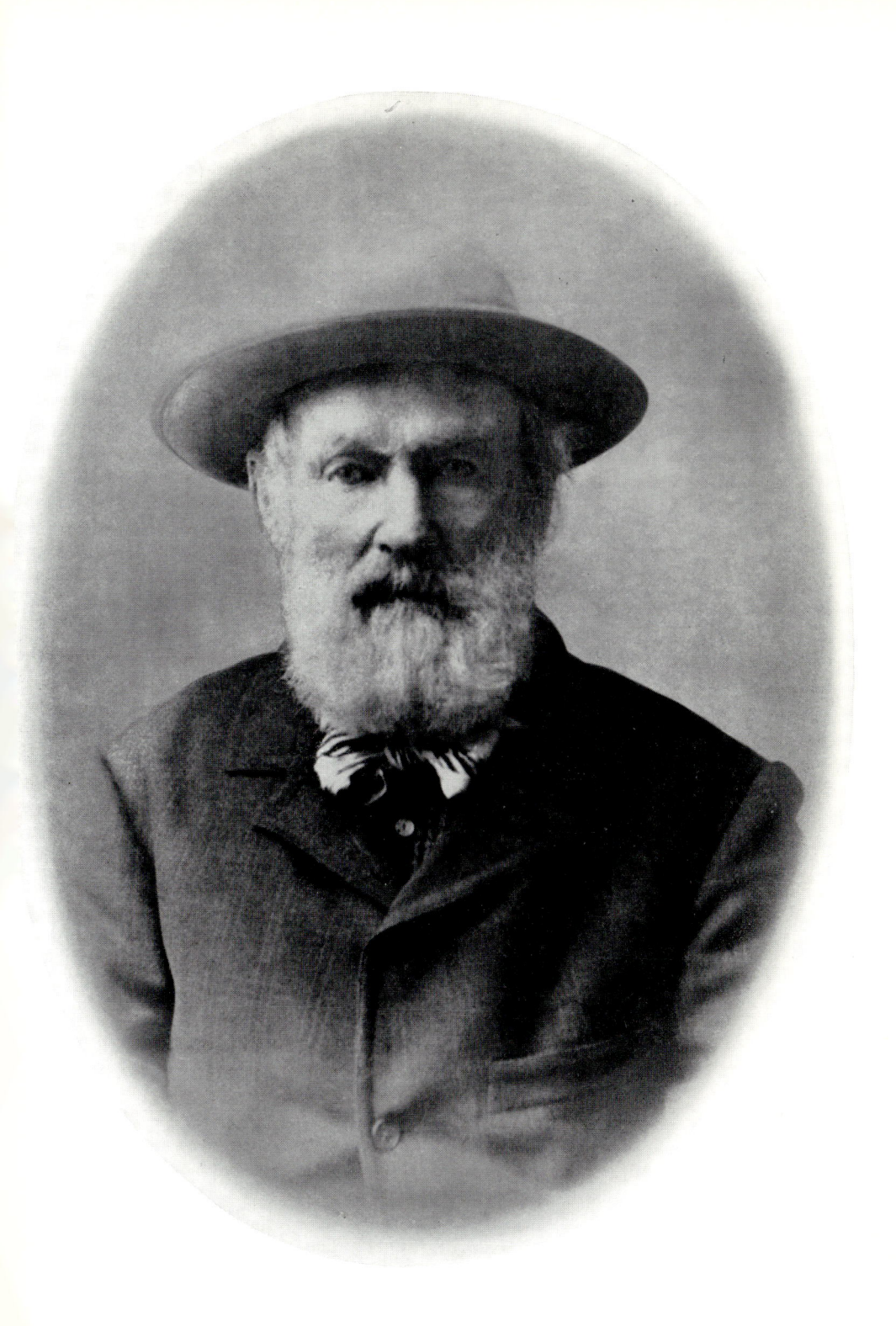

7 'Billy' Barker of Barkerville, British Columbia

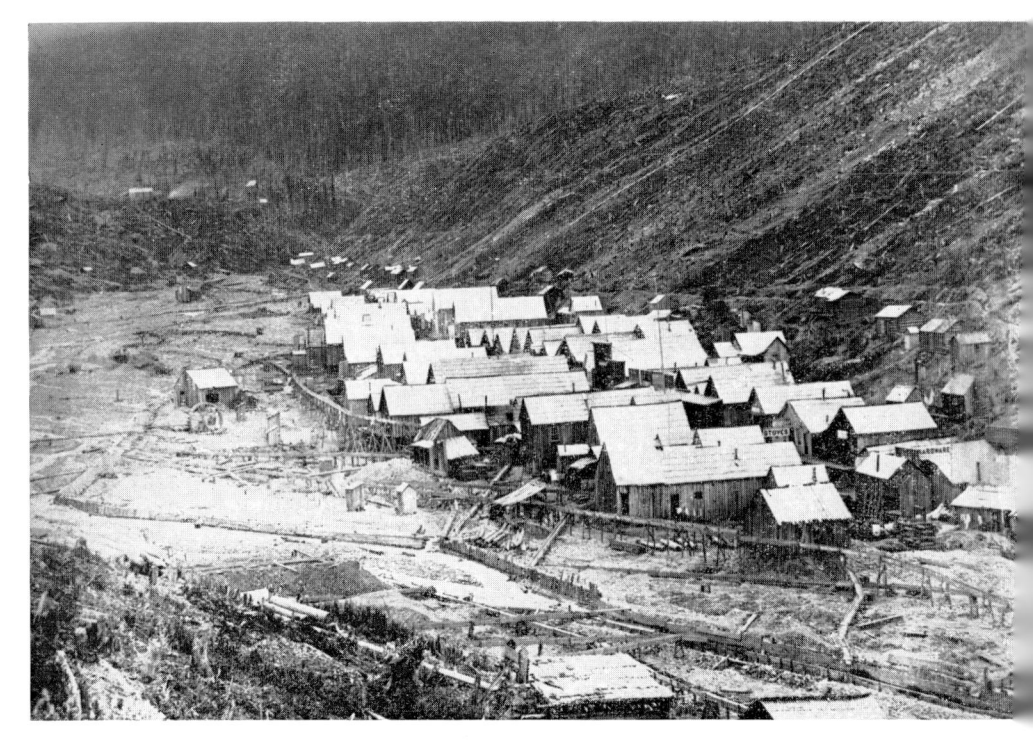

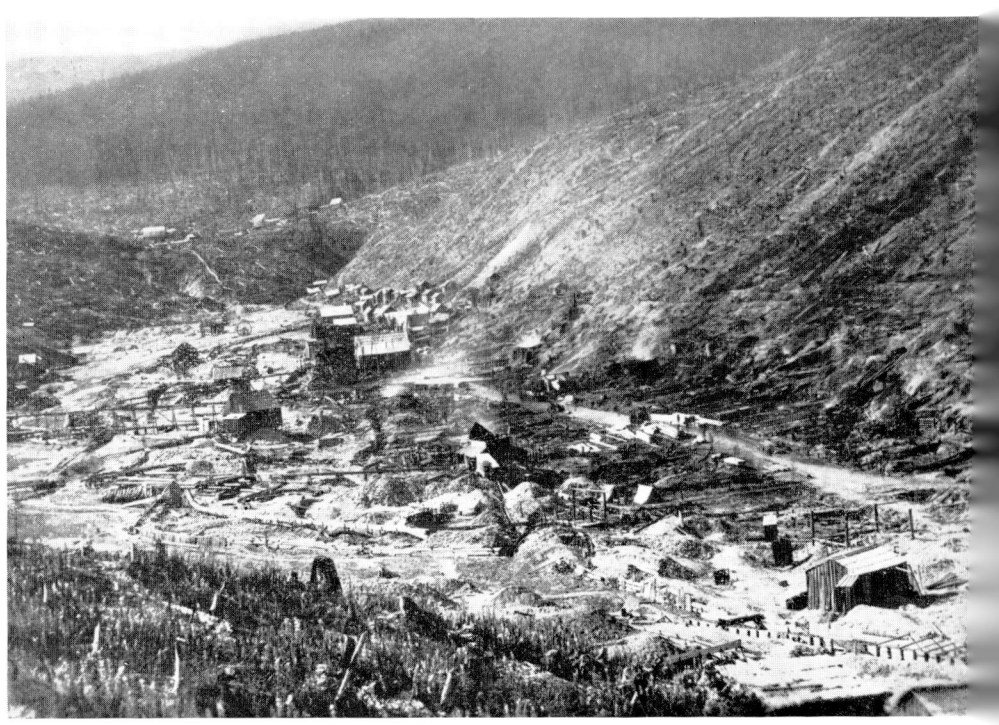

8 Barkerville, B.C., before and after the great fire of September 1868

of the Union was denounced by the 'Liberal' *West Briton* as territorial imperialism. Methodist 'Chapel' people were rather more pro-Northern than members of the Anglican Church. Lincoln's Proclamation of Emancipation did not evoke the sympathy which it did in the depressed cotton manufacturing districts of Lancashire, while as the fortunes of war turned against the South there was some tendency to stress the gallant fight the outnumbered Confederates were making to preserve their freedom and their way of life, however mistaken the latter might have been.

As for the Cousin Jacks in America a few may have fought in the Southern armies while quite a number enlisted on the Union side, but it seems that the majority tended to look on the war as a purely American quarrel that was none of their concern. It should not be forgotten that throughout the struggle there were Americans who looked upon the war as a disaster brought down upon them by scheming or misguided politicians, and 'copperheadism' did not diminish as the conflict dragged on and the carnage intensified. In Wisconsin, a generation later, it was difficult to acquit the Cornish immigrants of lack of excessive zeal for the Union by recording that two or three companies of militia were recruited among the lead miners and acquitted themselves tolerably well.[40] In November 1862, however, the threat of another militia draft led many miners in the Mineral Point district to claim exemption on the ground that they were British citizens, even though they had taken out their first naturalisation papers and voted in American elections.[41] Even a patriotic Cousin Jack of a later generation had to admit that they did not furnish more than their share of the forces recruited for the Union armies in the lead-mining region,[42] and they certainly did not flock to the army in anything like the numbers or with anything like the enthusiasm with which they had gone off to California in the early days of the gold-rush. Admittedly they could do the Union cause good service by working their lead-mines—which they did and made better returns than they had done for the past several years.

It is curious, however, to say the least, that many of the more substantial Cornish citizenry of Wisconsin twenty or thirty years later had, in the war years, pursued their mining vocation not in the lead region of the Upper Mississippi but in the mining camps of the Pacific Coast.

40. L. A. Copeland, 'The Cornish of South-West Wisconsin', *Wisconsin Historical Collections*, Vol. XIV (1898) p. 332.
41. *Mineral Point Tribune*, 25 November 1862.   42. Copeland, p. 332.

Naturally enough this seemed utterly reprehensible to fire-eating patriots, but of all the Celtic races the Cornish were possibly the least, just as the Irish were the most, ready to get into any fight that happened to occur, though many Irish actually returned to the 'Old Sod' in the autumn of 1861.[43] If Republican editors in Wisconsin misunderstood the Cousin Jacks in 1862, thirty-seven years later editors in the Old Country did the same when, rather than fight in a war they regarded as the result of the conflict of antediluvian Boer politicians, on the one side, and avaricious European and Semitic capitalists on the other, they left the Rand in hundreds to find other mining fields until the trouble was over. In the eighteen-sixties they preferred the hazards of new mining frontiers in Montana, Idaho, Nevada, and even Apache-ridden Arizona to being drafted into the army to fight in a war which they regarded as a most unwelcome and unnecessary obstacle to their earning a livelihood in the way in which they were accustomed. At all events, the 1881 history of Iowa County reveals that out of forty-five citizens of Cornish birth residing in Dodgeville that year, no less than ten had spent part of the Civil War years in western mining camps, and one of them had actually spent most of the war period in New Zealand. A Linden Cousin Jack, Thomas M. Goldsworthy, was unfortunate enough to be a prisoner in Andersonville, but he was the only one of the thirty-three Cornishmen treated biographically in that history with any record of war service, while of the forty Cousin Jacks of Mineral Point with biographical records in the same work, Samuel Hocking was apparently the only one who had served in the Union army.

In California there were a number of Southern sympathisers, while 'Copperheads' were numerous. The phenomenally rapid progress of the commonwealth into statehood and as a state since the gold discoveries had led to a form of State Rights sentiment, while there were also feelings of antagonism to 'Easterners', both North and South, caused first by the Congressional bickering over the admission of California to statehood, then by the failure of the East to provide transcontinental transportation facilities. As the war dragged on, it seemed likely to outsiders that the outcome might well be the disintegration of the Federal Republic not into two but into several sections. With Lincoln's apparent surrender to the Abolitionists at the beginning of 1863, the regions of the Union which detested slavery but which at the same time regarded the Negro, whether servile or free, as a threat to white standards of wages and living, began to feel that they had no moral interest in a

43. *W.B.*, 27 September 1861, quoting *Once a Week*, n.d.

conflict of abolitionists and slavocrats, and even that their material interests would not be advanced by continuing in alliance with Northern capitalists against Southern plantation owners. The family farmers and free miners of the Upper Mississippi Valley had had an interest in keeping the great river open to the sea, but they came to resent the insatiable demands of a 'centralised' government in remote Washington for money and men, and there were, too, increasing fears that cheap free Negro labour would be a threat to the maintenance of white living standards. Such resentments and fears increased with distance from the Federal capital, and were most intense in some of the mining camps of the Pacific Coast.

The threat of disintegration was very real in the dark days of 1863, and even after Gettysburg the shadows were not entirely dissipated. The Western mining communities grew more and more dissatisfied with the costly war, particularly since the decline in Californian gold output had been going on unchecked since 1855 and the gains both to labourers and to investing capitalists were diminishing. In the autumn of 1864 a Cornish miner wrote home warning fellow Cousin Jacks against coming out to Grass Valley 'to seek for riches'. Though wages were averaging three dollars a day, board cost at least a dollar a day; miners had to work a ten-hour day six days a week; it cost twenty dollars a year to keep a miner in boots alone, and all other items of clothing were dear; furthermore every miner in California had to pay six dollars a year poll tax and a further four dollars annual road tax. Besides these exactions, every foreigner was liable to a levy of four dollars a month for licence to mine in California. This tax had been introduced with the aim of checking Chinese immigration and was not being levied on the Cornish miners in Grass Valley in 1864.[44] It might seem that even with these high prices, taxes and so forth, a miner could save three or four hundred dollars a year, even if he allowed himself as much as fifty dollars for travel or other 'luxury' expenses. In fact, however, no miner could count upon being employed for more than eight months out of the twelve, unless he had influential friends, and the enforcement of the foreign miner's tax late in 1864 or early in 1865 cut a man's potential savings by fully a third. It was alleged that the miner's licence tax was enforced because Nevada County voted against the Government in the 1864 elections; thereafter the tax collectors appeared regularly at the mines in Nevada County, sometimes if not always with 'six-shooters'

44. *W.B.*, 30 December 1864: letter signed J. D. dated Nevada County, 16 October 1864.

swinging against their hips, and made sure that no foreigner escaped paying, no matter if he were in steady employment or not, or if a mine had only taken him on a few days previously.[45] Another Cornishman late in 1864 reckoned a miner lucky if he earned six hundred dollars a year and fortunate if, after the normal charges of board, taxes and travelling were deducted, he had a hundred dollars left for clothing and other 'necessary luxuries'; moreover, partly through high prices, partly through the currency depreciation caused by the war, the purchasing power of the dollar was no more than that of an English shilling.[46] It was little wonder that about that time several Cousin Jacks left Grass Valley for Cariboo in British Columbia while others went to the new 'strike' on the Boise River.

The declining value of the dollar was the main worry of Cousin Jacks in Upper Michigan during the Civil War. At best they were earning about forty dollars a month, at least half of which went to pay boarding expenses; they could, however, count on being employed more of the year than the gold miners of the Pacific Coast. As a rule during the war labour was scarce in Upper Michigan, and mine owners went to quite considerable trouble and expense in arranging for substitutes to serve in the stead of drafted men in their employ. For their part, many of the immigrants in the Northern Peninsula were no more anxious to serve in the Union armies than some of their fellows in Wisconsin, although a number were prepared and even eager to enlist; but Upper Michigan, in the Civil War period, was essentially an isolated frontier region. When the war began in 1861, navigation to Keweenaw and Ontonagon was still closed. Since South Carolina had passed its Ordinance of Secession stocks of copper had been piling up on the shores of the Keweenaw Peninsula while, expecting war, foreigners had been rushing in cargoes of copper to the Atlantic ports. When Lake copper reached the East, it found the markets glutted, and prices slumped to unprecedently low levels. A gradual recovery did not save the Copper Country from the pinch of hard times during the first war winter, and the early misfortunes of the Federal cause made matters worse. That there was some financial mismanagement was unquestionable, but even the best conducted mining companies were unable to purchase adequate supplies for their company stores to last through the winter of 1861–62; the competition of the armed forces for supplies and the Union's cur-

45. *Mining Journal* (London), 27 May 1865: letter of 'Working Miner' dated Nevada County, California, 31 March 1865.
46. Ibid., 7 January 1865.

rency troubles made matters still worse. There seems to have been plenty of whisky but a shortage of most other things in the Michigan copper-mining region that winter and the result was trouble, unrest and near riot. Some of the foreign miners seem to have been convinced that the United States was ruined and that they would soon have the opportunity to seize the country for themselves; life and property were no more secure in some of the Michigan copper-mining areas than in the roughest Pacific Coast mining camps a few years before. For their part, the mine officials and the steadier, more reliable workers organised a secret society that was to all intents and purposes the replica of the Committees of Vigilance organised in San Francisco in 1849 and 1856.[47]

As copper prices recovered and then soared to heights hitherto unheard of on Lake Superior, matters improved. Currency difficulties in the immediate locality were solved by the mining companies issuing their own paper money; it was unconstitutional, but necessity recognises no law, and these 'shin-plasters' served their purpose. Army drafts caused labour shortages which were aggravated by high copper prices stimulating the opening up of new mines. Nevertheless, there were no very appreciable increases of wages during the war years; improved methods, notably the substitution of underground tram-roads for wheelbarrows in some mines, enabled economies to be effected in the labour forces employed. The output of copper rose somewhat, but not sensationally for the new mining ventures did not prove immediately productive since so much preparatory work was needed before they could raise any ore at all. Attempts to bring in more foreign miners to meet the labour shortage were not altogether successful; some of the new immigrants refused to honour the contracts they had made with the agents of the mine-owners who had been sent to recruit them in Europe; still, it was comparatively easy to ensure that such troublesome newcomers, apparently mainly from Scandinavia, were drafted into the army,[48] tempted by a three hundred dollar bounty to go and fight for the Union. Whether these Swedes came back later to the Upper Peninsula is unknown. On the whole there does not seem to have been such a strong 'copperhead' element in Northern Michigan as there was in the quartz-mining regions of California, and the celebrations of the fall of Vicksburg were as sincere as the mass meetings held in the open air to

47. John H. Forster, 'War Times in the Copper Mines,' *Michigan Pioneer and Historical Collections*, Vol. XVIII (1891) pp. 375-82.
48. Ibid., p. 380.

express the Copper Country's heart-felt sorrow at the assassination of Lincoln.[49]

The development of the Copper Country was accelerated by the Civil War, but it had been a time of trial and tribulation. New mines had been opened up, but not all of them were to be successful. Methods of working and local communications had been improved, perhaps more quickly than they would have done had copper prices not reached a peak of nearly fifty cents per pound by the close of the war. Cornish miners continued to come to 'the Lakes', even during the War, but if they could still command a wage of thirty-five or forty dollars a month they were unable to send such substantial remittances home as they had done before hostilities began. Currency depreciation and increasing restrictions on transfers of gold made it difficult to send money to families left behind in the Old Country; as early as February 1863, shopkeepers in Redruth were complaining that families which previously were receiving at least three pounds a month from America were now getting less than two;[50] the situation about such remittances and the straits of those dependent upon them were to get worse still before the war was over.

And when the war ended, Michigan copper-mines had to adjust themselves to lower prices for their ores, although they were not to fall to the low levels of the summer of 1861. The fall, too, was checked by the dispute between Spain and Chile and the Spanish blockade of the Chilean copper ports. Some unwise speculative mines and a few badly managed concerns collapsed, but the discovery of the Calumet lode promised a supply of copper that might be worked and pay dividends even if the prices dropped lower than they had done in 1861. The Upper Peninsula, after the end of the Civil War, could look forward to the future with confidence. The prospect before the copper-mining industry of the Old Country, however, was soon to be one of gloom, hopelessness, even despair.

49. J. H. Forster, op. cit., p. 380. 50. *W.B.*, 6 February 1863.

7 Go they must— but where?

Hundreds, even thousands, of Cornish immigrants scattered throughout the United States, but concentrated in mining regions before the Civil War, suggested that conditions at home had driven them to a distant foreign land to seek a living. Their numbers indicated that their presence in America could not be explained as the result of individual foibles, traits of personal character, adventurousness, private discontents, and the like but that more general, fundamental forces existed forcing them out of the land of their birth. The sufferings of the Lancashire cotton workers in the days of the 'cotton famine' caused by the War between the States, perhaps even more than the Irish Potato Famine in the 'hungry forties' had made the British public more sensitive to the conditions under which the labouring poor lived and worked—or existed and slaved—than it had been in the past. Then in the summer of 1866 a letter appeared in the London press declaring that Cornwall had entered upon a 'terrible crisis', and that the position of the people of the south-western county was worse than that of the cotton workers of south Lancashire four years before since there was no hope whatever of recovery. Although fewer people were affected than had been in the far more populous northern textile manufacturing region, yet, the letter alleged, a quarter of a million souls were on the verge of starvation 'with no resource but external relief'.

The letter attributed the plight of Cornwall to the inability of its mining industries to stand against foreign competition. In recent years Cornish mining enterprises had been 'pushed quite to the verge of profit and loss, and occasionally even beyond it—so much so, that it has been said that Cornish ore, on the whole, costs more than its value to raise'. The writer foretold that 'at least' ninety per cent of the miners of West Cornwall would be unemployed by the next Christmas, and that the Cornish mines would be hardly supporting twenty thousand, instead of two hundred thousand, souls; he declared that, at that very time, no less than seventy-eight mines were being wound up in the

Stannary Court, and that more were being wound up by private agreements or were suspending operations through the fear of financial loss.[1]

Such figures were frightening, but they could not bear close scrutiny. The writer, Charles W. Merrifield, had only spent a short summer vacation in Cornwall, and had then communicated his views to the *Pall Mall Gazette*. The letter infuriated the mining interests, particularly Merrifield's statement that the Penzance workhouse was so crowded that temporary wooden huts were being used to lodge the rapidly growing number of pauperised miners and their families—a statement that was absolutely untrue; a few store huts may have been in progress of erection when—and if—he passed that way; his dismal estimate of the season's harvest was also wide of the mark. Furthermore, there were at that time about forty thousand hands employed in Cornish mines, of whom more than a quarter were youths under eighteen years of age; at the most less than a third of Cornwall's total population, then about 370,000, depended in whole or in part upon the copper- and tin-mining industries as the source of their livelihood.[2] The figures of the number of mines being wound up in the Stannary Court gave a false impression, for most of them had only been small concerns that had only employed a few men, and, the processes of law never being swift, some had probably ceased operating a considerable time back. Many miners, too, had already left Cornwall to find employment in the Welsh and Lancastrian coal-mines.

Allegations that Merrifield was neither a business nor a professional man but only a young clerk further discredited his case and opinions. Nevertheless, the mining industry of the south-western county was in an unsatisfactory, even in a critical, condition. Five months earlier, it had been revealed that no less than fourteen mines, in which an aggregate investment of £300,000 had been made, had suspended operations, and only nineteen mines were then paying dividends out of a total of three hundred.[3] The failure of these mines was a warning to those men with interests in Cornish mining ventures that falling metal prices, increasing poverty of ores, bad management, or even a simple run of pure bad luck could be fatal to other Cornish mines as well. Many did not choose to heed the warning.

In his letter, Merrifield had been mainly concerned with the declining condition of the Cornish tin-mines. At that time they were employ-

1. *Royal Cornwall Gazette*, 23 August 1866.   2. Ibid., 30 August 1866.
3. Ibid., 15 March 1866.

ing rather more than eighteen thousand hands, of whom a sixth were females and raising in the region of fifteen thousand tons of ore yielding about ten thousand tons of pure tin annually. High tin prices in 1860 had led to a marked increase in mining that metal by nearly fifty per cent; the increase, however, was partly owing to the declining deep Cornish copper-mines attempting to maintain their solvency by working the stanniferous lodes that lay below their formerly productive copper deposits. The American War brought some reduction in the demand for tin, and at the same time there was a marked revival of the ancient alluvial tin workings of the Isle of Banca in the Dutch East Indies. Deep tin-mines were far more expensive to work than alluvial surface diggings, quite apart from the fact that the coolie Chinese labour employed in the East Indies was content with a living standard reckoned to be the equivalent of a shilling sterling a week. By the mid eighteen-sixties, tin prices were governed by East Indian supplies which yielded the commercial interests concerned a good profit if, in European markets, the price stood at only £45 sterling per ton whereas, in 1860, the ruling average price of unrefined tin had been £71. 11s. The outbreak of the American Civil War had caused an almost immediate fall of about seventeen per cent in tin prices, and growing Eastern competition along with an uncertain demand for the metal, aggravated by the tendency of many Cornish mining 'adventurers' to keep up income by unloading still more of their produce on to falling markets, brought prices down to an average level of £55. 6s. in 1865, while in 1866 there were big tin sales at which the average price was only around £45 per ton. While it is true that what was a remunerative price to some mining concerns would not even meet the working costs of others, the Cornish tin-mining interest as a whole believed that a price less than £55 per ton would not keep the industry solvent, and, in fact, few tin-mining adventurers felt financially secure unless prices were well above that level.

The position of the Cornish copper-mining industry, too, was far from reassuring at the beginning of the eighteen-sixties. The peak level of production had been reached in 1855–56 and it had fallen away by a fifth in the year the American Civil War began; that lower level was maintained throughout the war years, but for the year ending 30 June 1866, the total output was roughly thirty per cent less than it had been ten years before; a decade later production was hardly a quarter of that of the mid eighteen-fifties. The record of prices for copper was still more dismal. In 1856 the price of copper had been hovering about £140 per ton, and through the years of the American Civil War an average of about £120 had been maintained which, however, compared none too favourably with the decennial average of £127 for the eighteen-fifties. The financial crisis of 1866 ushered in a series of lean years when levels of £120 were never reached, when there were some periods with prices below £100, and the average for the decade, 1866–75 was only £106. Declining output and low prices were bad enough, but in addition the produce of Cornish copper ores only averaged about 6·5 per cent. In some years when prices were low slight improvements in ore produce were only an indication that the mining adventurers were trying to reduce their financial losses by 'picking the eyes out of their lodes' and selling their best reserves—a policy that would be regretted when better times returned. In any event in 1867, the income from Cornish copper sales was barely forty-three per cent of what it had been eleven years before.

Statistics of production for later years show that the copper mines of Cornwall had largely been worked out; the best ores were exhausted what remained were too poor to warrant working. During the American Civil War the decline of Cornish copper mining had been checked to some extent by an increased demand for copper goods to be used in trading for the Egyptian and Indian cotton that was brought to Britain to fill part of the void caused by the cessation of American supplies. The day of reckoning, aggravated by a financial panic, came in the summer of 1866, and the fact that mining adventurers had been 'picking the eyes out of the lodes' of both tin and copper, albeit for radically different reasons, in recent years, made matters worse. Although Charles Merrifield exaggerated the calamity that threatened Cornwall, declining prices of both tin and copper made it impossible for existing wage standards to be maintained. In addition, the declining production of copper meant unemployment; furthermore, during the prosperous years, there had been heavy capital expenditure on labour-saving machinery, especially by the bigger mining companies which usually, though not invariably, were those best adapted to meet economic recession and survive.

Merrifield had blatantly asserted that the Cornish mining population could either emigrate or starve; before he had visited Cornwall many had already left for the coal-mines of Wales and Lancashire and, in a short time, a fairly considerable number 'emigrated' to Scottish collieries. Depression and distress did come to the Cornish mining districts in the winter of 1866–67. In some areas the cost of relief in money and in kind rose alarmingly, but most of those who sought relief were not able-bodied miners and labourers—but women and children, many belonging to families whose husbands or elder sons had emigrated or migrated elsewhere. Most of the men who left families behind in Cornwall sent remittances home as soon as they could do so but there were some who left their dependents destitute and abandoned. Those who stayed in the county had almost invariably to accept wage reductions but while some still managed to earn enough on which to live at the low standards to which they were accustomed, others suffered privation and want.

Neither in the gold-mining camps of California or Cariboo nor in the tin- and copper-mines of Cornwall was there such a thing as an 'average wage'. Tales of men averaging from ten to twenty dollars a day in Pacific Coast diggings or of making two hundred dollars in a day or two, had no more relation to the lot of the majority of miners than records of tributers at Devon Great Consols averaging £4 sterling a month or of two lucky miners making over £142 in two months, for in that same mine a group of four men only made £9 between them working a four- or a five-week month.[4] In the Helston district, miners' average earnings in the 1866 slump fell from £3. 15s. to £2. 15s.; some, however, were still making £4, while others were only getting £1. 10s. or even £1— and this at a time when the monthly flour bill of a moderate-sized family in Cornwall was about £2. 10s. In Breage the Rev E. Pridmore told a Committee of Inquiry into the distress of the labouring classes which had been set up by the county magistracy that he knew a case, which he regarded as typical and not exceptional, of

a man and wife and six children. Five of the eldest were upon one bed of straw. They had neither sheet nor blanket, the only thing covering them being half an old counterpane. There was another bed in the same room, in which the father, mother and youngest child slept, and the only covering they had was a counterpane. The father who was a steady, hard-working man, found it impossible to purchase bed or body linen out of his wages; the utmost he could do was to find barley bread for his family, his earnings being only fifty shillings a month.[5]

Cases of distress could be multiplied and the plight of the Cornish mining population was tragic. Responsible persons in the Helston district asserted that a third of the mining population, over a thousand families, were in need of clothing; local magistrates visited cottages in Tywardreath and found that forty families did not possess a single blanket, while forty others had scarcely any bedding at all.[6] The aver-

---

4. *Royal Cornwall Gazette*, 8 and 15 March 1866.   5. *W.B.*, 26 July 1867.
6. *Royal Cornwall Gazette*, 28 November 1867.

age working miner in St Just in Penwith was only earning 1s. 5½d. a week; a rather more fortunate family in the Truro area had a monthly income amounting to the large sum of £6 a month, but three-fourths of it were hardly enough to keep the family in flour alone.[7] Many miners went to work a ten- or twelve-hour shift underground taking no food with them save a thick slice of barley bread, while surface-working 'bal maidens' were subsisting on a diet of fried turnips.[8] In the Truro district, which included the mining parishes of Chacewater, St Agnes, and Perranzabuloe, a local committee set up to investigate cases of distress and to administer relief whenever possible, reported, in December 1867, that

> the great distress now existing, and they feared now increasing in the neighbourhood of Truro, came from the American war, when cotton rose to such a price that the poor were not able to keep up their stock of clothing, in consequence of the large mines in the place having stopped working, and the high price of flour.[9]

Gifts of blankets, bread and soup can only have been palliatives in those grim days. There was reason to fear lest privation breed epidemic disease; and that the incidence of miner's consumption, 'miner's con', was aggravated by the lack of adequate clothing and bed linen among mining families. It is hard, too, to avoid the conclusion that in those starving times boys were being born and reared who were destined from the very circumstances of their early environment to find premature graves in the copper region of Butte and on the gold-fields of the Witwatersrand about the turn of the century. Others came to their graves long before that because, in order to get a living wage in those times, they slaved as tributers for longer hours than usual in the deep mines that were still working and did so with far from enough food.

The Cornish mining industry had known bad times before, but now, in 1866–67, matters had reached such a pass that optimistic talk about waiting for copper and tin prices to rally from a blow which, certainly, could be attributed in part to general financial and economic slump, carried little weight. Statistics of output and prices could be studied and interpreted in many ways, but it was impossible to argue that the Cornish mines could again employ the labour force they had done in past years. The uninterrupted decline in copper production since 1856 and the opinions of many practical mining men indicated that copper mining

7. *Royal Cornwall Gazette*, 12 December 1867.
8. Ibid., 12 December 1867.   9. Ibid., 12 December 1867.

in Cornwall was a dying industry; not only were Cornish ores becoming increasingly poor in metallic produce but many mines had been practically worked out of payable ores; furthermore, the most productive mines were now not those of the Camborne–Redruth area that had been working for over a century, nor those around St Austell which began about fifty years before, but the mines in the East Cornwall and Tamar Valley region which had only been working for twenty or thirty years.

The tin-mining industry did not face an immediate problem of declining or near-exhausted reserves of ore; its trouble was foreign competition. The great tin markets were those of industrial Europe and the United States. Consumption of tin had not increased so rapidly as had that of other metals during the immediately preceding twenty or thirty years, and markets were subject to considerable fluctuations, which were intensified by the Dutch East Indian mining interests 'dumping' supplies irregularly upon the metal markets and upsetting the delicate balance of supply and demand. In Britain the marketing of tin had fallen into the hands of a small group of smelting firms, while the trade was virtually controlled by a few commercial houses, which not only facilitated the activities of speculators but even, at times, led to schemes of 'cornering' the tin supplies which caused wild fluctuations in the price of the metal. All through the American War the tin-mining interest had been complaining of dull times and had hoped that the restoration of peace in the States would lead to prices recovering the high levels of the late eighteen-fifties when over £70 per ton had been regularly realised by 'black' tin as against an average of just over £60 during the war years; during that period some of the mine-owners had been attempting to keep up aggregate returns by increasing output, while, after Appomattox, the hopes of the downward trend of tin prices being checked encouraged some new 'adventures' to start mining operations. In 1866, however, the general level fell considerably below £50; then, in the summer of 1867, rumours began circulating that 'a mountain of tin' had been discovered in Missouri.[10]

Missourian speculators and adventurers lost no time in declaring that theirs would soon be the richest tin-producing country in the world, and that 'Cornwall's glory has departed'. The successive vast discoveries of lead, copper, iron, gold, silver and other minerals in the United States within the past three decades had led to the belief that it would only be a matter of time before tin would be discovered somewhere or other within the extensive domains of the Republic. Tin was the only

10. *W.B.*, 30 August 1867.

metal of great commercial value and industrial utility that had not been found in that country, and it is possible that the nationalist fervour evoked by the Union triumph in the Civil War predisposed Americans to believe that their country 'had everything'. Demand for metals in the war years had stimulated prospecting and exploring in likely 'eastern' areas like the Ozarks, and the reports of the 'tin strike' in Missouri two years after the end of the war caused almost as much excitement as some of the earlier gold rushes. In the reputed stanniferous region itself, that summer of 1867, it was reported with picturesque exaggeration—or, rather, imagination—that

one half of the population own tin mines, and the other half are trying to own some. Everybody has a piece of tin-ore in his pocket, and there is scarcely a blacksmith's shop in the country where ladles and pans have not been coated with it. Blow-pipes protrude from pockets as frequently as 'bowies' do in Arkansas. Several thousand acres of land have recently been entered in Madison and Iron Counties upon which the owners hope to find tin—lands which have heretofore been considered as almost worthless because of their rocky character and remoteness from river and railroad communication. These lands have been entered and purchased by persons from various parts of the States and elsewhere; and as the explorers have three experienced Cornish miners—Mr. R. W. Dunstan being amongst their number—to guide them, it is unlikely that anything of value will be overlooked.[11]

Far from anything of value being overlooked, the reverse was the case. The whole affair seems to have been nothing more than a speculative swindle conceived for the purpose of deluding the investing public into purchases of worthless properties in one of the most barren districts of Missouri. Early accounts of the 'mountain of tin' were padded and crammed with descriptions of the region written by 'graduates' of mining schools, geologists, and 'professors' whose credentials would not have borne close scrutiny; even if some of them were genuine, it was possible for 'scientific gentlemen' to make mistakes. An early 'boost' of Missourian prospects significantly, if inadvertently, admitted that skilled Cornish tin miners would not have suspected the existence of tin from certain surface ores which were alleged to have assayed satisfactorily.[12]

11. *W.B.*, 30 August 1867.
12. 'The result of the assays thus far made, even in a crude manner and from surface ore, where, according to Cornishmen, little or no real tin should be expected, are very satisfactory, and the conclusion arrived at is that there exists beyond question an almost inexhaustible supply of tin ore in Missouri, of a quality that will pay a handsome profit for working it.' Letter of 'J.R.T.' quoted by the *West Briton* of 30 August 1867, from the *London Mining Journal*.

'Ee do knaw tin' was a phrase long current in Cornwall to describe a level-headed, practical man endowed with foresight and common sense, and the most charitable explanation of R. W. Dunstan's connection with the Missouri 'tin' discoveries would be to say that he did not know tin. He wrote home saying that, while he did not anticipate the newly located deposits would substantially reduce American imports of foreign tin, he had no doubt

> in the prospect of a New Cornwall springing up in Missouri, and that enough tin really exists to be worth the attention of the English mining capitalists, especially at this moment, when land can be secured at very low rates, while, at the same time, but few are acquainted with tin mining or with the formations of tin ores.[13]

The hopes of any unemployed Cornish tin miners, then in the Old Country, that they might soon find employment in Missouri were quickly blasted. Facilities for testing and assaying ores hardly existed in the rough mountainous country a hundred miles or so to the southward of St Louis, so half a dozen samples were taken up to Dodgeville, and promptly proved to be nothing more than a type of iron ore. The Missouri tin bubble burst, leaving a few land speculators and their dupes sadder, poorer, and possibly wiser men. Moreover, all later reports of tin discoveries in America came to be regarded in Cornwall and in English financial circles with scornful scepticism since, it was scoffingly reported, Missouri's 'tin mountain' had not produced enough tin to gild a single coat button.[14]

Had the so-called geologists and mining experts really discovered tin in Missouri, Cornish immigrants would have been attracted to that region just as surely as the copper-mines on Lake Superior had drawn so many Cousin Jacks from the declining copper-mines of the Old Country. The high copper prices prevalent at the end of the Civil War and the scarcity of labour on the Lakes led many more Cousin Jacks to go thither soon after the collapse of the Confederate States, but soon copper prices obtainable for Lake Superior metal fell below remunerative levels; before that happened, however, lured by the high prices of 1865, many speculators had bought mining properties and locations in a seller's market, and had installed costly machinery to work them. American copper mining, however, suffered from the slump of 1866 almost as much as did the Cornish industry; early in 1868 only eight out of a hundred and ten mining ventures in Upper Michigan were

13. *W.B.*, 1 November 1867.   14. *W.B.*, 21 October 1886.

paying dividends.[15] Mining interests on Lake Superior were admitting that they could not compete against the cheaper labour of Chile or even with the traditional skill of Cornish miners in those difficult times.[16] The situation, however, in America, was radically different from that in Cornwall: whereas Upper Michigan had vast, even incalculable reserves of rich ore, in the Old Country the signs of impoverishment and exhaustion of cupriferous reserves were obvious and inescapable.

The proverbial passion of Americans for quick results had been accentuated by the Civil War and the demands for metal which it had greatly increased. Thousands, even millions, of dollars had been spent on pioneering work such as clearing forests, providing roads and other facilities of transportation, installing machinery and the like. Instead of steady mining development in which costs were kept in some proportion to production, assessment upon assessment had been levied on investors, and although mines like the Minesota and the Pittsburgh had returned fabulously high profits, many others had been such a drain on their shareholders that they had given up in despair. The opinion was general among Cornish mining men on Lake Superior that many of the American copper-mining companies there had suffered severely from incompetent management.

These Cornish critics, however, forgot that the Americans on Lake Superior had not their generations of tradition and experience behind them. Some forty years earlier in a speculative boom Cornish copper-mining ventures had all but permanently 'blotted their copybook' so far as raising working costs from the investing public was concerned, and thereafter they had been forced to rely upon their own produce as the main source of operating capital, only making 'calls' or assessments upon shareholders in cases of the most critical necessity. The American mining ventures on Lake Superior only learned this hard lesson as a result of the trying times after the Civil War during which reckless expenditure, of blood and treasure, had been regarded as essential to the economic life and political survival of the Union. With the post-war decline in metal prices and post-war disillusionment copper-mines in Upper Michigan, like mines in the Old Country, had to be managed so that their metallic produce was the source of their future development; they had to be operated either efficiently or not at all.

Many Lake Superior mining companies had been badly organised and chronically mismanaged. Sites for mechanical equipment had been

15. *Royal Cornwall Gazette*, 22 October 1868.
16. Ibid., 27 February 1868.

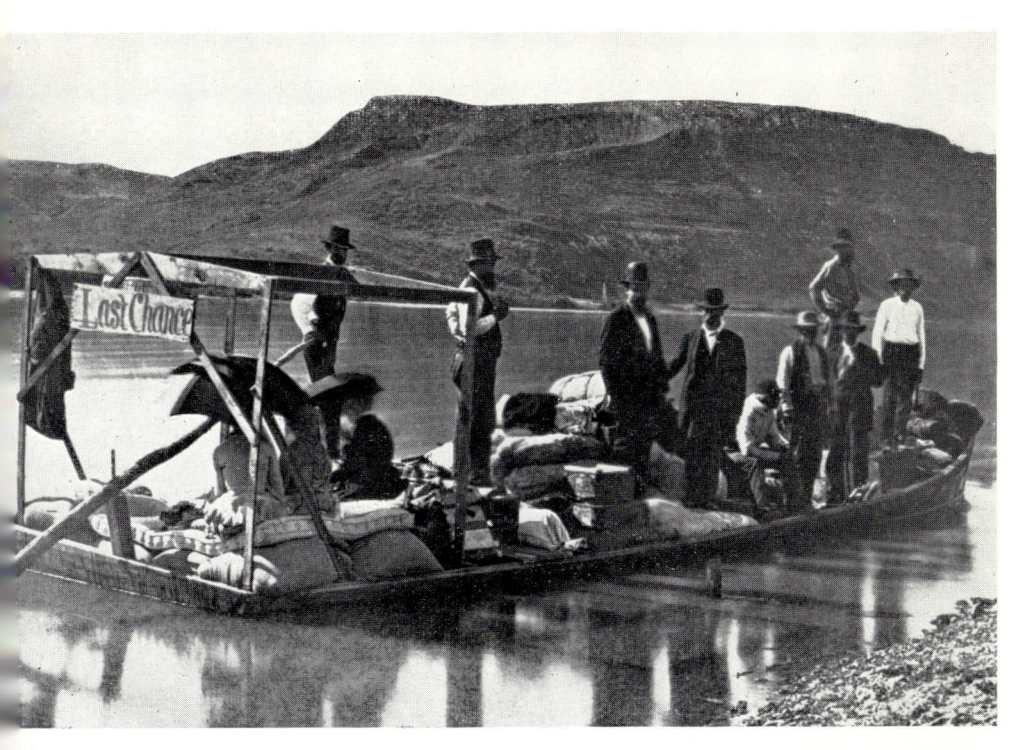

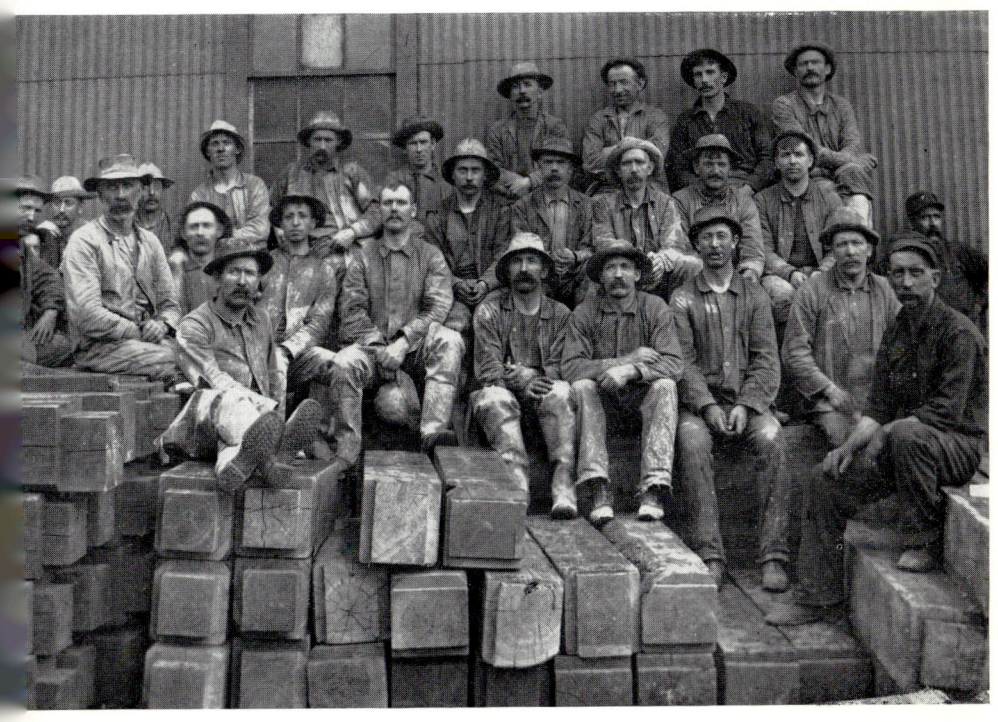

9 On the way to Last Chance Gulch, via the Missouri, about 1875
10 Miners at Speculator Mine, Montana, in 1890

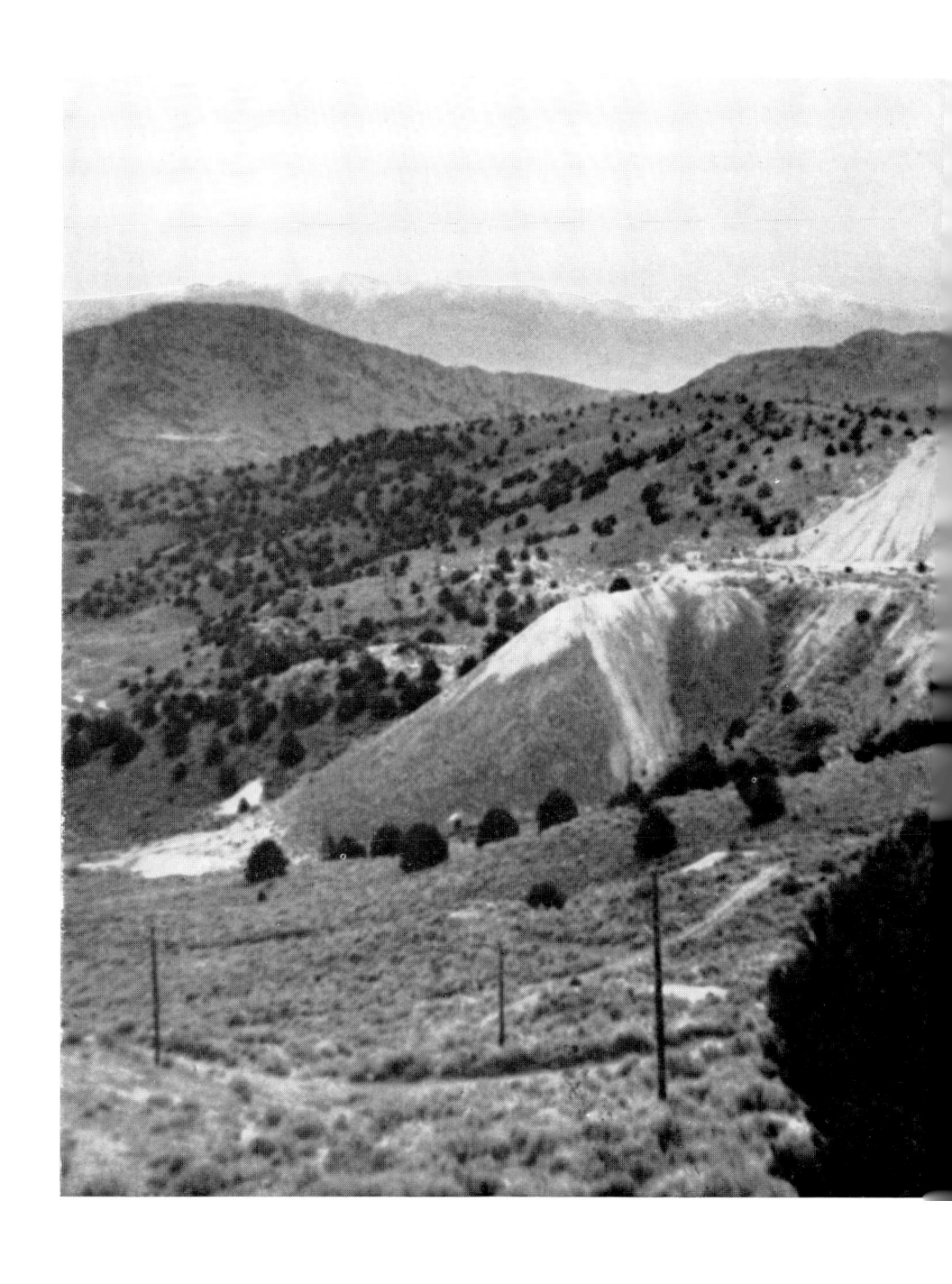

11 The Comstock country, near Virginia City, Nevada

selected without paying the least regard to accessible resources of water power, and steam engines were installed in places where simpler and cheaper equipment could have been effectively and cheaply worked by water power. Companies organised in Boston, New York, Philadelphia and Pittsburgh appointed managers and agents whose only qualification to conduct mining ventures was relationship or acquaintance with the company promoters. Managers of this ilk came to Northern Michigan and soon ran into difficulties, like the manager the Cliff appointed in 1852 who quickly made a dividend-paying mine a losing concern.[17] Some managers seem to have been afraid of tracing lodes deep underground; others were too inclined to regard the provision of more and more elaborate and dear machinery as the answer to temporary difficulties. On the other side, investors were avid for quick returns, and to satisfy them dividends were paid from profits which should have been devoted to further development of the mining properties with which they were connected.

Briefly, the Lake Superior copper-mines had been organised in 'boom' times of expansion and were unprepared for economic recession. Currency troubles and monetary depreciation made matters worse, particularly in those instances where, during the war years, mining companies had issued their own 'token' money; the later financial difficulties of such concerns entailed serious losses to their creditors, many of whom were working miners in their employ. Even before the worst slump came many such types of 'company money' had depreciated far below their initial face value and, furthermore, by the spring of 1867 a Federal 'greenback' dollar was worth barely seventy-five cents. This meant that the purchasing value of the remittances so many miners sent home to their families was materially reduced, while a great number ceased sending remittances home altogether because to them a dollar was a dollar and they had not sufficient grounding in elementary economics to realize that even in America the dollar was only worth three-quarters of its face value; they believed that the depreciation was simply an exorbitant tax levied somewhere on the way between Northern Michigan and Cornwall and, rather than submit to it, kept their paper dollars without realising that prevalent high prices in America merely reflected inflated currency.[18]

Currency inflation, too, gave the impression that wages were still high in the Michigan copper country. A miner's average monthly earnings on the Lakes in 1857 had been somewhat over $32, and victuals

17. See above pp. 80–1.   18. *W.B.*, 24 May 1867.

and lodgings cost him $12 a month; ten years later the wage average was $45, but it cost a man $19 a month in board and lodging;[19] the margin of $26 left in the latter year in terms of commodities and services was certainly worth no more than the $20 of a decade earlier. There is no doubt that many miners left Cornwall for Northern Michigan in the belief that all their wages above the cost of board and lodging could be saved, and they went out taking no account of incidental and unforeseen expenses or of the possibilities of their being 'laid off' through accident, illness, or unemployment.

The declining fortunes of so many Lake Superior mines benefited a few of the Cousin Jacks in the region, for some of the companies, including the previously extremely profitable Minesota Mine reverted to the tribute system as a means of carrying on till, it was hoped, better times returned; the abnormally high risks of fatal 'caves' occurring in working the soft 'country' rock of that Ontonagon County mine, however, in all probability deterred many from taking up tribute pitches.[20] Other mining companies struggled on in the hope of a protective tariff against Chilean and other foreign ores, but when at last Congress introduced such a measure it brought little relief. Salvation for the copper-mining industry of the Lakes, but ruin to many mining companies, only came when the Calumet and Hecla Companies managed to reach the extremely rich conglomerate lode and, by using the most modern and well-planned methods, began to raise copper at prices which could compete successfully against foreign rivals in the American market in 1868 and 1869. Within four or five years these two mines, soon to be amalgamated under one management, were producing nearly sixty per cent of the copper mined on the Lakes and were paying big dividends; other mines in Upper Michigan, however, had had their day and had ceased to be. Thanks to Calumet and Hecla it was still the 'Copper Country' where a Cousin Jack could hope to find more regular employment and higher wages than he could at home; nevertheless, even Calumet and Hecla were to experience their full share of labour troubles and the like in these difficult years.

19. *Royal Cornwall Gazette*, 22 October 1868.

20. *Mining Journal* (London) 21 May 1870. In the summer of 1870 the labour force of this mine had dwindled to eight or ten (Ibid., 24 September 1870); ten years earlier it had been employing over seven hundred hands, of whom three hundred were underground workers. In 1870, however, the Minesota in addition to its tributers was employing a few hands in exploratory work on a ridge to the south of the area the company had previously worked. The fall in copper production in Ontonagon County from 3,553 tons (shipped) in 1860 to less than 500 tons in 1873 was even more sensational than the decline in Cornwall during the same period. (Ibid., 2 February 1861 and 14 March 1874.)

GO THEY MUST—BUT WHERE?

In the spring and early summer of 1872 two Cornishmen, Grose and Vivian, were among the ringleaders of a strike which lasted for about four weeks. Beginning at the Calumet and Hecla it spread to the Schoolcraft, Pewabic, and Quincy mines. The main grievance was the long hours of labour. The men refused an advance in monthly wages of ten dollars from sixty to seventy dollars, and came out on strike demanding an eight instead of a ten-hour shift for underground workers. The sheriff of Houghton County asked for troops to be brought in to prevent disorder, and things looked decidedly ugly when, on the arrest of Grose and Vivian, about seven hundred miners gathered and forcibly effected their release. They were subsequently arrested a second time, but were released on bail. Despite a great deal of rash talk there was little actual violence, though a Cousin Jenny was reported to have killed one of the horses in the carriage in which the sheriff had driven to the scene of disturbance. A thousand miners were estimated to have accompanied Grose and Vivian when they were taken off in custody the second time. Yet there were no untoward incidents although, for almost the first time in American mining history, the Irish and the Cousin Jacks were on the same side, and the only workers that were inclined to accept the higher wage and longer shift offer were Scandinavian immigrants.[21] Grose had been a Methodist local preacher, and his talents for leadership and persuasion had, undoubtedly, been developed by that apprenticeship, while his connection, and that of so many of his fellow-strikers, with that church in large part accounted for the reasonably orderly conduct of the miners' movement of protest.

Shortly afterwards a financial panic in America had repercussions on Lake Superior. Several disappointed Cousin Jacks went home, and again there was talk of the 'almighty dollar' not being worth more than the 'old-fashioned shilling',[22] while many miners in Upper Michigan were counting themselves fortunate if they averaged a daily wage of a dollar and fifty cents.[23] Although some were content to work on, hoping that better times would soon return, there were many complaints about the cost of living. Some miners came back to Hayle early in 1873 after barely staying a year in the States, and the married men among them said that

> they did not find it better for themselves or families to be abroad, and not half as comfortable, for by the time the high rate of board is paid in America, and the fatherless family at home goes through its little sicknesses, and the dozen

---

21. *Cornish Telegraph*, 5, 12 and 19 June 1872: *Mining Journal* (London), 6 July 1872.
22. *W.B.*, 13 February 1873.   23. *Cornish Telegraph*, 18 June 1873.

necessities are paid for, which the father's presence would obviate, there is no gain... and after the balance sheet is drawn there is no real profit for the married man. If he takes his wife across, unless she is willing and able to rough it, to forego her quiet and comfortable house at home, it is still worse, for if she expects to live in the style in America, and dress herself and her children neatly, the cost will be more than her husband's earnings will provide. This is the married man's version. The young man says he might rough it and save money, if health was good, 'but it is rough work' at best, and money-getting is the only thing that would induce him to stay, for he has never yet seen a better country than 'old England'. He would not like to stay and marry an American wife, for if so, he would have to wait upon her, light her fire, and keep her house straight, and be himself the 'worse-half' and not the better. An American wife wants a rocking-chair, and plenty of candies, and plenty of chat to make her keep house in comfort. *Divorce* to her is a mere alteration and not a trouble. She says distinctly, 'she does not like the English working-man for a husband'; he is not sufficiently affectionate; he is too selfish, and he does not believe in 'women's rights'. American liberty consists... in being able to loll about loosely, chew tobacco, and spit in church, parlour, office, or in a superior's eye if occasion offer, and to carry a six-shooter in a little pocket under your coat-tails, or as a Yankee himself sums it up—'I guess this is a fine country, if you have no boots you can move about barefooted.'[24]

These Hayle migrants went home on the eve of the great economic slump of 1873. Their failure to make good in the New World, no doubt, in part accounts for the jaundiced account they gave on their return to West Cornwall. Other disillusioned men stayed on, and among the causes of disappointment not the least was that relations and friends who had preceded them to America, who had even perhaps advised them to emigrate and helped to pay their passages, had not, however, in many cases, shown any inclination to slaughter the fatted calf on their arrival. A man from Redruth wrote a warning to his 'fellow miners' in a local Cornish paper, saying they should not

> leave home supposing that you will make money enough in a few years to keep you for the rest of your lives. If you do you will be awfully deceived. Some here, I know, amassed 'heaps of money' in a short time, but such cases are the exception and not the rule. Do not come here supposing that you will always be able to walk in carpet slippers. You will find that in order to succeed you will have to elbow your way through thorns and thistles. All the 'vinegar of labour' isn't left in England. You will find even in this 'great American Republic' men every whit as mean as any you ever met with in Cornwall. I say it with shame, some of the meanest, most despicable and inveterate

---

24. *W.B.*, 13 February 1873.

enemies of Cornish miners are those who are born in dear old 'sunny Cornwall'. Some of them ... are always ready to play some trick on the greenhorn. Some ... have been the means of making their own countrymen work harder than workmen from other countries. ... I really believe that no class of 'bosses' act so meanly towards the men under them than 'Cornish bosses'. ... There are honourable exceptions, but the majority of them are too much like 'old country cappens'. So you need not be in a hurry to run forty miles from New York to work for a Cornishman.[25]

Nevertheless Cornish underground captains or 'boss miners' in Northern Michigan often used their influence and position to keep their fellow countrymen employed in hard times; in return they expected them to keep up the repute of the Old Country by working honestly and well. The rise of Calumet and Hecla and the failure of smaller mining ventures in the Copper Country, however, meant that, on the Lakes, there were no mines manned exclusively by Cousin Jacks, whereas in the gold quartz mines of Grass Valley in California some concerns were wholly Cornish. Indeed, it was said that there were mines in which old country provincialism was so strong that it was not merely a case of no Irishman, or Welshman, or German need apply for work, but even for anyone who did not come from the same mining village in Cornwall that was the birthplace of the superintendent in charge.

Whatever disappointments many Cousin Jacks suffered in the Old Country, there were on Lake Superior, even in the trying days of 1873 and 1874, Cornishmen who were convinced that men who had once worked there would, even if they went back to Cornwall, return to Upper Michigan, for they would never again be satisfied with the low wages that were paid at home.[26] In those depressed times, too, it was not only the Upper Peninsula of Michigan that seemed to offer better prospects to the enterprising miner than Cornwall. By this time, it is true, the lead prospects of Wisconsin had lost the appeal they had had a generation earlier, nor did the newer Wisconsin zinc industry attract many new immigrants from Cornwall. The most exciting days of Californian gold mining were over, but there were still quartz-mines and relations there who, if ties of kinship meant anything—and they usually, though not invariably, did—would welcome newcomers from

25. *Cornish Telegraph*, 4 June 1873.
26. *Mining Journal* (London), 7 February 1874. Thus the young Lelant miner, James A. Pearce, who may well have been in the party that returned to Hayle early in 1873 after a year or two in Upper Michigan, was back again six years later, and spent the rest of his days as a miner and metal worker in the Marquette iron-mining region. (Northern State Normal School Collections of Local Biographies, Marquette Historical Society).

the Old Country and help them find a place where they could earn a better living than they had been able to earn in Cornwall for many years past. The Comstock, too, for a time was under a cloud, but hardly a year passed without some new rich strike being reported from some place or other on the 'Pacific Coast'. Away in the North-west the Boise River had already, in the early eighteen-seventies, drawn many miners, but down in the Southwest, in Arizona, the Apaches were still barring the way against the rediscovery of 'lost Spanish mines'—mines whose very lostness enhanced rumours of their riches. Other Indian tribes were still causing trouble in the Dakotas and in Montana, and the bloody episode of the Little Big Horn when Sitting Bull proved a better general than the galloping cavalryman, George Custer, was yet in the future. Gold had been found and worked, and silver and more gold were still to be found, in Colorado. America, despite many blighted individual hopes, was still a land of opportunity but that could not be said of Cornwall.

The best opportunities round about the year 1870 were probably those offered by the northern Michigan Peninsula. Both the copper and iron mines wanted miners who were prepared to work for wages rather than prospectors and 'birds of passage', and the greater number of Cousin Jacks who came out in the late eighteen-sixties were predisposed to agree to such conditions. The Copper Country had survived the vicissitudes of boom and depression pioneering times, and if scores of grandiose mining promotions had failed a number of soundly organised and solidly financed companies offered employment. Even by the summer of 1873 the Lake Superior mines had recovered from the slump of the previous year; copper prices had rallied to thirty-five cents per pound, there was no surplus of labour, and miners could command wages of from fifty-five to sixty-five dollars a month, tributers might make up to a hundred dollars in a month and even hands hired by the day were being paid two dollars and fifty cents.[27] Furthermore it was

reported that there will be good times for two years to come, unless any sudden change takes place in the copper market. Miners are in great demand, and I can positively state that the influx of five thousand would not fill the mines from Marquette to Keweenaw Point, the Calumet and Hecla alone employs between three and four thousand men.[28]

The quick revival in the fortunes of the copper-mining industry of Northern Michigan provided a glaring contrast to conditions in Corn-

27. *W.B.*, 21 August 1873.
28. Ibid., 21 August 1873; letter of 'Red Jacket', dated Calumet, 6 June 1873.

wall. The rising copper prices of 1871–72 did not check the falling output of the Old Country mines which had sold only about 75,000 tons of low-grade ores in 1871 and which barely averaged an aggregate production of 50,000 tons during the five years from 1875 to 1879; a slightly higher metal produce in the ores sold in some of these years was not due to any lodes improving in quality but simply to the mine-owners selecting for sale the best accumulated 'parcels' of ore when prices were fairly high. In 1871 and 1872 the Cornish mines had realised about £300,000 from sales of copper ores, but in 1878 they got less than £200,000 and barely scraped up a monetary return of £150,000 the following year. Actually, these last two years saw extremely depressed economic conditions in Cornwall, but the dolorous statistics of the entire decade provided convincing proof that copper mining in the Old Country was a dying industry.

In this period the Cornish tin-mining industry had undergone vicissitudes of fortune; it had pulled out of a slump in the summer of 1868[29] suffered a recession the following April when abnormally large shipments of 'Banca' tin were dumped on the European markets from the Dutch East Indies, rallied but was in an uneasy state when the Franco-Prussian War broke out in the summer of 1870, and was comfortably prosperous in 1871. Then, however, the attempt of a group of speculators to corner the world tin market sent prices soaring to hitherto unprecedented heights only to drop to depression levels in 1874 and 1875. No real revival occurred until 1879, but 'uncertain' times were back by 1882, and the old story of depression was again told in 1883. Briefly, lean years were far more common than fat ones for the tin-mining interests; in fact, during the sixteen years from 1868 to 1883 the Cornish tin-mines enjoyed only three or four prosperous years, about the same period of trade revival, and for the rest of the time had been gloomy and depressed. If the Missouri threat had proved illusory, the East Indies had continued to provide formidable competition to the old Cornish industry while, in the early eighteen-seventies a new rival had appeared in Australia. The actual output of tin from the Cornish

29. English block tin prices had averaged £85 per ton during the last quarter of 1866, rose to £110 by the end of 1868, and touched £130 in April 1869. In December 1869, the price was down to £117, but early next summer rallied to £136. The outbreak of the European war caused a recession to £125, but for the first half of 1871 the average price was rather above £131. The corner attempt sent prices to £152 by the end of 1871 and to £163 by April 1872; there was then a slight decline to an average of £150. 10s. for the last six months of 1872 and the first quarter of 1873; in May 1873, the price fell to £135, remained 'uncertain' till November, when it plunged to £119; in March 1874 there was a sharp fall to £93, while in July 1875, a bottom level of £81. 10s. was reached. (*Mining Journal*, London, weekly reports.)

mines had hardly increased at all during the period, though there were variations from year to year, indicating that whether times were good or bad the industry could hardly be called progressive.

Some of the most able and go-ahead Cornish miners of the time were then managing mines in Michigan's Upper Peninsula. Captain John Daniell had left Chacewater and by 1873 was underground superintendent of the now veteran Cliff Mine in Keweenaw County.[30] Another Chacewater man, Captain Vivian Prince, occupied a similar position at the Delaware and Pennsylvania Mine, while Captain J. H. Moyle filled the same post at the Copper Falls Mine. Cousin Jacks were also managing the underground workings of the Quincy, Pewabic, and Franklin Mines in 1873.[31] Calumet and Hecla not only had a fair proportion of Cornish miners on its payroll, but a fair amount of its stock was held by men in the Old Country who had seen in it a more lucrative investment than the 'knacked' and deserted bals of their homeland.[32]

The success and prowess of these Cousin Jack mine captains on the Lakes encouraged others to emigrate in their wake, especially when Cornish mining was in an economic position that held out but poor prospects of promotion; this meant that emigration to Michigan and elsewhere was denuding Cornwall of its most enterprising and ambitious practical mining men. Their departure overseas was to be sorely felt in the homeland later when Cornish mining needed such men to check its declining fortunes; their emigration meant that, instead of the decline of the mining industry of the Old Country being checked, it was, in fact, actually accelerated.

Lake Superior, too, did not depend on its copper-mines alone. The immense iron ore deposits of the Marquette Range, although discovered in 1845, had only been developed on a large scale after the opening of the Sault Ste Marie Canal in 1855, and more particularly after the outbreak of the Civil War had created a great market for iron and steel and accelerated the industrialisation of the Northern states. It was not until the late eighteen-seventies that real underground mining began on the Marquette Range; up to that time the workings had been opencast pits and quarries, some of which were two hundred feet deep. With the driving of adits and levels and shaft-sinking, the Cousin Jack miner came into his own in the Iron Country, but before that time quite a number of Cornish immigrants had settled at Negaunee, Ishpeming

30. For Daniell's career see above, pp. 87–9.   31. *W.B.*, 21 August 1873.
32. *W.B.*, 15 April 1886.

and the other mining towns back of Marquette harbour. The later iron ore discoveries of the Menominee Range, which began to be worked about 1870, of the Gogebic Range a decade later, and, finally, over the Minnesota Line on the Mesabi after 1885, offered fresh opportunities which many Cornish miners seized although they were never to play such a predominating role in the iron-mining industry as they did in the working of copper in the Lake Superior region. Before 1880 the Cornish in the Marquette district included Angwins, Bennalacks, Pearces, Tippetts, Elliotts, Hodges, Martins, Pascoes, Trebilcocks, Tregembos, Trembaths, Urens, Vincents, Walters, and a number of others from the Old Country, most of them miners and labourers, who were content to make a modest livelihood and enjoy rather more comfort than they could at home in Cornwall. A few of them became quite prominent, and the Iron Country had its Cousin Jack 'Cap'ns' as well as the Copper Country, and it could be claimed, in 1889, that a Cornishman was managing one of the largest, if not the largest, iron-mine in the world.

Peter Pascoe, Superintendent of the Republic Mine, was born in Wendron in 1831; his schooldays ended when he was only ten years old and he went to work in the Gwennap mines. When he was twenty-three he emigrated to the United States, mined in West Virginia for a time and then was employed in railroad tunnelling work in Pennsylvania; his abilities got him a foreman's or boss's position on that job— the youngest employed by the construction company for whom he was working. The Cambrian Iron Works in Pennsylvania then employed him as 'captain' in charge of all the timbering in their mine at Johnstown, and in 1861 he came to the Upper Peninsula to occupy a similar post first in the Huron Mine, moving on first to become chief captain in the Calumet and then, in 1869, leaving for the Washington Iron Mine. He was first employed by the Republic Iron Company in 1873, becoming superintendent of their extensive works less than three years later. In that capacity he was said to have employed more of his own countrymen than any other mining captain on the Lakes, and the mine under his charge was attributed with paying more in dividends than any other on Lake Superior.[33] He won a great reputation for his ability to pick out and train promising young men, and his 'pupils' were to be found holding captains' positions in many American mines before 1890. Pascoe was justly held in high esteem in American mining circles for

---

33. That is, iron-mines; Calumet and Hecla had, by 1887, paid out thirty million dollars in dividends.

his ability, alert judgment, clarity of thought, affability, and practical knowledge.[34]

Others from the Old Country did not obtain as much prominence as Pascoe although making a significant contribution to the development of iron mining in the Lake Superior region. Cornish captains managed many of the iron-mines, particularly at Ironwood where, in the eighteen-nineties lived W. W. Stephens of the New Port Mine, John Tregumbo of the Pabst Mine, and John U. Curnow of the East Vulcan Mine.[35] Yet another Ironwood mining captain was William Trebilcock who, born in 1845, came out to Marquette County in 1866, tried his luck in the gold- and silver- mines of Oregon and Idaho in the unsettled early eighteen-seventies, though probably he would not have ventured so far West had he not been left a widower after four years of married life. Returning from the West, apparently having gained nothing more than experience, he had married a widowed Cousin Jenny and thereafter worked steadily in iron mines till he became mining captain of the Metropolitan Iron and Land Company. Trebilcock took a keen interest in public life, serving as Mayor of Ironwood no less than three terms, while another Trebilcock, James, was Mayor of Ishpeming in 1897.[36]

The careers of these Cornishmen were examples of what Cousin Jacks could do in the iron-mines of Michigan. What they could do in the copper-mines was most strikingly manifested by Captain John Daniell of the Tamarack,[37] but there were many other copper mining captains of Cornish nativity. About the turn of the century the Quincy was employing as captains Thomas Whittle, a Truro man, and S. B. Harris who had come out from the Old Country to work in the Wisconsin lead-mines before he was out of his teens, and had then come up to northern Michigan. At the Baltic Mine a latecomer, John Jolly, became captain when still in his thirties; even then he had seventeen years of American mining experience behind him. About the same time, in 1900, Edward J. Lord, who hailed from Liskeard, and who had started mining when ten years old, became captain of the Pewabic Mine.[38]

The list could be extended until it might seem that Cornish mining captains in northern Michigan were as common as colonels in Ken-

---

34. *W.B.*, 27 June 1889   35. See above, p. 85.

36. *Memorial Record of the Northern Peninsula of Michigan*, p. 388–9; *Iron Ore*, 24 June 1903, Commemorative Number for the Semi-Centennial of the Lake Superior Iron Company, p. 43.

37. See above, pp. 87–9.

38. Details of Whittle, Harris and Jolly are given in the *Biographical Record of Houghton, Baraga, and Marquette Counties*, pp. 75–8 and 85–6, and of Lord in A. L. Sawyer, *A History of the Northern Peninsula of Michigan*, pp. 785–7.

tucky. Most of them came to work mineral deposits which other men had discovered, but some Cousin Jacks in the Upper Peninsula attributed the discovery of the great Calumet Conglomerate Lode not to Edwin James Hulbert but to James Tregaskis, a native of the Old Country.[39] There is no doubt, however, that a Tywardreath man, John Wicks, discovered the Chaplin iron deposits at Iron Mountain, and earned the title of the 'Grand Old Man' of that town, for he was eighty-nine when he died in 1908, thirty years after his great discovery; his mining days had begun as far back as 1831, some twenty years before he first set foot in Michigan.[40]

These few random examples of mining captains in Northern Michigan give some indication of the way in which Cousin Jacks moved from mining camp to mining camp in the United States, although so many either settled in or returned again to the Upper Peninsula. Several had not come directly from Cornwall. To the Marquette district in the eighteen-nineties there came a number of Cornish people or their sons and daughters who, as a result of the slump of 1866, had sought a living and a home in the Furness iron-mining district of Lancashire, including Bowdens, Curtises, and Prowses. Others came to the Upper Peninsula from the Bruce Mines in Ontario, from the iron and coal mines of New Jersey and Pennsylvania, up from Wisconsin, or from small mines in New England or New York State.[41] Once the Cousin Jacks were in Northern Michigan, it was significant that the development of the iron-mines in the same region attracted not only those Cornish immigrants who had not found the mines of the Keweenaw Peninsula such lavish paymasters as they had anticipated, not only unemployed and half-employed tin miners from Cornwall, but some of those who had crossed the Tamar and had gone 'up country' in the late eighteen-sixties and had found northern England and other parts of the British Isles less homelike than Upper Michigan seemed to be from the many letters and accounts sent home.

Lake Superior, however, had its drawbacks and shortcomings, and it was hardly surprising that those unsettled by the very fact of their being immigrants and, still more, those who were natural rolling stones, should show that restless nomadic trait which had been so marked in California in the late eighteen-fifties, and be ready to listen to and to heed every

39. *Detroit Evening News*, 16 October 1899.
40. A. L. Sawyer, *A History of the Northern Peninsula of Michigan*, p. 1049.
41. Based on biographical sketches collected by Northern State Normal School in 1925-7, preserved in the Marquette Historical Society.

tale of new mineral finds, true or false, exaggerated or not, and take themselves elsewhere. Comstock and Pike's Peak, the Gila and the Boise, Last Chance Gulch and the Black Hills, all these and many more caused their flurries of 'excitement' wherever two or three miners foregathered. A row with a captain, a succession of light pay-packets, even a winter of chilblains and frostbite, all helped to keep miners from the Old Country on the move in the New World. With the Old Cliff and the Minesota showing signs all too much resembling the fatal creeping paralysis that had struck Wendron Consols, Great Wheal Busy, South Tolgus, Wheal Reeth, Hallamannin, and many other Cornish bals, it was little wonder that Virginia City, Butte, and Tombstone acted like magnets upon Cousin Jacks in Keweenaw and Ontonagon, and they hastened away to new places on the American mining frontier.

8 South-western desert empire

For some time after the first Californian gold discoveries the vast region stretching from the semi-arid high plains of the western Mississippi Valley, over the Rockies, and then across the great inland basin to reach the Sierra Nevada beyond which the new El Dorado lay, was regarded as ground to be travelled—nothing more. Trials and privations suffered by those who went overland to California underlined and gave new life to the old geographical myth of the Great American Desert. The first Californian 'Argonauts' had but the single idea of getting to the diggings as quickly as they could; it was only later parties and persons going back East by more or less the same trails that could stop long enough to look about them and conjecture about the mineral potentialities of the region that was to become the latter-day states of Colorado, New Mexico, Utah, Nevada, and Arizona. The greater number of the overland migrants of 1849, however, went by the more northerly route along the Platte and thence to South Pass before veering south-west to Great Salt Lake and the Humboldt Valley; comparatively few used more southerly routes across New Mexico and Arizona. The antagonism of the Mormons towards their American persecutors, flaring up to almost the point of war against the States in 1857 and marked by such bloody incidents as the Mountain Meadows Massacre in the fall of that year, did not tempt those going to California to linger long in Utah, while to the south-west, in Arizona, Apache and Comanche warriors were an even more deadly threat than the Avenging Angels of Mormon Bishop John D. Lee. Further west, nature herself made the most remorseless challenge of all to the overlanders—the Humboldt country and the alkali desert region which wayfarers cursed as the waste left by the Almighty on the sixth night of Creation, waste which the devil would not accept 'to fix up hell'. The fate of the Donner party of 1846 lost nothing in the way of horrors through countless retellings, and those who followed on their trail to the gold diggings were not inclined to linger on the way as they had done and be reduced to starvation and even cannibalism by the onset of an early winter.

In earlier times Spaniards and Mexicans had probed the region, but while the chronicles and legends of this extensive Spanish borderland were both fascinating and numerous, there is no gainsaying the conclusion that, so far as the American mining frontier was concerned, it was, at the beginning of the nineteenth century, an unknown territory wherein men believed there existed scattered gold- and silver-mines guarded by fierce Indians and haunted by the vengeful spirits of luckless prospectors and martyred priests who had been slain by the aboriginal tribes, a territory, furthermore, in which Nature herself was implacably hostile. Travellers' tales had become confused into a fantastic mizmaze; a few adventurers had reached the Grand Canyon, a few had stumbled into the unearthly region south and west of Great Salt Lake, some had wandered among the towering peaks of the Rockies, and, further to the westward, one or two may have stumbled into Death Valley. Petrified forests, alkali sinks, fantastically shaped and formidably thorned cacti, rattlesnakes, Gila monsters were all encountered in this region, and Nature herself warned, in turn, Conquistadores, missionary priests, prospectors, and white settlers that it wanted none of them.

Anglo-American penetration of the region began with Zebulon Pike's explorations of the great but ill-defined area which President Jefferson had bought from Napoleon Bonaparte in 1803. In the next quarter century, Anglo-American fur-traders swung down into the region beyond the Rocky Mountain divide from the northward, but hardly any of them had any eye for mineral prospects. In the eighteen-forties came the Frémont expeditions which revealed the more prominent topographical features of the Inland Basin, but again little knowledge was gained of the mining potentialities of the region. There followed the Mexican War, Marshall's discoveries, and in 1849 thousands of travellers were crossing a region wherein, up to that time, but few white men had set foot.

Of these thousands few were miners; most of them knew only the colour of manufactured articles of gold and silver but had not the least idea of the appearance of auriferous and argentiferous ores in the veins and rock-stuffs where they might reasonably be expected to occur. Only a mere handful of overlanders from Georgia, North Carolina, and Virginia had had any practical experience of gold mining, while some of the Michigan men had worked silver ores; like the rest of the forty-niners, however, they were all intent on getting to the diggings where gold had been found, and they did not mean to spend more time than they could help, to run short of provisions, or to get winterbound like the Donner party on the way. Had it not been for this hurry to get to the mining camps west of the Sierra Nevada, a Cornishman, Prouse, might well have located the Comstock lode ten years before it was actually discovered.[1] For the most part, however, men had to gain experience in prospecting in California, and provisioning points had to be established along the overland trails between Independence and Sacramento, before much prospecting could be done in the territory lying to the west of Missouri, Arkansas, and eastern Texas.

Under these circumstances, it fell mainly to returning miners and prospectors going back east from California to make most of the 'strikes' which marked out new frontiers of mining settlement in the late eighteen-fifties and early sixties, apart from those in Colorado which had been located and left behind by the Ralston party on their way to California in 1850, and which were 're-located' when dreams of quick fortune-making in California faded and when Americans were seeking fresh bonanzas nearer than the Fraser River. A minor gold find on the Gila, by a Texan homeward bound from California, in the fall of 1858 brought some adventurers to the vicinity of Fort Yuba, but the real sensation was the discovery of gold and silver in 'Washoe' in 1859. A new series of names began to be bandied about by prospectors and miners, by company promoters and by those who were ever ready to dash off to the rainbow's end for the proverbial —and usually mythical —crock of gold. Virginia City and Reese River, White Pine and Tonapah over the years were to make the headlines from Nevada Territory and State, and with romantic or original Indian names running out, last to come was simply Goldfield itself. In barely half a decade of the first mineral discoveries, Nevada, from being hardly an organised county of California, was admitted as the thirty-sixth State of the Union.

To many 'Comstock Lode' has been a synonym for Nevada, and the first swarm of immigrants into 'Washoe' settled around Virginia City and Gold Hill. Today Virginia City is a lively 'ghost town' living upon a tourist clientele, while Gold Hill has almost disappeared. Under the shadow of a towering mountain peak, set on a series of hill slopes, these Nevadan mining camps of the eighteen-sixties, seventies, and eighties drew a large but, mostly, transient population, including many Cousin Jacks. A number of them remained—in the graveyard overlooking the

1. T. A. Rickard, *History of American Mining*, p. 83.

steep, twisting valley running away southward and eastward from which so much gold and silver had been brought from the depths of the earth by their hard labours. Relatives and friends erected memorial stones to some of them on which the weathered legend 'Native of Cornwall, England' can still be read. For the most part they lived out barely two of man's allotted span of three and a half score years. Accidents took their toll; there was John Henry Carter who was killed in a mining accident in September 1879, leaving a widow and two children in St Agnes;[2] three months later another St Agnes man, James Prout, who had been in Nevada twelve years, perished in another mining accident;[3] yet another St Agnes man, Martin Tregellas, lost his life in a fire in the Gould and Curry shaft in July 1887.[4] Occupational diseases, especially miners phthisis, were particularly fatal in the dry dusty Comstock mines. And the Cousin Jacks were as likely to get involved in 'shooting scrapes' as those of any other 'nationality' in the boisterous western mining camps; in December 1874, Alfred Rule of Camborne and John Skewis of Chacewater quarrelled over a cribbage game at the Washington House in Gold Hill; Rule called Skewis a liar, and Skewis shot him.[5]

Gold Hill and Virginia City, however, were comparatively tranquil places compared with some of the Nevadan mining camps which sprang up overnight and vanished nearly as quickly. Pioche gained, and perhaps deserved, the reputation of being the toughest camp in Nevada; it was reputed to have buried twenty-five men who died in gunfights before a grave had to be dug for a man who died of 'natural causes'.[6] Rival Pioche mine-owners were said to have hired gunmen to fight for possession of rich mining claims, and to have brought them in at the rate of a score a day. Back in the Old Country a few who may have considered going to Pioche must have thought again when they read the following letter written by a miner to a friend in California which was, subsequently, communicated to the Cornish press:

Pioche is the county seat of Lincoln County, a mining camp a year old. It is on the Great American Desert, and situated between bare mountains overlooking a bare, dry plain. Water has to be brought eight miles by wagon, and is sold at six cents per gallon. There are about twelve hundred people here, half of whom have been in the State prison (stage and highway robbery, &c.), and the rest ought to be. Our graveyard has forty-one graves, of which

2. *W.B.*, 9 October 1879.   3. Ibid., 22 December 1879.
4. Ibid., 11 August 1887.
5. Ibid., 28 January 1875 and *Cornish Telegraph*, 3 February 1875.
6. *W.P.A. Guide, Nevada*, p. 173.

two were filled by death from natural causes. The rest all died with their boots on—shot mostly, some cut ... There is no law. Anyone feeling aggrieved seeks redress, generally with his pistol. It's been a close game for me several times. I got cut in the leg once, but am here yet. I sleep with a big bulldog, a Henry rifle, and a six-shooter. The mines (silver) employ about six hundred men, about a hundred are in business, and the rest are blackguards of the worst kind. Cattle thieves, renegade Mormons, and men who are banished from society for their crimes, and ready for anything. It's thirty miles to railroad or telegraph, we have three stages a week and one mail. I've done very well here, and would have made some money, but two of my partners in a claim were killed, and I can't get on alone. My life has been attempted twice by the party who killed my partners. I don't allow any man to scare me if I can have a show, but when it gets down to cases where you dare not sit by a window or open door after dusk, it's time to jump the game. If ever I get back to California, I think I'll stay there; anyway I've had enough of this kind of living. I'm tired of packing a six-shooter around night and day.[7]

A further lurid account of Pioche, written, apparently by a Breage miner, appeared in the Truro *West Briton* in July 1873, warning the writer's friends at home to stay there and not to be beguiled to Nevada by tales of wages of four dollars a day, for

gambling, theatres, balls, saloons, and, not the least, houses of ill-fame, drain the working man dry. There are many here in Pioche to-day well-known to you, but I shall not give names, who have been here on this coast for many years who have not got a dollar in their pocket, and not a change of clothes—Sunday and Monday all alike—and many Cornishmen are here to-day who are bumming around the saloons for a drink, and are half-starved.

He wrote on to give the impression that within the past week or fortnight three Cornishmen had died of starvation, and that

Pioche is three hundred and fifty miles from *everywhere*. There is a big desert to cross from Utah Territory to this place, and many fast boys who have run through the last dollar, and are not able to get work, attempt to foot it out. Some go right, others miss their way and are not heard of again—either killed by Indians, or die of starvation.[8]

Since this writer had referred to Pioche and California as being the two 'poorest States in the Union', he was not so much contradicted as corrected by a Redruth miner who caustically wrote from Pioche, on 18 August 1873, to the effect that, had the Breage man told the truth it would have been bad enough, but since he called a mining camp of barely two thousand population a 'State', he ought to refrain from

7. *Cornish Telegraph*, 15 November 1871. 8. *W.B.*, 17 July 1873.

writing letters.[9] Pioche, nevertheless, then ranked as the second mining camp in Nevada, yet, while forthrightly denying the poverty-stricken plight of the two western mining states, California and Nevada, the Redruther went on to say that he had

> never known any miners that attempted to walk from here to Utah. I have known men walk from here to White Pine, with their blankets on their backs, a distance of a hundred and fifty miles, and their pockets very light. The three men brought in (dead) were not Cornishmen, as stated by him. It is a fact that many men have been on this coast for years, and have not got a dollar, but I have never known men to go short of food, unless they were habitual drunkards or very bad men. Some men want to save all the money they can make, living more like hogs than men; but, taking men on an average, they are in better circumstances at home than what they are in this country. The fact is, there are twice too many here for the work, and many men would return (home) if they had the money. Most men think, when they leave home, if they can get to the States of California or Nevada, and get four dollars per day for their labour, that in a few years they can live independent of anyone, but they will soon find that they are counting their chickens before they are hatched—men are likely to be deceived. A man's life is not valued much in this country. There are four men in the gaol here now for murder.[10]

It should be emphasised that these letters were written in 1873, fourteen years after the first Washoe excitement, and at a time when the second great bonanza period of the Comstock had hardly begun; Nevada's silver mining industry had been shaken by the demonetisation of silver by the Federal Government, and the entire American economy was reeling towards the commercial panic of 1873.

For more than a decade, however, the mineral wealth of Nevada had been attracting hosts of immigrants from many places, including a fairly high proportion of Cornishmen from the Old Country, from the East and from California. At the first Nevada census, in 1860, the Territory mustered less than seven thousand inhabitants; in 1870 it was nearly fifty thousand and by 1880 had reached sixty-two thousand;

9. *W.B.*, 17 July 1873.
10. Ibid., 25 September 1873. The letter of the Breage miner had been sent to the *West Briton* by a person signing himself 'One who Remains at Home', who had, in an introductory note, vouched that the writer was 'a worthy representative of his class, and no doubt gives a faithful picture of the prevailing condition of things in the country now attracting so many able-bodied miners from this county, when their services are wanted at home'. It was published with the aim of checking the flow of emigration from Cornwall at a time when tin prices were still above £130 per ton, but the prefatory note certainly indicates the magnitude of emigration was alarming many in Cornwall, though their motives may not have been altogether altruistic.

there was then a falling away of more than twenty-five per cent until the Goldfield and Tonapah strikes in the early years of the present century boosted the census returns over the eighty thousand mark. Census returns, however, make no record of numerous transients of which class Nevada had an inordinate proportion, men who moved from one mining camp to another regardless of state or even international boundaries. Nor did the Federal Census ever have a record of places like Star City, about a dozen miles from Winnemucca, which boomed overnight to a population of over twelve hundred in 1863, but which, in less than five years, was deserted.[11] The short-lived camps of White Pine and its 'twin', Treasure City, however, were there to make census returns in 1870, but Story County, which was by far the most populous in the entire state that year, by 1930 could only muster 630 inhabitants, a figure that might be compared with an extremely conservative estimate that, in 1870, there had been at least two hundred and fifty Cornish people living in Gold Hill and well over three hundred in Virginia City. About seventy per cent of these were 'single' men, while of those with wives and families the recorded birthplaces of the children certainly indicated that they had been rolling stones; thus, two of John Spargo's children had been born in Maryland, while the three children of John Rosevear had been born in Michigan, Canada and Vermont respectively. Only three of the Cousin Jacks in Gold Hill were noted down as being 'unable to read or write', a fact which suggests that the charges of illiteracy levied against the Cornish immigrants were grossly exaggerated, but which also suggests that, since this was the year of the first British Compulsory Education Act, Nevada and other American states were attacting immigration by the better-educated miners of the Old Country, lending point to charges of mine-owning interests that the best workmen had left and were leaving Cornwall.[12]

The 1870 Census showed Cornish immigrants in other Nevadan mining centres. There were about thirty Cornish men living in White Pine, all either bachelors or with wives or families left in Cornwall or elsewhere. About fourteen or fifteen more were to be found in neighbouring Treasure City. The Cornish contingent in Pioche numbered at least a score, not including a number of Australian-born youthful

11. W.P.A. Guide, *Nevada*, p. 122.
12. See above, p. 172. An estimate of the illiteracy rate of the working Cornish miners round about this time would suggest that the percentage of those unable to read or write in the Old Country was as high as eight or ten, as compared to about one per cent among the emigrants from Cornwall in Nevada.

Cousin Jack miners. In Lincoln County 'English'-born hotel-keeper John Bowen returned three teenaged children born in South Africa, and three younger children born in Utah; since the sixteen-year-old wife of the Scottish teamster John McNeill, who made the next entry in the record was also South African born it is likely that she was also Bowen's daughter. At Meadow Valley, in the same county, an 'English' gardener, Harry Hodge had also come in from the Mormon Territory where he had married a Swedish wife. Lincoln County was obviously a microcosm of the American 'melting pot' of nationalities, but the Cornish element in the mines was numerous enough to make a fair contribution to the economic and social life of Nevadan mining communities. Up in the Reese River district there were certainly fifty, and more likely nearly a hundred Cousin Jacks at Austin, though only about a tenth of them had wives along with them. A year later, in July 1871, a Cousin Jack, William Craze, was fatally injured in a blasting accident in one of the Reese River mines; the local newspaper report of the tragedy not only mentioned that he had a wife and four children in Cornwall and that he had formerly worked in Grass Valley, but that another Cornishman, John Kinsman, who had lost half his left hand and a finger on his right in the same accident had a wife and children in New Jersey, and it also commended the care that 'Captain' Prideaux had taken of the dying Craze.[13]

For Nevada as a whole a guess can be hazarded that at least a tenth of the mining population, whether semi-domiciled or transient, was of Cornish origin in the early pioneering days of the eighteen-sixties and seventies. In some districts, particularly along the Comstock, in Virginia City, Gold Hill, and Silver City, the proportion may well have been considerably higher, while in other camps it was less, but in Nevada, as elsewhere, wherever a pit was found a Cousin Jack would be there digging away at the bottom of it. Somewhat later, in 1894, a visiting Cornishwoman reckoned that half the population of Virginia City, which had then fallen away to about six thousand, was Irish and a fifth Cornish.[14] It was probable that the proportion of Cousin Jacks had risen above what it had been in the first pioneering ten or a dozen years after which free-for-all prospecting for quick wealth had been largely replaced by wage-earning on the pay-rolls of capitalist corpora-

13. *Reese River Reveille*, 20 and 21 July 1871.
14. *W.B.*, 19 July 1894. (Article entitled 'A Visit to the Comstock' signed 'Cornish Girl in Virginia'.) The same writer, probably the wife or daughter of a visiting Cornish mining captain or engineer, also visited Grass Valley in July 1894, Butte City, and Upper Michigan (*W.B.*, 13 September and 20 December 1894, and 28 March 1895).

tions. Furthermore, the Comstock's second bonanza period coincided with the slump that hit Cornish tin mining in 1873; in fact, the old Cornish mining industry never was to regain the, admittedly artificial, prosperity it had enjoyed in 1872, and Cornish miners were ready to seek employment wherever it was offered; if some Nevadan mining ventures, often owned by British capitalists, were offering wages of four dollars a day, to Nevada they were prepared to go and 'work for wages' —a thing that did not come easy to individualist pioneers. The latter class, indeed, now took themselves off to newer fields in Arizona and Colorado, just as a similar development of capitalism and of wage-earning had made Californian pioneers ready to go to Washoe when the news of the first discoveries in 1858 and 1859 reached the western side of the Sierra Nevada. Where the trek of the forty-niners was chronicled over and over again, comparatively little has been recorded of the hundreds of men who left the Golden State to seek the precious metals elsewhere. As early as the fall of 1859 Californian editors were counselling their readers not to be lured into a second Fraser River 'madness' and, if they were resolved to go to Washoe, at least to defer their departure until the next spring, since

the whole mining region is a barren desert, destitute of timber and water, and it will not only require a large outlay of money, but great expedition, to enable miners to establish themselves there in even passable winter quarters, before the severe cold weather sets in. Besides, the season is at hand when crossing the mountains will be, in a measure, impracticable except for the stage with its relay of horses, and means at the command of the proprietors to push their way through under all circumstances.[15]

Time and time again the Fraser River disappointments, the 'humbugs' of Gold Bluff, Kern River, and other rushes after will-o'-the-wisps were refurbished to caution Californians and others to refrain from, or at least think twice before going to 'Washoe'. Those afflicted with symptoms of 'Washoe Fever' were warned that the lodes were limited in size, that provisions were scarce, and that they would find on the eastern side of the Sierra Nevada

a country socially and physically very inferior to California. Bloodshed, violence, and strife, it is to be apprehended, will be fearfully dominant in that region before long. Without law, courts, or authorities, filled with desperate and turbulent men, reckless of life, and excited by the strong passions of cupidity, frequent dissensions and quarrels, leading to personal collisions and

15. *Daily Alta California*, 2 November 1859 (Editorial).

deadly conflicts, may be looked for. Sufferings, hardships, and deprivations, too, such as even the mining pioneer of California were not compelled to undergo, will have to be encountered by those who go out to labour and prospect in these deserts. Fierce heat, fainting thirst, toilsome travel, exposure and even hunger, must be the sure lot of those who go there. In this Utah[16] there will be arid plains to cross, steep tablelands to surmount, and marshy lagoons to wade through, without a green tree to protect the weary traveller from the glaring sun, or a drop of water to quench his burning thirst. Toiling over the yielding earth, or flinty stones; sleeping unprotected from the night air, half famished with dearth, the strong limb will grow feeble, and the stout heart faint, and the hardy miner in the strength of his manhood will yearn like a child for the cool streams that danced by his cabin amongst the woody hills of California. He will find, when perhaps it is too late, that he has come a long way, to reach a lonely, inhospitable, and unfruitful region, of savage aspect and dubious wealth—a land abounding with bitter waters and blistering rocks, a basin filled with mephitic pools and ponds of lye, thickly strewn with lava, basalt, slag, and cinders, the apparent vestiges of a pre-existent system—a primitive wilderness so scorified, saline, and sulphurous that it would seem to have been rained upon fire and brimstone, and afterwards sown with salt. Here, without shelter or guide, the miner will be exposed to unwonted deprivations and dangers, and it is to be feared many a stalwart and intrepid man, overcome with fatigue and thirst, will perish miserably on these solitary deserts, with no shroud but his gray blanket and no sepulture but the drifting sands.[17]

This description was applicable to the Humboldt Valley—or to California's own Death Valley—but not to the whole region lying east of the Sierra Nevada: had men been disposed to heed such writers there would have been no Virginia City, no Bingham, Globe, Bisbee, or Leadville, hardly even any Denver, Helena, Butte City, or Boise, no States of Nevada, Arizona, Colorado, Montana, or Idaho: indeed, had such counsels been heeded but a few short years before there would not have been any State of California.

It was the men who refused to listen to such cautionary advice who pushed back the mining frontier from California to the valleys of the Truckee and the Carson, to the Comstock, and then, in the course of a few brief but eventful years to Pioche and Bodie, to Reese River, Star City and White Pine. By March, 1860, 'Washoeites' were passing through Placerville on the Californian threshold of the high sierra at the rate of

16. The Washoe region, for the most part, lay in the Utah Territory in 1859.

17. *Daily Alta California*, 17 March 1860, article by a special correspondent, H. de Groot, who had been sent to report on Washoe.

seventy-five a day,[18] and five months later a Californian wrote back from Gold Hill that there had gathered in that camp 'subjects of all nations', and that 'the experienced miners of Mexico, Chile, and Peru, the hardy gold-seekers from California and the sturdy copper miners from Cornwall, one and all agree that this region is without a rival in the richness and variety of its metallic deposits'.[19]

For many years, however, the trails eastward over the Sierras from California did not provide comfortable travelling though certainly not unexciting, judging from the account given by a Crowan miner, Thomas Couch, who went from Grass Valley to Treasure City in 1870. He left by the stage that ran regularly from the Californian mining region to Virginia City. For a long time the route was uphill, till a summit was reached, from which, far below, a river was seen and something over it that looked like a log but which the stage driver assured his passengers was the bridge that they would cross. The horses at the crack of a whip plunged away at the spanking rate of ten miles an hour for a three-quarter mile descent of which, Couch wrote, there was

not a tenth straight, the road is entirely zig-zag, from the upper peak to the base of the mountain. As the leaders pass the ox-bow turns, they seem to be plunging head foremost against the thither bank, but as their noses touch it they spring quickly and with certain bound to the centre; round come the wheels in fine style, and the vehicle follows as smoothly as on the best race-track. The driver's face occasionally wears a conquering smile, and he says, with a slight impatience at our timidity and in an assuring tone and manner, 'Perfectly safe, boys, driven here a long time and no accident has happened; I guess you'll get down all right.'[20]

At the bottom the travellers got the driver to stop the stage for them to get off to take a drink from the stream and to shake off the road dust that had covered them while riding up and down the mountain. Couch remarked that the red dust was like that around the Wendron tin mines in Cornwall, while the water tasted like that of the springs trickling out from the granite rocks around Crowan Churchtown. Then, aboard the stage again, they went over the bridge and up the other side of the gulch, the track being

just wide enough for a single vehicle. In many places, where it is blasted out of the solid rock, there are not six inches of leeway. The hubs of the left-hand

18. *Daily Alta California*, 31 March 1860.
19. Ibid., 15 August 1860, letter of 'H.C.P.' dated 4 August 1860.
20. *W.B.*, 31 January 1871, letter dated 31 December 1870.

wheel revolve close to the perpendicular rock, and the others almost jut over the edge of the precipices, some of which go down nearly straight from a thousand to three thousand feet. After a fifteen or twenty mile drive the summit is reached. It offers one of the grandest views of mountain scenery which the globe affords. As far as the eye can see (and the vision sweeps many times further in this clear atmosphere of the Pacific shore than anywhere east of the Mississippi) mountains succeed mountains, peaks are piled on peaks; gorges, ravines, canons divide them, serving to throw into shadow the steeps as the fleecy clouds go scudding athwart the bluest of heavens. Gazing away for scores of miles the earth's surface seems nothing but mountains. We wonder where the plains are, the fields waving with grain, the vine-clad hills, the orchards, the villages and towns. Apparently we are in an endless region of mountain waste, and doubt if it be possible to find our way out again to civilization, even with a compass, unless by the skill of this champion reinsman of the world. He was probably born amid the granite hills of the New England States, and there doubtless imbibed some of the spirit which led him to drive over this wild but charming route, to the summit of the Sierra Nevada mountains.

Many travellers in this region since Thomas Couch went that way have marvelled how men ever found their way into the Sierras, found gold and silver lodes among them, and then managed to find a way out again. The men who worked in Cornish mines, especially those on the steep slopes of the Tamar by Gunnislake or on the cliffs between Cape Cornwall and Gurnard's Head had probably acquired a 'good head for heights', but even they must have wondered at times if they would safely reach their destinations along these mountain trails.

Couch did not describe the later part of his journey on from Virginia City to White Pine, but a wayfarer from Marysville who travelled to Star City in 1863 described his experiences along the road in much less rhapsodical terms than the Crowan miner. Accommodations along the stage route from Nevada City to Virginia City were roughish, although no-one bound for the realms of silver and gold was inclined to quibble at being charged a dollar for a meal on his way, which was usually mountain lake fish or tough beef steak, or at being asked seventy-five cents for a bed consisting of a straw mattress, sheets and blankets. Even these comforts were rare east of Virginia City. Nevertheless

the house at which the stage stops at Nevada (City) deserves a passing notice for the accomodating disposition of its inhabitants, who are careful to keep several pairs of heavy boots proceeding up and down the stairs and through the passages all night occasionally treading on a dog's tail by way of variety—

all of which is very useful to travellers who must get up at three o'clock in the morning or lose their discharge.[21]

Beyond Virginia City this traveller proceeded on horseback, and as far as Ragtown found the road tolerable, but then he came on to desert where

for some seven or eight miles the laboring horses must drag their load through the heaviest kind of sand and it takes from six to ten hours to get through with a load, as they were obliged to stop every two or three rods.

The route then, for a further seven or eight miles, was over black sand or gravel, half an inch thick, which seemed to the traveller a very desolate country 'fit to commit murders or suicide in', and it was strewn with the bones of dead oxen. The surrounding mountains instead of being rough and jagged, however, seemed to be

rounded, soft, and mouldering, as if they were reduced to ashes by the heat of the sun which pours down its rays with fierce intensity on this desert track.

Sixteen miles from Ragtown, at a brackish well where the stage company kept changes of horses, there was a store run by a one-legged man. His prices were high—twenty-five cents for water for a horse, eight cents per pound for hay, and eighteen cents per pound for barley.[22] By the store a crudely painted sign announced that the 'best of cigars and liquors' were for sale, but the cripple had smoked and drunk his entire stock the day before the traveller for Star City passed through. The route then traversed mile upon mile of white alkali flat, glittering in the fierce sun, and there he came upon and had words with a Dutchman who was using an iron chain as a whip upon his wagon team of five horses that had come to a standstill with the wheels sunk, through being overladen, full eighteen inches deep in the sand. Later on he passed two women driving along, apparently unattended by any male escort, in a buggy, and he philosophically reflected that wherever men went, women were sure to follow. At length, after being deluded by a mirage of a lake of fresh water, he reached the 'Slough', and thence

21. *Daily Alta California*, 18 June 1863.
22. At the West Australian gold camp of Coolgardie in 1894 water cost 9d. (18 cents) per gallon, but in parts of this desert goldfield 5s. was charged for a gallon of water, and a case was recorded of a man paying £5 for water for his three horses for one night. (*Cornish Telegraph*, 15 November 1894.) The prices of roughly $9 per hundredweight for hay and $20 per hundredweight for barley in Nevada may be contrasted with the then prevalent English prices of $1·04 per hundredweight for hay, and $2 per hundredweight for barley. *Journal of the Bath and West of England Society*, Vol. x, 1862, p. 294.

pursued his way to the mining camp of Unionville, or Dixie, which had become half-deserted through so many of its inhabitants going off to the Star City excitement.

The pioneer mining camps of Nevada, indeed those of all the American mining frontier west of the Mississippi, had much in common. Almost invariably men rushed in ahead of supplies and had to pay fabulously high prices for the necessities of life and the implements which they had not had the ability or the forethought to bring with them. The first shelters could only, by a compliment, be even called ramshackle—tents, brushwood huts, even 'coyote holes' dug in the ground. The population, predominantly male, was ever-changing with new arrivals appearing and earlier comers drifting away. Saloons, gambling-hells, and houses of ill fame quickly appeared. Life was violent, law unknown, crime common, but for the majority of men there was nothing much save heavy work and very little gain from it. A prospector struck a lode; the news quickly spread—sometimes even before he struck it—and was exaggerated: men poured in from all points of the compass, but it was unwise to ask many of them where they had come from or even their names. In a few days there might be three or four saloons and a store or two doing quite good business, even though their roofs and walls might be canvas and their counters and bars packing cases and barrels. If the camp looked like lasting some mail and stage facilities were arranged, and some crushing machinery installed to expedite the garnering of the precious metal from the ores. The Cousin Jacks, for the most part, only appeared on the scene when a lode looked like lasting a fair while and was of such a nature as demanded their particular Old Country skills of hard-rock mining—drilling, blasting, tunnelling, and shaft-sinking. A fair number of the Cornish, too, came in under contracts arranged by British-owned mining companies or by British firms connected with 'Eastern' mining enterprises. This was probably the case at Battle Mountain in 1870, when the Cornish mining superintendent, W. S. Nancarrow, was living in close neighbourhood with about a score of Cornish 'quartz miners'; in Lyon County, the same year, there was a colony of at least seventeen Cornish miners, only three of whom, including the superintendent, John White, were married or had their wives with them. The same Federal Census returns, listing a colony of at least thirty Cousin Jacks in the White Pine district, reveals that the most prominent of them, Superintendent John Treglone, must have 'struck it rich', for he possessed $15,000 in real and $15,000 personal estate; hardly any other Cornishman in that camp had more than a thousand dollars worth of any sort of estate, and not one of them had a wife with him.

The great disproportion of males to females and the very few families in the Nevadan mining camps was not peculiar to the Cornish immigrants but general. Indeed, as late as 1910, men still outnumbered women two to one throughout the state, while the proportion of school children was but forty per cent of the average for the entire Union. When the gold- and silver-mines were booming, as they were intermittently right up to the outbreak of the First World War, probably a third, if not a higher proportion, of the people in Nevada were 'birds of passage' who had no intention of settling permanently in the Sagebrush State. Yet even the transients who worked in Nevadan mines helped to build the state and contributed much to its economic and social life, even being partly responsible for such features as its minimal residential qualifications for franchise and divorce; the risks of mining life led to Nevada being one of the earliest states to introduce workmen's compensation and minimum wage legislation, and it can be argued that the disproportionately high number of men may have caused the early introduction of women's suffrage in order to attract female immigration.

The very size of Nevada, its extensive tracts of mountain and desert country, and its aridity were features making for nomadism, and in its first half-century of statehood the strikes of gold and silver, some rich, others mere 'flashes in the pan'—and especially the latter—accentuated this. False hopes were buoyed up by the very richness of the first discoveries east of Mount Davidson. Second and greater Comstocks were announced time and again; the Pioche, Reese River and Eureka districts, were highly productive and remunerative camps in their booming heydays, and the last-named region was estimated to have had a Cornish population six hundred strong in 1881,[23] when it was nearing the end of its fourteen-year bonanza period during which it had produced forty million dollars worth of silver and about half that value of gold.[24] It is well nigh impossible to assess how much of that wealth could be attributed to the efforts of the Cornish colony in Eureka; perhaps a fifth of the total product might not be an unreasonable guess. At all events the Cousin Jacks in Nevada that year, 1881, were reckoned to be making money wages seven times as great as they could have made had they stayed at home,[25] the cost of living in the Sagebrush State may have been twice as high as in the Old Country, but not much

23. *W.B.*, 16 June 1881.   24. W.P.A. Guide, *Nevada*, p. 255.
25. *W.B.*, 16 June 1881.

more even in places where water cost a dollar for forty gallons; two months steady work in Nevada would repay their fare thither from Britain; if they were provident—admittedly a mighty big 'if'—Cousin Jacks could save over and above their living expenses as much in a nine-month working year in Nevada as they could hope to save in as many years in Cornwall. In the high altitude camps fiercely cold winters curtailed the mining season, but 'short time' was by no means unheard of in Cornwall as many immigrants knew full well. Even if the best bonanza lodes petered out after a few years working, any man who worked in a good Nevada mine for four or five years had the chance to lay by a reasonable competence for himself, as much, in fact, as any man who spent his entire working life in a Cornish mine.

The chance of something more than a good wage or a competence, however, and the tendency of many good lodes to 'pinch out' or deteriorate served to accentuate nomadism among the miners. Strikes and rumours of strikes drew men over the mountains and across the deserts to such camps as Star City, Bodie, and White Pine. Even if these districts produced some millions of dollars worth of precious metals, the returns to the majority of the miners who flocked to them hardly recompensed them for the costs and privations encountered in their journeying to these short-lived bonanzas; a host of gullible speculators, too, lost all the investments they made in companies working or supposed to be working in these camps. At one time White Pine stocks had an aggregate value of seventy million dollars on the stock exchange, but the total output of that camp over the years was less than a third of that sum.[26] Such company promotions increased the intensity of the 'excitements'; genuine claim-holding companies hired miners to go and work their properties, while the flamboyant prospectuses of genuine and fraudulent companies alike served as propaganda inciting men to go to these 'greater Comstocks'. Hardly one in a hundred of those who set out for White Pine knew that it was ten thousand feet above sea level, exposed to bitter winds throughout a longish winter, and that the nearest spring of water was miles away. Water had to be piped in to the main mining centre at Hamilton and the miners had to pay dear for it; then, in 1873, a store-keeper turned off the water-valve just where the pipe entered the town, set fire to his premises in order to collect insurance, and half the city was burnt to the ground.

Through such acts of arson, but more often through pure accident or sheer carelessness, many western mining camps suffered severely

26. W.P.A. Guide, *Nevada*, p. 253.

from fires in their long or short lives. What was left of Hamilton after the 1873 fire, and the rapid decline of the mines after that date, was burnt out in 1886. Pioche had a fire in 1871, which, catching a powder store, cost thirteen dead and many badly maimed. Virginia City had severe fires in 1873 and 1883, while as late as 1923 fifty-three city blocks were swept away by fire in Goldfield. White Pine district's early days were also marked by a smallpox epidemic, and in that high altitude rheumatic complaints were rife, especially among miners who had in time past suffered limb injuries in the course of their calling; pneumonia, too, was a swift and common killer.[27] One Grass Valleyan in Treasure City wrote back in April 1869, to a Californian friend that he had been laid up with rheumatism for a week, and that he had found that it did not

pay to be sick here. A prescription that would cost a dollar and fifty cents in Grass Valley costs here five dollars. So you see it's very pleasing to a party that happens to be sick—*it's so cheap*! There are now nine cases of smallpox in one house on Main Street, right in the principal part of the town. People are going in and out of the house every day and mixing among the people in the crowded saloons every night; so that if it is true that the disease is contagious it would not surprise me if one-half of the population would take it, and in this country it would prove fatal in almost every instance on account of the scarcity of accomodations.[28]

To complete the catalogue of possible disasters, Nevada, despite its general aridity at times suffered grievously from flash floods, Austin being devastated in such manner in 1868 and again in 1874, while Eureka experienced bad floods and cloud-bursts during its chequered history which, too, included a fire which did three-quarters of a million dollars worth of damage 'covered by about $150,000 insurance' in August 1880.[29]

Fire and flood were hazards to be encountered anywhere, and so were epidemic and other diseases. Nor could the extremes of summer heat on the deserts or of winter blizzards in the mountains deter men who were born miners like the Cousin Jacks or the men of all races who could not resist the gambler's chance of quick riches, or any men who, in Cornish phrase, had 'any sprawl in them'. The same man who wrote home to Grass Valley about his rheumatics, the high price of medicines, and the smallpox epidemic, wrote again, six weeks later, in May 1869,

27. *Grass Valley Union*, 3 and 6 April 1869.
28. Edward McSorley to John Patterson, 1 April 1869, quoted in *Grass Valley Union*, 6 April 1869.
29. *Cornish Telegraph*, 8 September 1880.

from Main Street, Treasure City. He had found the only sure bonanza in a mining camp, not a silver lode but a saloon. But he had every confidence in the future of the White Pine district, writing confidently that

it is impossible for the bottom to fall out of this district, as there is ore enough now on the surface, awaiting crushing, to keep twenty mills going for one year. This is no place for a working man to come to unless he has a few hundred dollars in pocket, which are necessarily required to build a cabin and to purchase tools and grub while prospecting. In this district the mines are not sufficiently developed to give employment to a large number of men, and all those who come here with the expectation of getting immediate employment will necessarily be disappointed, and become disgusted with the country, and, as a matter of course, will damn White Pine to all eternity, or until the shipment of bullion puts the 'Riaree' on them. My private opinion is that this is, and will prove to be, the best country in America, but it is no place for a man sick in mind or body. Times are quiet now, and may remain so for a month, but there will be a reaction that will put all croakers to silence forever.[30]

Ed McSorley, in his saloon, was 'on to a good thing,' but it might be asked how many of those who travelled the trails to Eldorados and Silverados had a few dollars in their pockets, let alone a few hundreds. The survey of a day's traffic passing through Austin on the way to White Pine, recorded by the *Reese River Reveille* reporter in late February, 1869, suggests not many. He wrote that he had

noted casually the following persons and things moving in that direction. Three ox teams, each hauling two wagons loaded with machinery; one mule team hauling mill machinery; a Concord wagon, containing one man, two women and a child; two men on horseback, carrying rifles, and followed by a wicked-looking bulldog; three small farm wagons—one containing two men and a lot of household furniture, another a man and woman and a cat, and the last the driver with packages of provisions; four mule teams, each hauling two large wagons filled with barrels of liquor; a little old man under a blue blanket coat, mounted on a tall mule; a Concord wagon containing six Chinese men, their goods and chattels, seven mule and horse teams, hauling lumber, shingles, and merchandise; a cart drawn by a single yoke of antique and bony oxen, containing a cooking stove, several boxes, a few pieces of 'side meat', and a very long and heavy rifle—the sole attendant of which was the driver, a tall, weather-beaten, iron-looking man of apparently sixty years of age; a queer looking vehicle, a sort of improvised Rockaway, or German state carriage, in which were many trunks and boxes, and three persons—a man driving, an elderly woman of consequence holding a cat on her lap, and a

30. *Grass Valley Union*, 14 May 1869, letter dated 10 May 1869.

small Indian girl; nine wagons drawn by oxen, horses, and mules and fitted with traps; thirteen men of all nations, afoot packing their blankets and bottles, and followed by fifteen ill-looking hybrid dogs; three wagons containing the remnants of a Californian livery; a substantial covered wagon, in which were three ladies of remarkably free ways and speech, and one man driving—one of the ladies, who was pretty, saw us looking at her, and said something to us which we could not repeat; five strapping boys from the 'ould sod' afoot, without blankets, each carrying a bit of a stick, and roaring with fever as they passed; three Mexicans mounted, and driving one mule packed; two wagons containing a saloon and fixtures—

assuredly some inkling of White Pine's water problems must have reached the outside world, for the *Reveille* man went on

eleven teams of twelve mules each, and each hauling two first class wagons piled up with casks, barrels, kegs, boxes, and baskets, having a strong suspicion of spirits; two men on horses, with grave bearing, and looking like evangelical parsons; five colored citizens on white and grey horses, dressed in fancy blanket shirts, smoking cigars, and talking louder than white folks; more wagons; six teams; one wagon;—and so the living flood set in the direction of White Pine until night closed the scene.[31]

Such hegiras were to be seen on many Nevada trails in the decades following the Comstock strike, and were, indeed, to be seen in many parts of the Southwest in those times, and notably—in both directions—to the Cherry Creek camps in the Pike's Peak region of Colorado in 1859 and 1860. Thither most immigrants came in from the eastward, thousands of them by the famous Smoky Hill Trail from Sharon Springs in Kansas to Denver, along whose two hundred weary miles water and game were so scarce that it acquired the name 'Starvation Trail'.[32] In Colorado, as in Nevada, a succession of strikes of precious metals sent miners and prospectors scurrying hither and yon among the mountains and ravines—first in Gilpin County that was quickly to challenge the claim of the district round Chacewater in the Old Country to be considered the 'richest square mile on earth'; then to Tarryall, Fairplay, Buckskin Joe and other camps of the South Park; to Leadville and California Gulch when rich carbonate silver-lead ores were discovered in 1880; past the Continental Divide to 'King Solomon's Mines' at Creede; and, in 1891 and 1892 to Cripple Creek, which, briefly, enjoyed the repute of being the richest gold camp of all with, however, rather more reason than a host of other camps which, in Colorado, Nevada,

31. *Reese River Reveille*, 27 February 1869.
32. W.P.A. Guide, *Colorado*, pp. 144, 271–2.

and elsewhere on the American mining frontier have become 'ghost towns'—even if their site be as much as remembered less than a century after their brief days of fame.

The mining pioneers of Colorado, like those of California and Nevada, came from 'all over', but in the early days of Gilpin County, which was the first major mining district of the Territory, the number of Cornish hard-rock miners was very noticeable. In the successive Federal Census reports of Gilpin County for 1870, 1880, and 1890, fifteen, twenty-five, and twenty-three per cent of the population were recorded as 'English' born; a large proportion of these were Cornish,[33] possibly a total of five or six hundred in 1870, and well over a thousand in 1880 and 1890. The first Cornish arrivals were Cousin Jacks from the Lakes and Wisconsin, but from 1865 or 1866 the numbers coming directly from the Old Country became more prominent.[34] Gold was the lure, but in the early eighteen-sixties some came to Colorado from both south and north of the Mason–Dixon line to evade personal involvement in the War between the States. Gilpin, too, was the most easterly of the great auriferous regions and, for that reason, seemed easier to reach than the Pacific Coast mining camps. On the other hand, it is unlikely that the Probus man, C. P. Holten, and his friend, Alfred Hill, were the only Cousin Jacks who enlisted in the service of the Union in Colorado Territory,[35] but most of the Cornish miners who came to Colorado before 1870 were 'birds of passage', since the Census of 1880 indicates only three or four Cornish families with children of more than ten years of age born in Colorado.

Though in Federal Census returns the Cousin Jacks usually gave 'England' as the country of their nativity, typically Cornish mining family names suggest that, in 1880, Gilpin County had at least seven hundred Cornish residents, of which at least three hundred were 'single' men mostly living as boarders in the homes of families to whom they may have been related; a few, however, in groups of two or three, seem to have fended for themselves in 'bachelor quarters'. The remaining four hundred were married couples, a quarter of whom had young families born in Colorado, another quarter had children born in both Colorado and in England, about a fifth were childless though mostly in age-groups that might yet expect to have children, and the remainder had families born in one or more of the other states, or in one of the

33. Lynn L. Perrigo, 'The Cornish Miners of Early Gilpin County', *Colorado Magazine*, Vol. XIV (1937), pp. 92–93.
34. Ibid., p. 92.   35. *W.B.*, 20 February 1863, and see above, pp. 145 ff.

other states and in England with, perhaps, the youngest one or two children born in Colorado. The main states from which Cornish families had come on to Colorado were New Jersey, Pennsylvania, Michigan and Wisconsin, but among the infants of Cornish parentage were children who had first seen the light of day in Massachusetts, Vermont, Connecticut, Iowa, Illinois, Virginia, Tennessee, Missouri, Nebraska and California. The eldest born of Edward Tippett and Seymour Reynolds had been born in Michigan, in 1863 and 1864 respectively, but whereas Tippett had a child born in Colorado in 1869, his youngest, aged four, had been born in England; Reynolds' second child, aged twelve, was born in Illinois, and his third, aged six, in Colorado. The first-born of George Hill had been born in Michigan in 1873, but two younger children, aged five and four years, respectively, had been born in the Old Country. Of the five children of John Prouse, the eldest, aged eleven, had been born in England, the second, aged six, in New Jersey, the third and fourth, aged three and two, in Pennsylvania, and the youngest, one year old, in Colorado. The largest family, that of John Simmons, numbered eight English-born, two Pennsylvanians, and one Coloradan; the ages of the eldest eight ranged from eighteen years down to seven, while the American-born were aged five, three, and one year old respectively.

Analysis of the ages of these immigrants and their families reveals certain significant features. There were just over four hundred 'children' of less than twenty years of age returned in the family groups bearing Cornish surnames, and about three out of every four of them was under ten years old. This high proportion of young families underlines the argument that the Old Country was losing the flower of her youth and manhood by emigration, and this is still further emphasised by the ages of the 'single' men. Of the latter, nearly three hundred in all, only about a dozen were over forty-five years of age and less than a score still in their teens; a further breakdown of the age-group of twenty to forty-five years, reveals that rather less than half were between twenty-five and thirty-five, that slightly more than a quarter were between twenty and twenty-five and that slightly less than a quarter were between thirty-five and forty-five years of age. Hardly ten per cent of the entire immigrant group, born in 'England' was over forty-five years old, but at that time through the incidence of occupational risks and diseases a similar age distribution pattern was to be found in mining parishes in the Old Country; in Gilpin County, however, there was a great disproportion of numbers between male and female, there being roughly nineteen men to every seven women in the age-group of twenty to forty-five years, while even in the group above forty-five years of age, males outnumbered females nineteen to fourteen whereas in Cornwall the women would have outnumbered the men to an even greater extent. Of course, in Gilpin County, in 1880, the general excess of males and pioneering conditions seriously affected the sex proportions in adult age-groups, but the Colorado mining frontier was no less hard than any other on womenfolk, and many aged prematurely and died young. Nevertheless, pioneer conditions here were less lethal to women than were the occupational risks of mining to men, and the former were to be more quickly ameliorated and mitigated than the latter.[36]

Colorado mining camps had their share of fatal mining accidents, but, as a rule and as in Cornwall itself, there were comparatively few incidents in which more than one or two individuals lost their lives. Late in 1869 John Moyle, who had lived at Creegbraws in Redruth, was killed in the Briggs Mine, leaving a wife and three children destitute.[37] Five months later a Roche miner, James Roberts, was killed in a blasting accident at Georgetown.[38] In the early summer of 1877, William Pope, who had gone out from Camborne, was killed, and though but twenty-three years of age, left a widow and three children in Colorado.[39] One of Colorado's worst mining disasters occurred in Central City on August 29, 1895, when miners working on the Fiske vein holed into some long disused mine workings that had become flooded; the torrent that poured into the Americus and Sleepy Hollow mines drowned eight Cornish and six Italian miners. Months passed before the flooded workings could be pumped out and the bodies of the victims recovered.[40] The list of such accidents and casualties could be extended, but Colorado escaped such fearful disasters as the Osceola Mine fire in Upper Michigan that, in September 1895, claimed between thirty and forty victims,[41] the Levant Mine disaster at Pendeen in the Old Country with thirty-nine fatalities,[42] and the Speculator, and Granite Mountain Mine fire in Butte in June, 1917, when a hundred and sixty-four lives were lost.[43]

36. The above paragraphs are based upon the Federal Census returns for Gilpin County for 1880.
37. *Cornish Telegraph*, 12 January 1870.   38. Ibid., 6 July 1870.
39. Ibid., 26 June 1877.
40. *Boulder Daily Camera*, 30 and 31 August 1895, and Caroline Bancroft, 'Folklore of the Central City District, Colorado', *Californian Folk Lore Quarterly*, Vol. IV (1945), pp. 324-25.
41. *W.B.*, 19 and 26 September 1895; *Colorado Chieftain* (Pueblo), 12 September, 1895.
42. *W.B.*, 23, 27, and 30 October 1919.   43. W.P.A. Guide, *Montana*, p. 73.

Accidents did not occur only in the mines. At least one St Just man was among the twenty-five victims killed in an avalanche early in 1883;[44] now and then flash floods claimed victims,[45] and in the early days many prospectors vanished into the wilderness never to return. Troubles with Indians were not as bad as they were in Arizona and Dakota. 'Shooting scrapes' were not unknown in Colorado, though none of the camps round Pike's Peak or elsewhere in the state, gained such notoriety as did some of the 'roaring' camps of Nevada and Montana for violence; the shooting of Stephen Pearce, of St Ives, by a Redruth man, Terrill, was perhaps the most sensational incident of this nature in which Cousin Jacks were involved in Colorado, and it is noteworthy that it seems to have been a case of deliberate murder, not what Westerners regarded as 'fair gunplay'. Terrill, who was working a tribute pitch near Georgetown in the fall of 1877, had been accused by Pearce, who was working nearby for wages, of taking some silver ore that he, Pearce, had taken out, and had hotly denied the charge, declaring that if Pearce repeated it he would be 'done for'. The accuser repeated the charge, and Terrill then went off to his home. After telling his wife that he was leaving in the morning for a distant and more promising location he went back after dark to the mine with a revolver and shot Pearce who died two days later.[46]

Miners' occupational diseases, however, claimed more victims than accidents and homicides. The dreaded phthisis claimed several victims, though, again, 'rocks on the chest' or 'miner's con' killed far more in the mining districts of Nevada and Montana than it did in Colorado. In the high altitudes of Colorado, especially in such places as Caribou—the place where the wind was born—bronchial complaints brought suffering and death to many men with mine-dust afflicted lungs; pneumonia, too, was particularly deadly, but it is likely that the incidence of serious illness in the mining camps of Colorado was rather less than it was in those elsewhere in the South-west where there was less family life. 'Cousin Jennies' looked after their menfolk, while a married man living at home with his wife and family was less likely to waste his earnings in saloons and brothels, nor had he any need to neglect and starve himself in attempts to save up money enough to pay the fares of his wife and family out from the Old Country to join him in the western mining camp where he had located.

44. *Cornish Telegraph*, 22 February 1883.
45. C. Bancroft, *Californian Folk Lore Quarterly*, Vol. IV, p. 321.
46. *Cornish Telegraph*, 27 November 1877.

Opportunities for making a fortune or, at least, a comfortable living by mining in Colorado were very similar to those in other western mining states. The first gold discoveries were alluvial deposits which could be worked by 'free miners' with very little capital at their disposal, but soon quartz gold and silver lead mining developments called for capitalisation, while many of the Coloradan ores proved to be of an abnormally refractory character which required complicated and costly reduction and extraction processes. It was not long, therefore, before the normal miner was a wage-earner, but wages, at first, were comparatively high, although successive waves of Irish and Austrian immigrants tended to depress them and caused considerable racial friction. Wages and employment, of course, were adversely affected by the slumps in silver after 1873 and again in the last decade of the century, but the widely scattered mineral deposits of the state and the organisation of large companies to work them by Eastern and European capitalists kept up the demand for labour. A fair amount of mining was done on the Cornish tribute system which gave the ordinary working miner the opportunity of benefiting from bonanza strikes. Furthermore, the very richness of several ore discoveries enabled the underground workers to resort to the practice of 'high grading',[47] with many miners hiding the best specimens they came across during the course of their day's work, smuggling them up 'to grass' when they left work, and selling them to shady ore dealers and assayers. Any mine manager who attempted to check high-grading was liable to find himself without any hands at all, but in several instances the ore bodies were so rich that the capitalists were content with just that part of the ore their workmen saw fit or could not avoid turning over to them. In any case, many working miners in Colorado gained considerable 'nest-eggs' by high-grading.

Wages in Colorado were rather low in comparison with those that were paid in some more western mining regions, but compared with the cost of living and current wages in Cornwall an average daily wage of $2.50 to $3 was reasonably good. It represented wage levels just high enough to ensure comfortable, if 'careful', family living. There was trouble when, first Irish and, later, Austrian and Italian immigrant labourers arrived who were used to lower living standards and who, therefore, were prepared to accept lower wages, yet even in these circumstances it was hard to induce the Cousin Jacks to become loyal members of the Knights of Labour in the mid eighteen-eighties and of

47. W.P.A. Guide, *Colorado*, pp. 247–248; Marshall Sprague, *Money Mountain*, pp. 203 ff., 312.

the far more militant Western Federation of Miners a few years later. By the time these wage, labour and union issues arose, however, a fair proportion of the Cornish who had come as mining immigrants to Colorado had either become mine managers themselves, or had left mining for various trades and callings particularly in the rapidly growing metropolitan centre of Denver.

The mines of Colorado and those of the Pacific Coast generally had the benefit of experience in organisation and technology gained in Britain and in the older mining regions of the East. Over the Old Country they had the advantage of being new ventures exploiting hitherto untapped mineral resources whereas the Cornish mines were generations old and, of course, the metals raised in Colorado were gold and silver, with copper occasionally appearing as a 'by-product'. Eastern capitalists, however, were reluctant to invest heavily in Colorado mines after the wildcat boom of 1864 when a horde of companies had been promoted in New York, Boston, and Philadelphia; these concerns bought up mining claims in Colorado at exorbitant rates without much preliminary investigation of either their mineral potentialities or even their legality; they had then despatched agents to Colorado who knew nothing about practical mining and who quickly involved their employers in lavish purchases of costly machinery, especially unproven and ineffective devices for amalgamating gold. All this expensive and heavy equipment was brought into Colorado overland from 'ports' on the Missouri, some six hundred and fifty miles distant at the enormous freight charge of five hundred dollars per short ton. Much of this machinery was never used by the companies who brought it to Colorado, and five or six years later could be bought at scrap iron prices.[48] Although such practices ruined many of the early Colorado mining ventures organised in the East, others failed because the free-milling gold found in surface outcrops was traced in lodes down below the water line where the ores were unoxidised and shut up in pyritic matrices which required reduction methods that were very much in an experimental stage at the time of this first Colorado mining boom.[49] Tales of single hundred-foot claims producing over a quarter of a million dollars were enough to start a stampede of capital to Colorado in which it was reckoned that Easterners promoted companies capitalised at over £150,000,000 when the aggregate gold production of the Ter-

48. *Mining Journal* (London), 26 March 1870, letter of Thomas Jennings, dated Central City, 26 February 1870.
49. Ibid., 4 May 1872 and 10 Aug. 1894.

ritory for the six years beginning in 1859 had not exceeded ten million dollars.[50]

Such reckless speculation coupled with mismanagement by agents, made later investors reluctant to risk anything on Colorado mines, nor did Federal and, later, State legislation curbing the rights of ownership of foreigners help the development of the state's mining industry. Litigation over disputed mining claims checked the development of the English-owned Terrible Mine, one of the best run and richest mines in the Territory, and threw several men out of work in 1875.[51] Cornishmen who had come to Colorado to mine were disgusted by Eastern mismanagement and by political maladministration, one of them writing home in 1875 that

> the mining interest of Colorado is in its very infancy, and is likely to remain so for some time. Notwithstanding the untold mineral wealth hidden in her everlasting hills, the little development of which has proved to be bountiful and inexhaustible, eastern capitalists as well as foreign, cannot be induced to speculate, not because they doubt the existence of gold and silver, but because so many have been fleeced by bogus companies. The instances are not rare in which companies have been organized and thousands upon thousands of dollars expended in stamping mills, machinery, and other mining apparatus, then some 'broken-down' professor put in as agent, who knows as much about practical mining as a child does about astronomy, but having plenty of rhetoric at command to produce in the most fascinating language the probable results of the veins of gold and silver not yet discovered. Months and months of prospecting, and then, and not before, were the shareholders satisfied in the deception practised upon them, the 'professor' being again liberated to prey upon other innocent victims, the machinery locked up awaiting a purchaser, and the victims wiser by experience. But this is not all. Oft-times when gold does exist, and the claims have been purchased of supposed reputable parties, who have received their half-a-million more or less, other claimants come forward and demand another half-million, and the company is obliged to pay it, or an injunction will be issued; often a long and doubtful lawsuit follows. . . . But a brighter era, an era of prosperity and plenty such as she has never known before, awaits Colorado. She has long been governed by the minions, popularly termed 'carpet-baggers' of President Grant, her people having little or nothing to say in the management of their public affairs. But another year or so will see her undoubtedly admitted into the Union as a sister and sovereign State, when her citizens will have the power to make their own laws and regulate their internal affairs, and then confidence will be restored, and the attention of monied men will be attracted,

50. *Mining Journal* (London), 13 January 1866 and 4 May 1872.
51. W.B., 27 May 1875.

and the melodious sound of hammer and pick will be heard on every hill, thousands will seek homes within her borders, and a little time will see her as independent as the great States of New York and Pennsylvania.[52]

Such lavishly capitalised ventures, although unfortunate to luckless shareholders, were not wholly detrimental to the mining development of Colorado. The first mine adventurers might go bankrupt, but those who bought up their plant cheaply were able to develop mining properties profitably. Much of the machinery installed was similar to that in use in Cornwall, and until the end of the nineteenth century Cornish pumps were used in Colorado mines to deal with large quantities of water.[53] Hoisting gear and stamping mills, however, by that time, had been vastly improved in America, although in pioneer mining localities the most up-to-date methods might be employed at one mine and the most primitive devices in an immediately adjoining one. In Colorado, as in other American mining regions, companies were compelled to develop labour-saving machinery and methods on account of fierce competition and, in places, scarce and dear labour. One big saving of labour was effected by using tram-roads to move ore-stuff underground, while the Americans made more economical use of water power than mines in the Old Country; thus, before 1892, one mine in the Guston district, in Ouray County, was entirely worked by water brought from a distance of five miles.[54]

Colorado was admitted to statehood in 1876, and there is no doubt that the influence of immigrant miners and the pressure of mining capitalists had hastened this political consummation. The second great boom in Colorado mining which came four years later, must, however, be attributed to the Leadville silver discoveries rather than to political factors. On the other hand, in the neighbouring western territory of Utah, really large-scale and successful mining operations got under way only after the Mormon commonwealth had been admitted to full statehood in 1896, while the effort in the early days of the Civil War to curb the political and social power of the Latter-Day Saints by encouraging prospecting had not been conspicuously successful. Utah mining, too, got a bad reputation in financial circles because of the notorious transactions in which the Emma Mine became involved, transactions in which Grant's 'carpet-bagging' cronies were implicated. Quite apart from these considerations, however, men on the spot were

52. *W.B.*, 27 May 1875, letter of 'Sac', dated Denver, 29 April 1875.
53. Ibid., 4 February and 17 November 1892.
54. Ibid., 4 January 1892, letter signed 'Cousin Jack', dated Guston, Colorado, 8 January 1892.

far better qualified to assess the potentialities and needs of western mining locations than 'absentee' capitalists and politicians who were too prone to listen to the rhetoric of 'broken-down professors', and to send out as agents men who had no more practical sense than to drive drainage adits parallel to nearby lodes or to sink shafts within a few feet of each other at heavy cost.[55] The Cousin Jacks reckoned that many more Colorado mines would pay dividends if they were only managed by 'thorough, practical Cornish mine agents',[56] and men who had come from Cornwall with some experience of mine management, like Thomas Jennings, criticised 'Yankee companies' for leasing off mining claims on 'tribute' instead of developing them in a 'systematic, miner-like manner'.[57]

Despite all the adverse criticism the gold- and silver-mining industries of Colorado made great progress and to this progress the large numbers of Cornish immigrants contributed. A group of four Cousin Jacks found the silver lodes near Mount Lincoln in the fall of 1869, giving them the name of Hecla, possibly from Michigan connections, but they had not the capital to develop the property.[58] Several Cornish miners played a part in the rapid rise of the Caribou Mine, twenty miles from Boulder and nearly ten thousand feet above sea level where, in less than two years from the first discovery in 1870, eleven shafts were sunk, the deepest going down two hundred and ninety feet, ten levels and a well-placed drainage adit driven, and a great reduction mill installed with every modern feature available including a hundred and fifty horse-power engine.[59] Boom times in Caribou lasted barely a decade, and the Cousin Jacks dispersed to other mining centres both within and outside Colorado; Rocky Mountain weather soon erased wooden dwelling houses and other buildings; a few scars in the earth now remain as evidence of lively human occupation less than a century ago, when many Cornishmen laboured and made a living and a home, however temporary, in the 'place where the wind was born'.[60]

55. *Mining Journal* (London), 4 May 1872; letter of H. B. Grose of St Austell from Central City.
56. Ibid., 17 January 1874, letter of C. S. Richardson, dated Central City, 25 December 1873.
57. Letter from Central City, 14 March 1870, published in the *Mining Journal* (London), 9 April 1870.
58. Letter of H. B. Grose, dated Central City, 7 December 1871, printed in the *Mining Journal* (London), 6 January 1872.
59. Ibid., 28 December 1872.
60. David H. Stratton, 'The Cousin Jacks of Caribou', *Colorado Quarterly*, Vol. I (1952-3), pp. 371-84.

Practically all the Cousin Jacks of Caribou were 'unknown to fame,' but one Cornishman made a contribution to Colorado and western American mining almost as great as that of Richard Trevithick in the Old Country. Richard Pearce, the son of a Dolcoath mining captain, was born at Baripper, near Camborne, in 1837. When a youth, he attended a small mining school at Truro, but had hardly been there three months before he was appointed assistant to the principal. When this school closed down to be replaced by a smaller one, Pearce continued teaching there for half the year and spent the rest of the time giving lectures in the various Cornish mining centres. He seems to have been largely self-taught, picking up his vast technical knowledge of mining more from direct observation and first-hand experience in the Carnmenellis area where he had been reared than from scientific manuals although, in 1859, he attended chemical and metallurgical courses at the London School of Mines. He gave up teaching and lecturing in 1865, when he became manager of the silver and copper works of the Williams and Foster Company at Swansea. Pearce paid his first visit to Colorado in 1871, when he was sent out by some British capitalists to report on the prospects of certain mines in which they were interested in Clear Creek County. A year later he came out again, bringing his family with him, to settle at Georgetown, then and until the Leadville discoveries the most important silver-mining camp in Colorado. In 1875 Pearce became metallurgist of the great Boston and Colorado Smelting Company which had erected the first smelter in the Territory at Blackhawk seven years before, in 1868, but which had only been crushing silver and copper ores for shipment to Swansea until 1873 when it built its own separating works. These works were moved to Denver, where coal was cheaper, in 1878. Subsequently Pearce became superintendent and finally, in 1887, manager of the Boston and Colorado Company. In 1885 he had been appointed British Vice-Consul for Colorado, and in 1890 his standing in the mining world was signalised by his election to the presidency of the American Institution of Mining Engineers.

From his mining forebears Richard Pearce inherited a 'nose for a vein,' and in Colorado he 'pointed out the existence of a number of mineral species which were not known before to occur in this country'.[61] As early as 1864 he had advanced the theory that some of the richest Cornish tin lodes were not fissure veins but had been caused by meta-

---

61. *W.B.*, 21 August 1890: article based on an article in the *Magazine of Western History* (n.d.): the phrase 'this country' refers to Colorado or the Western States.

morphism, and when he came back to the Old Country on a visit in 1896 he told the Dolcoath people, who since Trevithick's time had taken a pride in being progressive, that they needed far more up-to-date mining methods and appliances to exploit a hitherto untouched rich deep tin lode. This, along with some other pointed remarks about Cornish and British backwardness and inertia in mining matters, roused some controversy, though Pearce realised that the Cornish tin-mining interests had been compelled to resort to the most stringent economies to meet the East Indian competition, and had nothing but praise for the modernisation plans the far-sighted Dolcoath managers had already adopted. Back again in Denver, however, he could give only slight comfort to the silver-mining interests which had been hopeful that the bi-metallic currency movement in Britain might influence the American Federal Government towards a similar policy. A strong case could be made out that the silver currency in the East Indies gave 'Straits' tin a great advantage over the Cornish metal, but with leading Members of Parliament so ignorant, according to Pearce, as to believe that silver cost sixty cents per pound and not sixty cents per ounce to produce,[62] the silver-mining states of the West could hope for little help from Britain.

Pearce's visit to the land of his birth hardly roused much interest in American mines in the Old Country. He told the *Denver Republican* that he had been asked several times 'if Cripple Creek was a great mining camp, and if there was anything of moment there'. Stung by numerous fraudulent mining promotions in Colorado, Montana, Utah and elsewhere in America, injured and penalised by anti-foreign discriminatory legislation and litigation in many of the states, British capital had become far more interested in the South African and West Australian gold-fields than it was in Colorado's last great gold strike, while before long, too, Canada's Klondike was to steal the sensation mining headlines in the popular press. To the working miner of Cornwall, in the mid eighteen-nineties, Coolgardie and the Rand were the new El Dorados, and comparatively few immigrants came to Cripple Creek from the Old Country till the outbreak of the Boer War in October 1899, closed the Rand for some time.

Another Cornishman who became fairly well known in the south-western states was the Truronian, Thomas V. Keam. He went to the west coast from Cornwall in 1862, intending to work in the gold diggings, but soon enlisted in a Californian regiment of the Union

62. *W.B.*, 8 October 1896, quoting interview of Pearce in the *Denver Republican*, n.d.

Army, in which he served as adjutant, and after the war joined the Federal Indian Service. In that work Keam travelled extensively in the all but unknown wilderness areas of New Mexico, giving his name to a canyon in the Navajo country of north-eastern Arizona. He became widely known as a true friend and wise counsellor of the Indians among whom he spent many years of his life, and was credited with having prevented more than one Indian uprising in the turbulent Indian frontier of the south-west, no mean achievement in times when such redoubtable warriors as Cochise and Geronimo were 'plowing up hell' in that region.[63] Only rarely did Keam go back East, although he maintained at the remote Indian trading post in 'his' canyon the habits of a man of culture and refinement. In the course of his duties which were an amalgam of those of a trader, a colonial administrator, and a missionary teacher, Keam travelled long and dangerous trails. What he called 'a very pleasant journey among the Indians' in 1883, he described to a Truro friend. He made the trip to attend

> one of their great dances, called a *Ya-bit-shy*, at the solicitation of some of their principal chiefs. It was quite a novel spectacle. The Indians were dressed fantastically, wearing hideous masks, having on silver belts, coral, and other beads. My journey there was perilous in the extreme. I started alone and too late in the afternoon. The distance to the scene of the dance was thirty good miles. I did not leave till p.m., hoping to reach my destination before dark, on my best and favourite horse 'Ute'. I failed, however, darkness overtaking me when I was about five miles from the place. Four of those miles were down the side of one of the deepest and worst canon in the country, the sides of which are solid sandstone rock, and eighteen hundred feet high, which to the uninitiated would be appalling in daylight. At night it is like looking down into a dark abyss, where one false step on the part of myself or horse would have sent us a bruised and mangled mass more than a thousand feet below. Determined to proceed, however, I slowly felt my way to the bottom, having to lead my horse the entire distance from the commencement of the descent. What made it more dangerous was that it was six years since I travelled down it, and the trail had looked altogether different in daylight. At this place also the two sides of the canon almost meet, so that in the day-time it looks like being boxed up in an immense rocky chasm—about eighteen hundred feet high, the sides almost perpendicular, showing just one narrow strip of sky. On arriving at the dance the Indians welcomed me with surprise to find I had come by night on the trail alone. I remained two days, enjoying the grand and imposing views in the canon and partaking of Indian hospitality.[64]

63. *W.B.*, 26 February 1883, quoting *Eureka Herald* (Kansas), 2 November 1882.
64. *W.B.*, 10 December 1883.

By gaining the friendship of the Indians among whom he worked, Keam was able to make contributions to archaeology, anthropology and kindred sciences. The Navajo who lived in the vicinity of his trading post on one occasion brought him scores of pieces of primitive pottery, relics of sacrificial offerings to native gods near the principal spring of a group of cliff dwellings.[65] Another time the Indians told him of some strange footprints embedded in red sandstone rocks in the wild, desert country of north-western Arizona. Keam, thereupon, set out upon a seventy-five-mile journey into a region of broken mesas or tablelands separated by deep, precipitous-sided valleys. To bring away the stones in which the footprints of some long-extinct vertebrate animal were embedded, he had, with the labour of a party of Indians, to blast a road down into the valley where the prints were from the top of a mesa, and it then took him twelve days to cut out the pieces of rock bearing the footmarks, which seemed to be those of antelope, a primitive type of horse, and of man, with hammers and chisels. Water had to be brought to the site from a place twelve miles away, but Keam reckoned the time and labour well spent, brought back the 'fossil' footprints, and later presented them to the museum of the Royal Institution of Cornwall.[66]

Many other Cornishmen came into the country where Keam's official duties took him, but most of them were either miners or men associated with mining ventures. New Mexico and Arizona were the territories into which Spanish colonial adventurers had penetrated, many failing ever to return, and the entire region was rich with legends of fabulously rich 'lost mines'. Spanish-American traits prevailed among many of the people in the southern districts, who were antipathetic to the 'Gringo' way of life, while the political instability and weakness of Mexico made the border country a refuge and a lair for a motley array of political refugees, malcontents, sheer ne'er-do-wells and outright banditti. It was, therefore, little wonder that when, in this region, rich mineral discoveries were made lawless and violent hell-roaring camps sprang up of which Tombstone was to win notoriety's darkest crown.

Several scattered rich strikes were to be made in New Mexico during the half century following the American annexation. In Spanish days gold had been worked in the vicinity of Santa Fe, and under Mexican rule lode gold was found at the Ortez mine in the rather over-hopefully

65. *W.B.*, 17 August 1893, quoting *American Anthropologist*, n.d.
66. *W.B.*, 26 September 1895. Keam spent his last days at home in Truro, where he died in November 1904; he left about £4,500 to the Truro Royal Institution—and in all a fortune of about $45,000. (*W.B.*, 17 April 1905).

named Sierra de Oro. Later there were to be several silver discoveries in the Territory, especially in the Socorro Mountains in the late eighteen-sixties, and valuable gold and silver deposits were found near Hillsboro, but Indian troubles deterred the immigration of both capital and labour into the rich silver and gold districts round Silver City and Taos until nearly the very end of the nineteenth century. The copper mines of Santa Rita had been fairly extensively worked in Spanish colonial times, but the Apaches delayed the effective modern development of that area until the early eighteen-seventies. No Comstock and no Cripple Creek were ever discovered in New Mexico, whose mining 'excitements' were in a rather minor key. Cimarron is said to have had several gold-mining booms without ever having had a really important strike, and it is doubtful if the British company which bought up the Maxwell Grant for gold-mining purposes in 1870 ever saw much return for the $1,350,000 it paid.[67] The Cochita 'rush' in the spring of 1894 attracted out-of-work silver miners from far and wide, but it was only a seven days' wonder. Many of the mining camps of New Mexico, like others in Nevada and Colorado, became 'ghost towns', notably White Oaks, Kingston, Elizabethtown, Glen Woody, and Bingham, and it is noteworthy that silver discoveries nearly made Socorro the largest city in New Mexico in the eighteen-eighties.[68]

Although New Mexico was very much the frontier territory of the Spanish Empire, yet it may be surmised that during two centuries of quasi-military and missionary occupation the Spaniards would already have located the most accessible rich deposits of precious metals as they had done to the southward of the Rio Grande. In the last generation of Spanish rule, however, when industrial and commercial developments in the Old World caused acute currency difficulties in a period when bullion was becoming increasingly scarce, there was fairly considerable 'prospecting' activity in this frontier area, particularly in the Cerrillos region to the north of Santa Fe; the same, period, too saw the beginning of copper mining at Santa Rita, convict labour being used. Subsequently in American times, mineral discoveries and the pressure of mining capitalists not only led to an influx of 'white' immigrants but compelled the Federal Government to subjugate the fiercely hostile Apache and Comanche Indians, although it was not until the very end of the nineteenth century that Eastern speculators were prepared to risk investing any money in ventures in the north-eastern part of the Territory.

67. W.P.A. Guide, *New Mexico*, pp. 271–72.   68. Ibid., pp. 252–53.

Silver and gold strikes in the Mogollan Mountains brought Cornish miners direct from the Old Country in 1871 to the Silver City district in fairly considerable numbers.[69] Two years later a Truro newspaper, the *West Briton*, may have turned the thoughts of miners who had notions of emigrating from Cornwall towards New Mexico by reprinting an article on the mineral wealth of that Territory from the *Chicago Tribune*.[70] This mentioned silver ores running as rich as four thousand dollars per ton in the region around Silver City; the gold placers of Pinos Altos; the Santa Rita copper-gold ore which was so rich that it paid to pack it a thousand miles on mules to the coast for shipment to Europe for separation; an iron mountain with ore so pure that it could be hammered out without smelting; masses of rich gold, silver and copper ores in the Burro Mountains; millions of tons of silver ore at Ralston; and scattered indications of, among other things, marble, kaolin and coal. The writer of this article, however, had to admit that the Pinas Altos placers were short of water for working, and that for the time being it was impossible to work the Ralston silver deposits owing to the lack of both wood and water. Still, he was confident, the railroads which were then penetrating the region would solve many supply problems, and, that before long too, New Mexico might be mining its own coal. No Cousin Jack was likely to be deterred from mining by lack of water, nor likely to pay much heed to the concluding sentence in the article which admitted that

Settlement in this country has been greatly retarded by the hostility of the Indians, but Cochise, the great chief of the Apaches, having made peace, and his people having gone to their reservations in Arizona, it is believed there will be little further difficulty of this kind, although it is still deemed advisable to travel in company and armed to insure freedom from molestation by roving outlaws whom the chief may not be able to control.[71]

The New Mexican silver-mining districts felt the 1893 depression keenly, and the Territory was mentioned along with Colorado and Montana as 'States' in which there were 'many hundreds of Cornishmen anxiously awaiting the results of the special meeting of Congress' which, in August 1893, discussed schemes for 'adjusting the relationship between gold and silver coinage'.[72] Yet of these 'many hundreds' it seems unlikely that more than a few score were scattered among the mining camps of New Mexico. That Territory never attracted anything like as much English capital investment as did California, Nevada,

69. *Cornish Telegraph*, 15 March 1871.   70. *W.B.*, 12 June 1873.
71. *W.B.*, 12 June 1873.   72. *W.B.*, 7 September 1893.

Colorado, and Montana, and it was capitalistic British mining companies that were responsible for the establishment of veritable 'colonies' of Cousin Jacks in some of the more famous mining districts of Old Mexico. The still pronounced Iberian character of the society living in the region around Santa Fe indicates that this territory might well be considered as the Mexican rather than as the American mining frontier. and the immigrant mining classes most likely to be attracted thither were technicians, engineers, managers and skilled artisans rather than the common run of ordinary working miners. New Mexico possessed a large available supply of cheap Mexican and Indian labour, while, in addition, the bottom had dropped out of the silver market before its argentiferous ores came to be exploited extensively by modern methods. No predominantly Cornish settlements like Mineral Point, Grass Valley or Butte were to appear in New Mexico, nor, for that matter, did Socorro, Hillsboro, and Silver City attract as many Cousin Jacks as did Globe, Bisbee, and Jerome in the neighbouring territory of Arizona. Some did come at various times, but it may be taken as an indication of their comparatively small numbers that when, in the early part of 1895, William Heather of Penpons, near Camborne, lost his life in a New Mexican mining disaster he was but one out of twenty-five victims;[73] had that disaster occurred in certain mines in Butte or Upper Michigan a dozen or more of the casualties would have been of Cornish birth or extraction.

It seems probable, indeed, that Cornishmen contributed more to the development of mining in Utah than they did in New Mexico. Latter-Day Saint missionary activity in the Old World brought many convert immigrants to Deseret, although comparatively few Cornish people joined the Mormon Church. There was in Cornwall as in other parts of Britain a scabrous interest in Mormon polygamy, while the Mountain Meadows Massacre became ten times more bloody in anti-Mormon propagandist telling and re-telling.[74] Nevertheless, a few Cornish people joined the Latter-Day Saints, notably the hymn-writer, Penrose, and members of the Moyle family.

The pursuit of Mammon rather than the creed of Mormon, however, brought several Cousin Jacks to Deseret. There were probably some Cornish miners serving in the western regiments of the Union Army which Colonel P. E. Connor sent out prospecting in the Mormon Terri-

73. *W.B.*, 15 April 1895.
74. An account of the Mountain Meadows Massacre, written by 'Two North St. Just Men', probably miners travelling in Utah, was published by the *Cornish Telegraph* on 7 April 1880.

tory under his military command in the time of the Civil War. Connor, there is no doubt, hoped that there would be enough rich strikes to attract mining immigration into Utah on such a scale as to swamp the Mormon population. For their part, the Latter-Day Saints had hoped to build up a peaceful agrarian commonwealth remote from their Eastern persecutors, and elders of the Church had proclaimed that they had no wish to see 'peaceful Deseret, the home of a people who fled for religious freedom and quiet to the mountain solitudes, converted into a rollicking, roaring mining camp'.[75] By 1858, however, Brigham Young himself encouraged the establishment of mining and foundry works— to work lead and cast bullets for use against the Federal forces President Buchanan was sending to subjugate the Mormon commonwealth. The Mormons, furthermore, wanted metal for farm implements, and felt that if they could provide their own, independent of the Gentiles, it would be so much the better for them.

Silver-lead ores were found in Bingham Canyon in 1863, and later massive deposits of low-grade copper in the same place. At first both the Mormons and the Union Army had more use for the lead, but, with the realisation that Deseret could never be independent of the States and with the end of the Civil War, the silver ores became a greater attraction to mining adventurers; subsequently the still more mundane copper ores came to be more important in the development of Utah's economic life, especially after the completion of the Union Pacific Railroad in 1869 ended Deseret's physical isolation and provided the transportation facilities needed to take Utah produce to the outside Gentile world. It also brought carpet-bagger Eastern capitalists, the men associated with the promotion rather than the development of the Emma Mine who succeeded in tainting the early history of Utah mining with one of America's most scandalous financial stigmas.

As a mineral property the Emma Mine was rich, and it should be admitted that even the propaganda of its utterly amoral promoters served to stimulate active and effective interest in Utah mining and to bring in mining immigrants, though not in such numbers as Colonel Connor had hoped would sweep out the Mormons from the Territory. Among those who came were a number of Cornishmen, and within two years of its first discovery at least forty Cousin Jacks were working at the Emma.[76] Among them was a young St Agnes miner, Stephen

75. Statement of Oscar F. Whitney, W.P.A. Guide, *Utah*, p. 120.
76. *Cornish Telegraph*, 13 March 1872. The Emma vein was first located by J. B. Woodman in 1869.

Davey, who wrote home early in 1872 that he had himself in the course of a single shift broken six thousand dollars' worth of silver ore; a month or two later news came to St Agnes that he had been killed by a fall of ground.[77] Other mines were opened up close by the Emma in Little Cottonwood Canyon, and a Pendeen man, in August 1876, wrote that the Prince of Wales Mine was 'worked in the Cornish style and chiefly by Cornishmen'.[78] At the end of the century, Daniel Jackling, by employing vastly improved metallurgical techniques and mass-production methods, initiated the profitable working of low-grade copper ores in the Bingham Canyon, and made Utah prominent among the copper-producing states of the Union. Furthermore, the example of the great Utah copper-mine in exploiting low-grade ores was to be profitably followed by mining companies in Montana and Arizona mining districts to which Cousin Jacks had migrated in far greater numbers than they ever did to the country round Great Salt Lake.

Indeed, many Cornishmen did not like Utah. Mining conditions in the Emma, Flagstaff, and other silver-lead mines were extremely deleterious to health. The Pendeen miner who wrote home in the summer of 1876 had little that was good to say about the Mormon country, apart from the fact that its mineral veins had yielded some of the richest silver ores ever found in the world. The climate was exaggeratedly described as 'nine months winter', but possibly in the highest mountains a night might not pass without 'some accident in connection with snow slides'. It was easy to obtain work in the mines, but only because they were so unhealthy that

men care only to work in them for a short time. The first question when you engage is 'What hospital can you go to when you are sick?' As if that was a natural occurrence!... Strong healthy men, whose mining experience would command the highest pay, have to work (amid giant powder smoke and arsenic fumes, where candles will scarcely burn ten hours) for three dollars! And happy is he who can save enough to pay his fee to the hospital when he gets 'loaded'. The best way to avoid disease common to these mines is to remain at a distance.[79]

The Pendeener concluded his letter with a forthright condemnation of the theocratic tyranny of the Mormon Church. Newly arrived convert immigrants were lodged in the tithing-house yard to sleep on 'the soft side of a plank'. Like other unfriendly critics, he had to say his say about

---

77. *Cornish Telegraph*, 26 June 1872.
78. Ibid., 5 September 1876, letter of 'W.A.', dated Salt Lake, 3 August 1876.
79. Ibid., 5 September 1876; W.A.'s letter of 3 August 1876.

polygamy and the Mountain Meadow Massacre, though that incident had taken place nearly twenty years before at a time when it looked as if the Federal Government was going to send an army into Deseret to destroy the Latter-Day Saints by force.

This Pendeen sojourner in Utah was obviously a strait-laced member of some Methodist denomination or other, for in a further letter he castigated Brigham Young for advising that children be taught how to dance, alleging too, that the Mormon leader believed that too much education was bad for 'the lambs of his flock, as there are so many Apostles and Elders to direct them safe to the other world'. And the spectacle of the ancient Mormon patriarch accompanied by many of his church leaders, along with fifteen hundred of their following patronising a travelling circus shocked this narrow-minded Cousin Jack to the marrow. Dances and circuses hardly seemed to him to be 'the latter day dispensation—the fulfilment of the everlasting covenant'.[80] Rather they smacked of the pit and of antichrist. The Mormon Zion was no place for godly Cornish people. Nor, on the other hand, was it a congenial environment for unregenerate Cousin Jacks whose ideals were high wages, full purses and quickly made fortunes. For them the Mormons could keep their Salt Lake City and Bingham Canyon—Pioche, Tombstone, and Butte City were far more in their line, more like warmed-up versions of St Just, Redruth, and St Blazey on Saturday nights.

Colonel Connor's dream of crushing Mormondom by mining immigration remained a dream. Not even the Emma Mine, before the scandals broke, brought in six hundred and forty immigrants in a single week as did the Mormon missionaries in 1876.[81] Two years before, another Cornish traveller, William Copeland Borlase, reckoned that there were then a hundred and fifty thousand Mormons in the whole Territory and that convert immigrants were coming in at the rate of four or five thousand a year.[82] Nevertheless, miners did come into Utah and the theocratic state was compelled to receive them. It had to modify its own ways somewhat, but from an outcast territory it became a sister State in the Union, although tardily, in 1896. Deseret could easily have been a backward pastoral state like Oom Paul Kruger's South African Republic, had not its pastoral and patriarchic pioneer religious and

80. *Cornish Telegraph*, 31 October 1876, letter of 'W.A.' dated 22 September 1876.
81. Ibid., 5 September 1876; W.A.'s letter of 3 August 1876. This unfriendly critic was unlikely to exaggerate the numbers of converts.
82. Ibid., 9 October 1877; Borlase visited Salt Lake City in February, 1874.

social ways been modified by the inherent progressive Yankee traits of many of its leaders aided by the immigration of hard-rock miners among whom there was a fair sprinkling of Cousin Jacks. Furthermore, once polygamy was abandoned, there was much in the ascetic Mormon creed that evoked a sympathetic response from the stricter Cornish dissenting sects.

Neither Mormon Deseret nor Spanish New Mexico attracted as many Cousin Jacks as did the territory of Arizona to the south and west of them. While American and Cornishman alike had to accept, and not always with good grace, co-existence with Latter-Day Saints and Spanish-speaking 'greasers', there was only a declining Indian race to encounter in Arizona, and to Apache and Comanche the whites saw fit to offer but two alternatives—subjugation or extermination. Yet the Indians put up a bitter, prolonged resistance which delayed the real start of extensive mining enterprise in Arizona for nearly twenty years after the first American effort to mine gold and copper in the territory. White settlement was also hampered by natural disadvantages. It was a country of deserts, mountains and canyons. There were great problems of water supplies and communications. It lay far to the south of what had become the main routes of overland traffic and trade. The ill-success of early pioneering mining ventures, besides bogus diamond 'discoveries' in the early eighteen-seventies, gained the territory disfavour and notoriety. This deterred miners from coming into Arizona with labour and capital until the Indians had been subjugated, substantial mineral deposits located, and communications improved.

Arizona's mining history began in the fall of 1858 when a Texan gold-miner returning home from California camped near Fort Yuma on the Gila River. While his teams were being grazed and rested, he spent his leisure in haphazard prospecting and found placer gold deposits. Through rumour and publicity, the strike was exaggerated, and a small-scale 'excitement' ensued; several Texans bound for California stopped on the Gila, while a party some seventy-five strong came from San Francisco as soon as the news reached the 'Coast'. The deposits, however, were not as extensive as the first-comers thought, while it was hard to work them since water had to be brought from the river half a mile to the diggings. An average daily return of four or five dollars per man seemed small to Californians, and the Gila gold-field was soon almost forgotten in the furores caused by the Washoe and Pike's Peak discoveries. Washoe was much nearer the by now overcrowded Californian gold diggings, while the finds in Gilpin County were far more accessible to Easterners than the Gila Valley. Early travellers, too, painted no very flattering picture of Arizona as 'a wild country, chiefly destitute of water, and presenting to the traveller dry-river channels, rocky wastes, shadeless trees, and a brickyard appearance of the soil, much like some of the inland of Australia, and not very inviting'.[83] Even the most enthusiastic early account of the Gila gold deposits, after warning prospectors and miners that the Apache were uncomfortable neighbours, went on to state that

the climate in December was delightful. It is said only to rain in August, but the country, for miles around, is one great region of brown hills and valleys. As the soil is impregnated with alkali, and there are no means of irrigation, agriculture will not flourish. Except in some few places, there is no timber, and what there is consists of cottonwood or mezquit. It is, however, a rich mineral country. Provisions are therefore, dear and are brought from California, chiefly from San Francisco, up the Gulf of California to the mouth of the Colorado, and thence to Fort Yuma by steamer.[84]

The cost of provisions and of transportation also hampered the development of early copper mining in western Arizona, and a Californian company faced with freight costs of a hundred dollars per ton sent inquiries to England about the possibilities of a newly patented traction engine to reduce such charges; something radical was needed to meet with a situation in which transportation reduced the profits on even fifty or seventy per cent ores to vanishing point.[85] Silver discoveries and prospects were located about the same time in southern Arizona, but the New York backers of the Trench or Sonora Company and of the mining promoter, Colonel Titus, were probably more interested in prospects in Sonora and Sinaloa. At the time the political condition of northern Mexico was chronically anarchic, and there were grounds for suspecting that the so-called Arizonan 'Emigration and Colonisation' societies of Henningsen, Lockridge and others were filibustering ventures designed to deprive Mexico of still more of its northern provinces.[86]

During the next decade mining made some progress in Arizona. In the eastern part of the Territory the Vulture Mine produced a fairly considerable amount of rather low-grade gold ore, but some interested British speculators were warned in 1871 that its riches might not hold out in depth, while the Big Bug Mine, near Prescott, was regarded as an even more questionable speculation.[87] Arizona's greatest mining swindle,

83. *Mining Journal* (London), 2 July 1859.   84. Ibid. 5 March 1859.
85. Ibid. 5 March 1859.   86. Ibid. 5 March and 2 July 1859.
87. Ibid. 25 February 1871.

however, was the reputed discovery of diamonds in the Territory in the early eighteen-seventies. The fabulously rich strike of diamonds in the Kimberley district of South Africa in 1869 had set prospectors looking for gem-stones in all sorts of likely—and unlikely—locations. In Pinal County in Arizona some prospectors picked up stones which they thought might be 'diamonds in the rough', but on examination they proved worthless. A correspondent of the *Arizona Miner* described one of the stones which a Pinal prospector had handed to him as so exactly resembling an egg in colour and shape that

> the shrewdest hen in the country might hatch out her very existence in the effort to bring a chicken from its crystal depths, and never discover that she had been 'sold'. Upon breaking this stone we found within the semi-transparent crust, which is about an eighth of an inch thick, a cavity literally filled with beautiful pieces of rock crystal, all pointing towards the common centre and each terminated by a pyramid of six sides. The perfect transparency of their lustre in the sun's rays is quite sufficient to suggest the idea of a nest of diamonds to the inexperienced observer. We understand that the Governor sent three bushels of these ovaliths to San Francisco to be there examined and their value determined. They should be of some value as ornaments.[88]

Obviously this Arizonan newspaperman was sceptical about diamonds being found in the Territory, but round about this very time two rough-looking prospectors called on Ralston, the cashier and real directing spirit of the famous Bank of California in San Francisco, and asked his opinion of some pebbles which, they reckoned, might be diamonds in the rough. Ralston was sceptical but interested enough to get them examined by Le Conte, the Berkeley professor of geology who said they were diamonds; Henry Janin, a San Franciscan mining expert, and some jewellers agreed with Le Conte. Ralston then contacted the two prospectors, William Arnold and Isaac Slack, and, with the backing of a syndicate which included some of the most prominent men in San Francisco's financial circles, agreed to pay them a million dollars for their claims if they took a party of experts to the site of the discovery to verify that diamonds existed there.

The episodes that followed were more like a melodramatic cloak-and-dagger Ruritanian conspiracy than a mining promotion. Arnold took the geologist, Janin, and two other prominent San Francisco mining men down to the Oakland ferry in the half-light of dawn, got them by train into 'Wyoming Territory', and thence by coach southward for some distance to the alleged diamond fields. Accounts vary, and it was

88. *Arizona Miner*, 4 November 1871, quoted in the *Grass Valley Union*, 19 November 1871.

subsequently reported that the prospector blindfolded his companions for part of the journey; what is certain is that during a few days on the ground the party dug out a number of rough diamonds and brought them back to San Francisco. Ralston and his associates paid Arnold and Slack their million dollars and then organised a three-million dollar company to develop the Arizona 'diamond field'. The attempt to keep all the shares of this company within Bank of California circles naturally tended to exaggerate rumours about the Arizona diamond finds, particularly as it coincided with the period of frenzied speculation which preceded the collapse of Jay Cooke's financial empire.

Despite the secrecy the location of the diamond 'discoveries' became known, and there hastened to the scene, Clarence King, who, some years before, had conducted a geological survey of the region for the Federal Government. King had declared that diamonds could not exist there, and his scientific reputation was at stake. Meanwhile, Arnold and Slack had left the 'Coast', the former to start farming in Kentucky with a capital of nearly half a million behind him, the latter to vanish 'somewhere in Europe'. King found the site, examined it carefully, noted tell-tale markings in a gravel bed in which there were a few diamonds, and came to the conclusion that the gemstones had been fired into the soil with a shotgun. In fact, King had exposed the most sensational 'mine salting' episode in American mining history. More experts examined the diamonds, and found that they had been brought from South Africa and planted in Arizona. Ralston and his partners managed to live down the ridicule they had incurred, and it was later discovered that Arnold and Slack had been hired by a shady financier to buy up diamonds and rubies in London, to smuggle them into the United States across the Canadian border, and to sow them in a locality which was topographically similar to the Kimberley region where Slack had previously been. The duped San Francisco capitalists tried to get back the million dollars they had given the 'prospectors', but it is doubtful if they recovered much from either Arnold or Slack. The former was later killed in a 'local quarrel' at Elizabethtown in Kentucky; of Slack, nobody ever heard again.

The episode certainly served to arouse interest in Arizona's mineral wealth, although it must have deterred the immigration of the capital which was needed to develop the real mining resources of the Territory. In the course of the next decade the gold and silver of Tombstone and Globe were to become prominent, and by the time of the Federal Census of 1880 a number of Cousin Jacks had come into Arizona, some of them direct from Cornwall, others from older mining centres of the American West. Mayhem and murder at the time and the imaginative perpetrations of the later motion-picture industry have obscured the very real significance of Tombstone as a mining camp of considerable importance. There were forty, and possibly more, Cousin Jacks living in Tombstone when the 1880 Census was taken; about ten or a dozen were residing at a single boarding-house on Tough Nut Street, but there was hardly a single Cornishwoman in the camp. Other immigrants from Cornwall were scattered throughout the Territory. Captain J. L. Tremayne was one of the 'boss' miners of the by then well-established Clifton Copper Mine. James Hunt was superintending a mine at Seymour in Maricopa County; in the Cedar Valley Mining District of Mohave County, William H. Gillis returned his native country as 'Cornwall', and working at the American Flag Mine in the same county were John Penberthy, William Prisk, and John Corin; another Penberthy, James, was mining at Mineral Park, another Mohave County location. The Hermosa Mine in Pima County had a number of Welsh besides Cornish employees. There were already a few Cousin Jacks in Globe, Prescott, and Phoenix, centres to which many were to come in later years, while, by 1880, other Cornish miners had found their way to the mines of the Huachuca Mountains, to Pinal City, to Tip-top in Yavapai County, to the Big Bug Mine, and to the Canon de Oro Creek among other places. In all probability at least two hundred Cornishmen were living in Arizona around midsummer, 1880, but as scarcely half a dozen of them had wives and families with them it seems likely that many of them had little intention of making a permanent home in Arizona, or, at least, not until the prospects were more assured and living conditions more secure.

Tombstone, either because of, or in spite of the activities of the Earps, settled down to being a reasonably tranquil mining camp in the early eighteen-eighties. The bonanza prospects of the first strikes, however, faded. Some promising lodes deteriorated in depth, while by the early summer of 1884 Tombstone's deeper mines were confronted with serious drainage problems if they wished to continue operations or to tap the ore bodies which were believed to exist below the water levels.[89] Massive pumps were installed by the Contention and by the Grand Central Companies, that of the latter being boosted as the second largest pump in the world and as costing no less than $130,000 to install, although, actually, it had been bought second-hand from the Colorado

89. *Silver Belt*, 3 May 1884; *Daily Alta California*, 1 May 1884.

mining venture that had originally intended to use it for $95,000 in San Francisco.[90] At the beginning of 1886 there had been a revival in Tombstone's mining fortunes, and such mines as the Grand Central, Contention, Tough Nut, Emerald, and Lucky Lass were being worked with confidence, but it was symptomatic of declining fortunes that one of Tombstone's original 'seventy-niners', Walter R. Coleman wrote back from South Africa to the editor of the Tombstone *Epitaph*, in February 1886, that there was an opening for two hundred goldminers in the Transvaal gold-fields, besides good prospects of employment for mill-hands and assayers.[91] Up to the end of 1886 the Contention mine had paid over two-and-a-half million dollars in dividends, the Grand Central paid three-quarters of a million before it was destroyed by fire, and the Tombstone Mill and Mining Company paid dividends aggregating a million-and-a-quarter dollars to its shareholders. These dividends, however, only represented about a fifth of the value of the capital stock of these concerns;[92] others did not do as well, and by the summer of 1887 Tombstone had declined so much that its famous newspaper, the *Epitaph*, ceased daily publication and became a weekly. Some mining, however, still went on, and every now and then in Tombstone, as in Cornwall, there was talk of a mining revival. The boom days of 1879–1881, however, never returned, and many Tombstone pioneers who did not get located in the notorious Boot Hill Cemetery scattered to other mining camps all over the world. Walter Coleman, who had gone to the Transvaal had actually paid a visit to Australia and had returned to Tombstone on a brief visit before going to Barberton; a number went from Tombstone to Cripple Creek,[93] but it is likely that the greatest exodus was that which went to nearby Bisbee.

Boot Hill, however, must have been the last dismal resting place of a few unfortunate Cousin Jacks, unlisted though they are by the Guide published nowadays for the benefit of tourists and unmarked by any plain headboard or any other memorial. Possibly one or two got involved in 'shooting scrapes' with the hardly distinguishable forces of law enforcement and misrule in the turbulent mining camp, but for the most part they were victims of the accidents and diseases to which miners were particularly prone. Within six months of each other, Edward Rich-

90. *Daily Alta California*, 19 November 1885.
91. *Tombstone Epitaph*, 14 February 1886.
92. Ibid., 28 March 1886; letter of Coleman dated 1 February 1886, from Barberton, De Kaap Goldfields, Transvaal.
93. *Tombstone Prospector*, 5 and 6 February 1896.

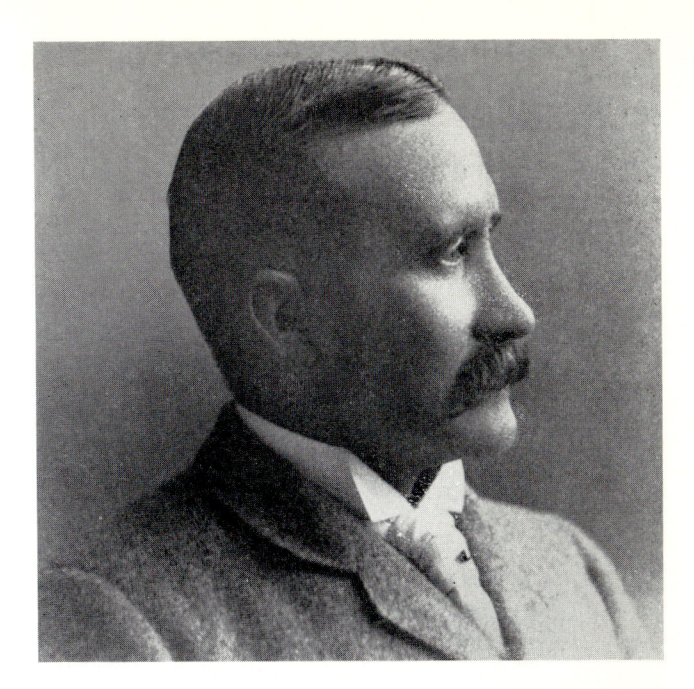

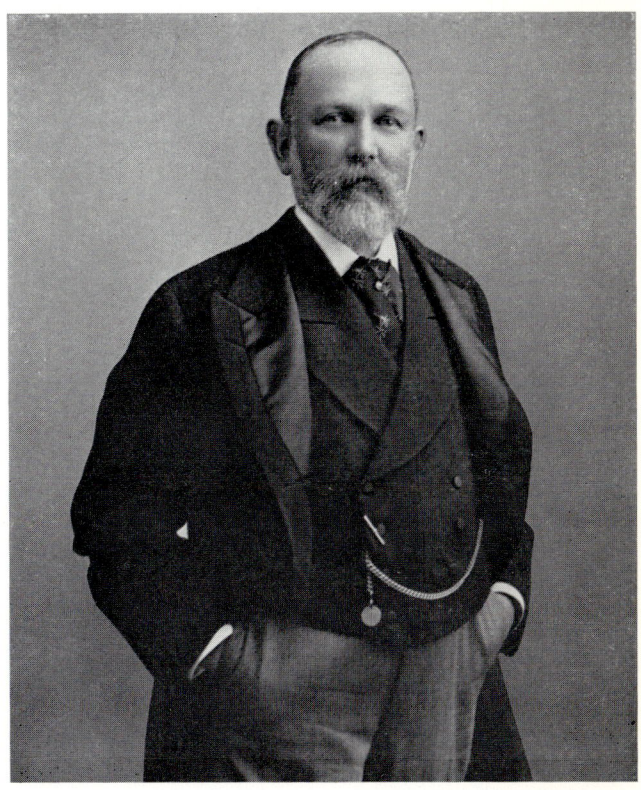

12 Two Cornishmen who were successful in North America:
Thomas V. Keam (top) and Richard Pearce

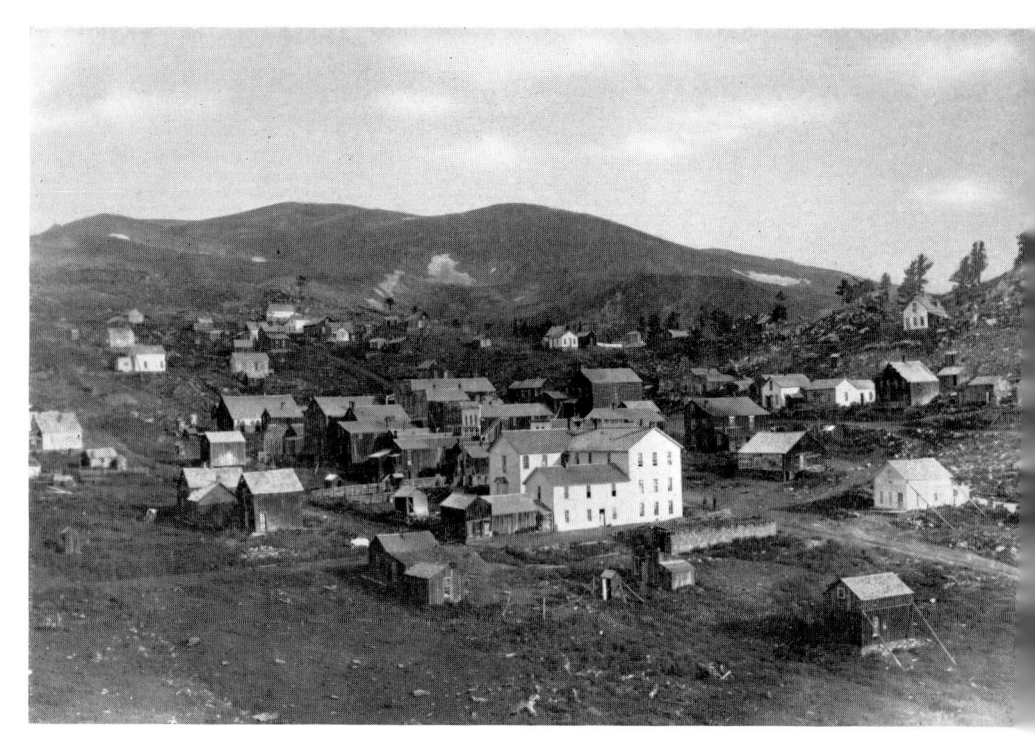

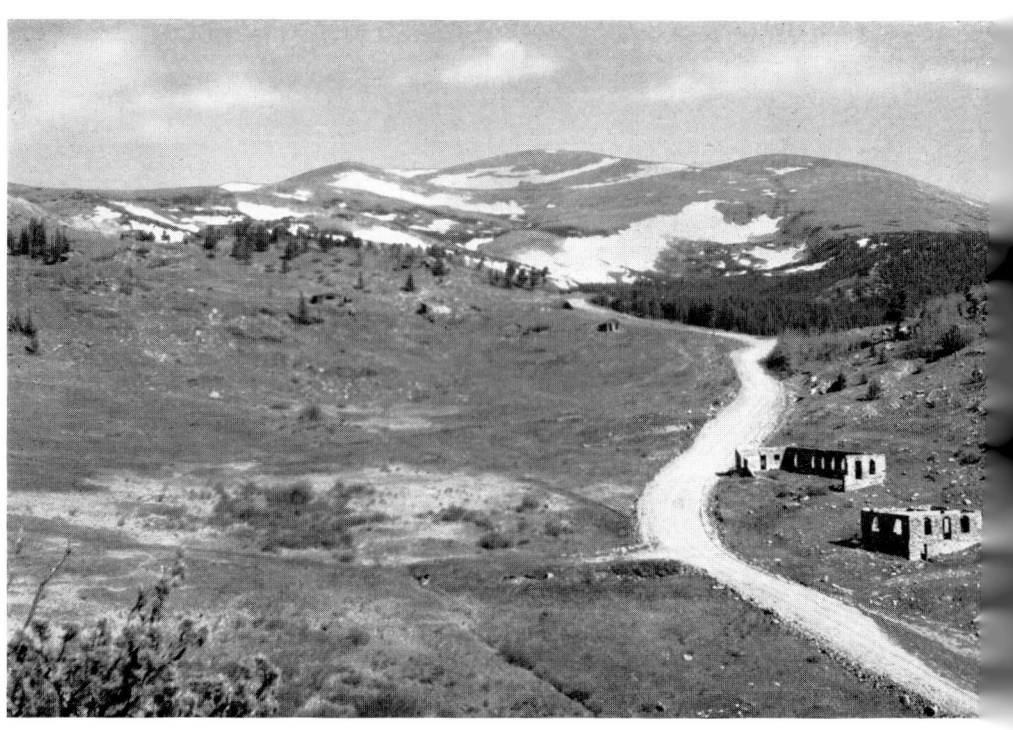

13 Caribou, Colorado, in its bonanza days, and the same site today

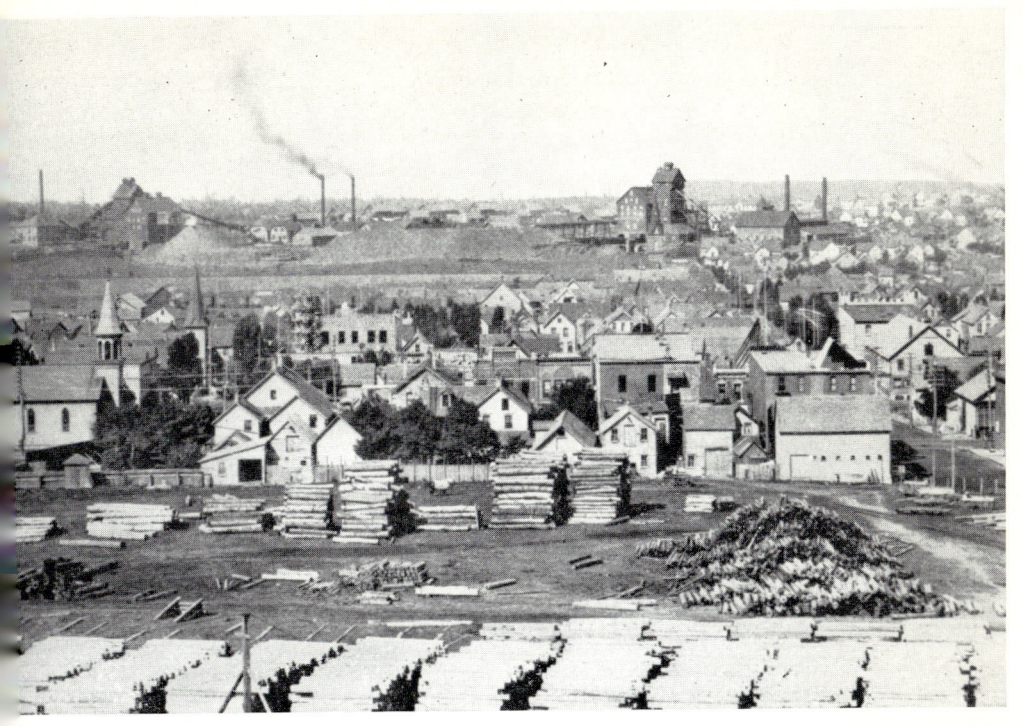

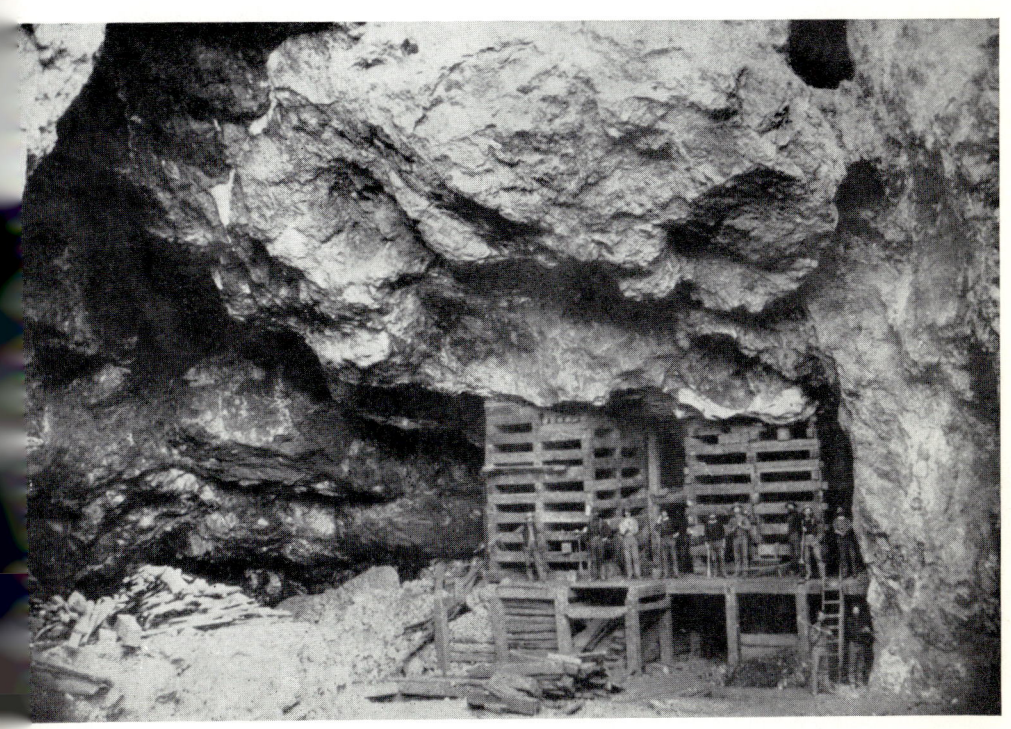

14 The Tamarack Location, Calumet, about 1880
15 Copper Queen Mine glory-hole, Bisbee, Arizona

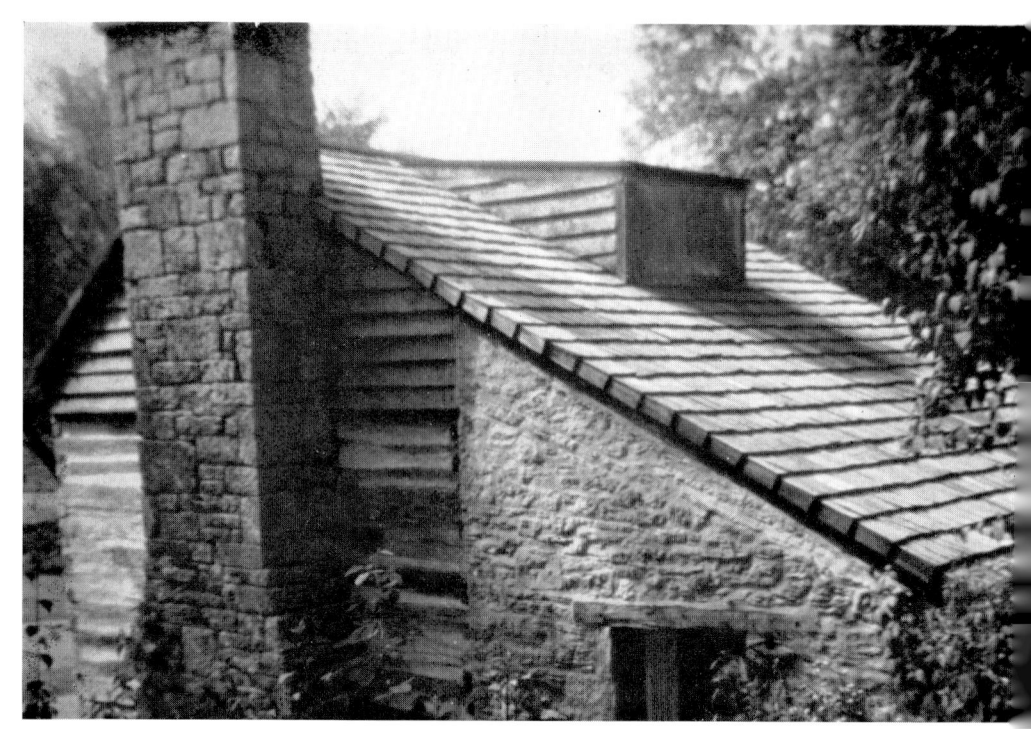

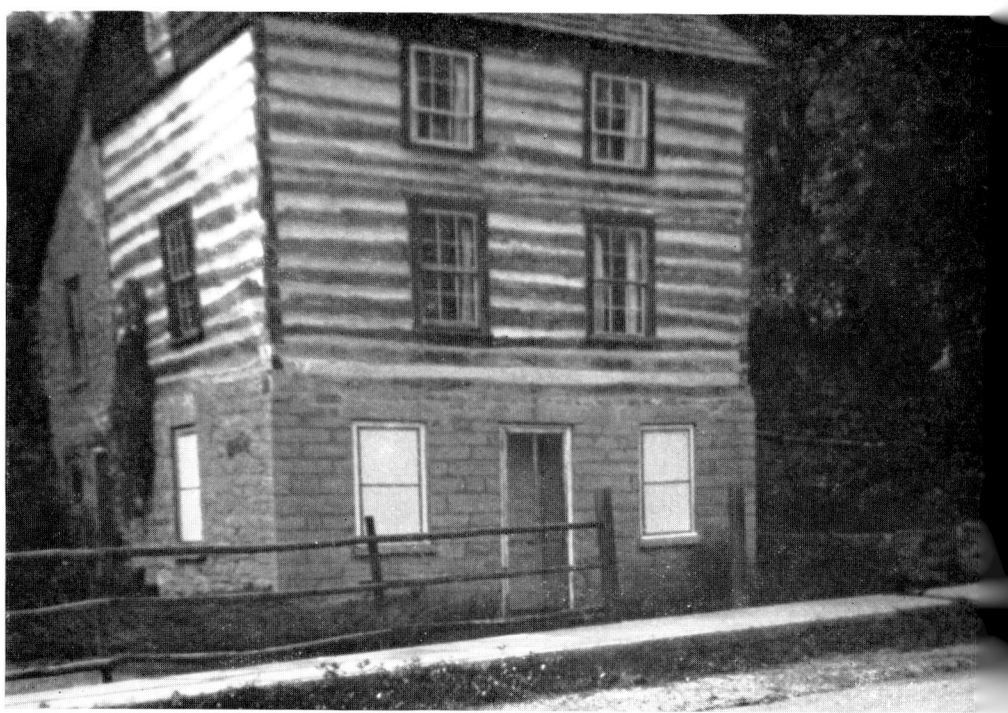

16 Cornish miners' homes, in the Copper Country of Upper Michigan (top), and at Mineral Point, Wisconsin

ards of Breage, John Bennetts of St Blazey Gate and Thomas Martin of Halsetown were killed in mining accidents in the fall and winter of 1886–1887.[94] At the inquest held on Martin in Tombstone on 16 March 1887, eight of the nine jurors agreed on the verdict that he was killed by the falling of a cage on him while he was working in the shaft of the Silver Thread mine, and specifically blamed the mine engineer for carelessness. The engineer, James M. Leggett, had admitted that the cage had only been secured by a single safety pin, and that that pin was by no means as safe as it could have been; he also admitted that he had not told the foreman, John Crago, that it was unsafe. The inquest had attracted interest of a kind that might well provoke trouble, for the saloon-keeper, John Martin, had doubtless already told many of his customers what he testified before the jury—that the dying man's last words had been 'Jack, boy, I'm gone: that man has killed me,' and that before he had gone to work the dead man had expressed the fear that Leggett would kill him. John Martin said that Leggett was 'particularly down on Cornishmen; that he had spoken disrespectfully of them as a class in Mrs Coyle's restaurant'. Leggett denied that he had had any personal grudge against the dead Cornishman, although he had on one occasion reprimanded him for washing his feet in a pan of water he and his partners kept for washing their hands and faces; furthermore, miners always tended to blame engineers for any accident which might happen.[95] Had Tombstone been the turbulent camp of its first days, and had the Cousin Jacks been less law-abiding, Leggett instead of testifying at an orderly mining fatality inquest would have been the central figure in a lynching bee. It is evident, too, that Cornish clannishness had evoked some antipathy among the other races which made up the polyglot population of Tombstone.

Three months after this incident another Cornishman died suddenly in Tombstone, James H. Stevens, an employee of the Tough Nut Mine. He had been visiting relatives in Michigan, had come back to Tombstone, was stricken with pneumonia, was apparently recovering when he had an unexpected relapse which carried him off within an hour. Mine-dust afflicted lungs made pneumonia especially deadly among miners, and Stevens was one of many to die in the prime of life, being only forty-four. The brief obituary notice in the Tombstone paper recorded that he died at the house of his sister, Mrs Whinner,[96] which

94. *W.B.*, 28 April 1887.
95. The inquest on Martin was reported in the *Tombstone Epitaph* on 17 March 1887.
96. *Tombstone Epitaph*, 8 June 1887.

indicated that already, a mere eight years after discovery of its gold deposits, Tombstone had become sufficiently well established as a mining town for Cousin Jennies to come thither and set up homes for their menfolk.

Unlike other mining camps Tombstone never lived down the Boot Hill phase of its history, which was, however, nothing more than the violent growing pains of its brief hectic pioneer and discovery days. As early as January 1886, the most controversial topic in the town was not the prowess of a gunfighter but how funds could be raised to keep the community's schools open. With about two hundred and fifty youngsters of school age in the place, the salaries of half a dozen teachers and a janitor already months in arrears, and the mines not nearly as prosperous as they had been, taxpayers began to complain. Miners in short employ resented the fact that teachers' salaries averaged out at a dollar per hour while their wage came to a bare thirty cents, and several Tombstone miners thought that no twenty-day working month of five-hour days really merited a hundred-dollar payroll. Women, however, had a different opinion when there was a threat that the schools would be closed, and the Cousin Jennies in the town certainly sympathised with the harassed 'Mother', who, in a letter to the *Epitaph*, declared:

> If the schools are closed, our children must be thrown upon the streets for their education. We have no money to send them to private schools, and we cannot educate them at home, for we must bend every energy to supply the simplest needs of life—food and clothing. Let us dispense with our clergy, let us mortgage our churches, let us abolish our expensive city government if need be, but in the name of God and humanity give us good public schools and short vacations.[97]

As the year 1886 wore on issues of this type were those of most concern and interest to the residents of Tombstone. In the early spring there was some agitation against the introduction of cheap Chinese labour which threatened the maintenance of white living standards and wages,[98] while the summer was to bring complaints about Tombstone's odoriferous and insanitary state,[99] but it is unlikely that either issue attracted as much interest from the Cousin Jacks as did, next year, the wrestling match organised as part of the town's Fourth of July celebrations. The names of the organisers, referees, and contestants, some two dozen in number, read like the report of any assembly in Cornwall itself. The first three prizes were won by B. Martin, H. Pope, and D.

97. *Tombstone Epitaph* 20 January 1886.
98. Ibid., 17 February and 19 March 1886.   99. Ibid., 5 June 1886.

Harris; the less-successful wrestlers included W. Sampson, J. Martin, B. Phillips, R. Hancock, W. Cocking, F. Broad, Joe Toy, A. Allen, and W. Martin. Charles Hancock, and G. Martin were the committeemen, and the three referees were J. Bluett, H. John, and G. Carlyon.[1] Truly, Martins seem to have been as plentiful and prominent in Tombstone as were Pengellys in Looe, and as Puckeys, Libbys, and Olivers in Polperro.

Briefly, within a decade of the first discoveries, Tombstone had become a town wherein communal life was following patterns which reproduced those of any Cornish community in any country, old or new. Mishaps and accidents often led to manifestations of over-hasty readiness to attribute outspoken—but not always undeserved—blame for their occurrence; there were grumbles about the squandering of taxpayers' money; there was criticism about the way local government was conducted; there were wrestling matches and other amusements. The racial antipathy to the Chinese was not to be found in the Old Country, but judging from the anti-Irish riots in Camborne in the same decade only because Cornwall, unlike the Pacific Coast, had no Oriental immigration.

What happened in Tombstone was typical of developments in other Arizonan mining camps where mineral lodes were extensive. In other places bonanza strikes petered out and ghost towns resulted; everywhere, as Jerome and Tombstone manifested, mineral once mined was gone for ever. The more permanent mining centres were Globe, Prescott, and Bisbee, and it was to the first and last of these that the greater number of Cousin Jack immigrants to Arizona came. Beginning as a silver camp, Globe in later times became one of the premier copper-producing regions of North America, and mining names like the Silver King, the Rustler of the West, the Blue Jay, the Cumberland, the Keystone, the Old Dominion and a number of others were as well known to Cousin Jacks as Grass Valley's Empire, North Star, Eureka, Idaho-Maryland, and Allison Ranch, as Michigan's Calumet and Hecla, Quincy, Minesota, Osceola, Tamarack, and Cliff, and even as the Old Country's own Dolcoath, Levant, Wheal Vor, Crinnis, Gonamena, and East Caradon. What Calumet and Hecla was to northern Michigan, the Copper Queen came to be to Bisbee. There were other important copper-mining developments at Williams, Clifton, and elsewhere in the Territory, and the soaring figures of production from less than a quarter million pounds weight of copper in 1883, and below thirty-five million pounds in 1890

1. *Tombstone Epitaph*, 28 and 30 June and 6 July 1887.

to over a hundred and fifty million pounds by the end of the century, valued in the region of two-hundred million dollars indicated the rapid transition Arizona was making from an Apache territory and battleground to statehood.[2]

The mining development of Arizona had had to wait for two things—the solution of the Indian problem and the provision of adequate transportation and other facilities for dealing with the copper ores of the Territory. As late as 1885 an Apache scare led to the hasty evacuation of over half of the population of Willcox and over the New Mexico line the mines and ranches around Lordsburg were deserted and 'as quiet almost as a graveyard'.[3] Five years later the London *Mining Journal* expressed the hope that a treaty between Mexico and the United States for joint action to suppress brigands, red, white, and intermediary in colour, might, at last, put an end to a state of society and affairs which, though romantic and picturesque, had rendered life and property insecure and had been a great obstacle to systematic mining development.[4] 'Bad roads and lack of coke' forced Bisbee's great Copper Queen mine to close down for a few days in the late summer of 1887,[5] but in his annual message at the end of that year the Territorial Governor, Zulick, boasted that Arizona now ranked third only to Michigan and Montana as a copper producer among the states and territories of the Union, and that

three thousand miles removed from the eastern seaboard, she has successfully competed with Lake copper.... With increased facilities of railway communication, cheapening of coke and transportation, and a fair price for copper, this especial industry must assume large proportions, giving employment to thousands of miners and adding greatly to the wealth of the Territory.[6]

The extent to which Arizonan mining was hampered by transportation problems in pioneer days may be gauged from a report that in March 1886, the Old Dominion Company had 'on the road from Willcox to Globe ... a hundred and five wagons, conveying 591,000 lbs. of coke, and at the furnaces there is four hundred tons of copper bullion awaiting transportation.'[7] Until the railroad came, this slow and costly method of transport meant that coke prices in Globe were about $65 per ton, which made it unprofitable to work copper ores of less than

2. *Tombstone Prospector*, 9 March 1901.
3. *Daily Alta California*, 7 December 1885.
4. *Mining Journal* (London), 13 December 1890.
5. *Tombstone Epitaph*, 3 September 1887.   6. Ibid., 10 December 1887.
7. Ibid., 18 March 1886, quoting *Silver Belt*, n.d.

eight per cent produce. Those advocating the building of railways to the copper-mining camp reckoned that such facilities would reduce the price of coke by half and make it possible to exploit the vast reserves of lower-grade ores.[8] Before railways were provided too, teaming operations were sometimes hampered by winter snows and summer storms. A heavy snowfall and the subsequent thaw damage to the roads cut Globe off from the outside world for some days in March 1881, and again in December 1884,[9] while floods cut off the mail for nearly a week in October 1896, when the mining interests were getting feverishly interested in the progress of Bryan's presidential election campaign.[10]

The drainage problems of the Tombstone and Bisbee mines were of a nature familiar to Cornish miners from the Old Country. At Globe the Old Dominion in its early days gained a notorious reputation for administrative mismanagement, but Arizona mines on the whole were not so plagued with absentee directors' 'pets' as were those of Colorado. The Territory had, by its late start, the advantage of the experience of Cornwall, of Michigan, and of the older Pacific Coast mining districts, but the early period of its rapidly increasing copper-mining development coincided with a time of general economic depression and labour unrest in the United States, and first the Knights of Labor and later the Western Federation of Miners were the centres of industrial conflict. In Tombstone the Knights of Labor fell into disrepute of 1887 through the association of some of their leaders with the anti-Chinese movement,[11] while in Globe, in June 1896, the attempt of the Old Dominion Company to reduce wages provoked a strike which was successful in its main immediate aim, though the Company refused to recognise the Miners' Union and asserted its right to employ whom they pleased.[12] Cousin Jacks seem to have been involved on both sides in this dispute, for the most criticised 'boss' was a Superintendent Parnell, and one of the chief spokesmen for the Union was a man called Bennett —both names associated with mining in the Old Country. Company stores and company-owned houses do not seem to have caused much friction in the early days in either Globe or Bisbee, but in the former town the Cousin Jacks were involved in events leading to the precipitate departure of a company-hired doctor from the district in 1883.

8. *Silver Belt*, 19 August 1882.   9. Ibid., 13 December 1884.
10. Ibid., 15 October 1896.
11. *Tombstone Epitaph*, 12 February 1887 (Editorial).
12. *Tombstone Prospector*, 6 June 1896: *Silver Belt*, 11 and 25 June and 9 July 1896.

The South Pioneer Company had hired Doctor Blackwood as 'medical attendant to the mines' at the reasonably generous salary of a hundred and fifty dollars a month, but his charges for attending the miners who asked for his services over and above this was regarded as exorbitant, and the mine superintendent, Comstock, had told him that his services were no longer required and dismissed him. Blackwood, however, did not leave, and shortly afterwards there was an accident in the South Pioneer mine. Four men were descending the shaft in a bucket when the brake on it failed and it plunged nearly three hundred feet; luckily the mine foreman managed to stop it before it reached the bottom of the shaft; the four men in it, however, sustained minor injuries and shock. Blackwood came on the scene, and told one of the men whom he examined, Henry Sparnon, that his back was broken and that he only had a few minutes to live, whereas he was, in fact, only badly shaken and bruised on the hips. Luckily another doctor arrived and was able to treat Sparnon successfully and to alleviate the sufferings of the other three injured men, Tom Desmond, John Jenkins, and John Laity. Sparnon's friends then sent Blackwood a note telling him to leave the camp by the next stage, and when he did not go, two miners called on him with another note to the effect that if he did not leave at once it would be the worse for him. This time the hint was far too broad to be ignored, and the so-called doctor left. It was not the first accident in which two of the men had been involved; Jenkins had previously lost a finger and an eye in a powder explosion, while a few years before Desmond had escaped with a few bruises when he had ridden his horse into one of the Silver King shafts, the horse being killed. Working in mines and living in mining districts had its hazards in Arizona as well as in the Old Country.[13]

The Cornishmen who came to work in the silver- and copper-mines of Arizona made fair but by no means high wages. The strike at Globe in 1896 only secured an average daily wage of three dollars, and in Globe and still more in the more southerly mining districts there was a risk of wages being depressed or kept low by cheap immigrant Mexican labour. This meant that comparatively few immigrants from the Old Country made quick fortunes and returned home, but it also predisposed many, especially those who had already wandered far afield, to move elsewhere on the news of better prospects. American and British companies seeking to work mining concessions in Sonora and elsewhere in Mexico drew upon Arizona's mining population. A fair

13. *Silver Belt*, 17 March 1883.

number drifted off to the Witwatersrand, but it is noteworthy that when the Boer War broke out there was a scheme to set up a Boer colony in either southern Arizona or northern Mexico.[14] David M. Neil, a Tombstone miner, went off to Alaska early in 1896, while a Winslow man, W. H. Morris, found his way to Circle City, nearly up in the Arctic Circle, in search of gold.[15] The Leadville silver discoveries in Colorado drew a number of Globe miners to the mountain mining camps of the high Rockies in 1879, but several of them came back disappointed; indeed William Beard returned grumbling about the snow and the high cost of living at Leadville and declaring that if only a quarter of the money invested in Colorado was put into Arizona mining, Colorado would never be heard of again.[16] Colorado was heard of again, and Cripple Creek attracted several Arizonan miners in the eighteen-nineties.[17]

Like Arizona, however, Colorado was used to the coming and going of miners. Whether and wherever they stayed for a month, a year, or the rest of their lives, they contributed to the transformation of the wild frontier to modern industrial life and society. That society, of which the United States is often held out to be the supreme exemplar, is far more nomadic and unsettled than is generally realised. The Cornishmen wandered further and more often, perhaps, than other miners, and thereby gained themselves a reputation in America and elsewhere. Their contribution to the social and economic development of the modern American Southwest is no mean one. And there were other mining camps to the northward in which they played a no less important part. Some of those who trekked to Pike's Peak and Washoe in 1858 and 1859 were on the move again by the summer of 1862 when the news reached Virginia City, Sacramento, San Francisco, and Denver that new bonanza strikes of gold had been made on the Salmon River in Idaho Territory. Soon there was talk of these new mining fields being richer than anything which either California or Colorado had known.[18] By the next fall the excitement in mining camps was intense, and Denver's *Rocky Mountain News* began uttering warnings like those the San Francisco press had been 'croaking' about Fraser River a few years before, and declaring that many Colorado men were

setting out upon the long journey without an adequate conception of what is before them. They expect and prepare for a journey of only a few hundred

14. *Tombstone Prospector*, 13 February 1901.   15. Ibid., 10 and 23 April 1896.
16. *Silver Belt*, 20 November 1880.   17. *Tombstone Prospector*, 5 and 6 February 1896.
18. *Daily Alta California*, 28 October 1863.

miles, whilst it really is more than a thousand. Almost the entire distance is through a mountainous country, where the winters are usually severe and the snows deep. Ox-teams cannot make the trip to Bannack City in much less than ten weeks, under the most favourable circumstances. Before the end of that time the mountain passes may be impassably gorged with snow.[19]

But it took more than a few inches—or feet—of snow to cool the fires of gold fever. Dangerous mountain trails and the menace of hostile Indians were no deterrents, and in the early eighteen-sixties the miners pushed the frontiers of settlement into Idaho and western Montana, and then, a little later, into the Black Hills of Dakota. As for the Cousin Jacks, if they were rarely the first on the scene of a new discovery they were certainly both prominent and early in the ranks of the second-comers.

19. Quoted by the *Daily Alta California*, 28 October 1863; date of the *Rocky Mountain News* not given, *circa* mid October.

9 Gold in them thar hills—perhaps

The exodus of miners from 'all over' to the Boise Basin in the early eighteen-sixties marked only one stage in the progress of settlement in the north-west. Before the gold discoveries of 1848, Oregon had been the first promised western land of eastern agriculturists who lacked the knowledge and the means to farm treeless, tough-sodded prairies whose few sluggish, mile-wide, inch-deep rivers were not comparable to the swift-flowing streams and brooks of the heavily timbered lands between the Mississippi and the Atlantic. The development of the Californian placers and mines had then given Oregonians a ready market for livestock and farm produce in the gold camps. Men who travelled the trails between the Columbia and the Sacramento had to cross many rivers with bars that looked similar to those wherein deposits of alluvial gold had been found in the south, and either came back to work the bars themselves or told the Californian diggers that they might strike it luckier if they came to the northern rivers. Before the stampede to the Fraser took place, gold had been found in fair amounts along the Klamath, Rogue and other rivers; the belief that the mother lode or lodes were high up these rivers encouraged enterprising prospectors to strike directly eastwards instead of following the tortuous course of the Columbia up from Astoria and Portland; the tales of fur-traders and Indians led others to penetrate the region south of the now well-frequented Oregon Trail. Within a decade of the first discoveries in California the deposits of Owyhee had been located, and with an ever-increasing number of prospectors coming into the region, the deposits on the Boise River were discovered and made known to the world by the early eighteen-sixties.

The newspaper publicity given to reports of gold strikes in the Oregon country in the late eighteen-fifties and early sixties had ulterior political motives. Oregon Territory, eager for admission to statehood, wanted to attract immigration to swell its population to such numbers that would make its claims irresistible. Furthermore, her southern neighbour,

California, was anxious to see the Pacific Coast more adequately represented in Congress.[1] In the Territory itself, an influx of miners was regarded as the best means of providing local markets for agricultural produce and also as the quickest way of terminating the Indian problem in a region in which the conflict of redskins and whites had been more bloody than on most American frontiers of settlement. Apart from these provincial considerations, national patriotism was eager to check the drift of miners up into the British possessions, and this led to the boosting claims that the Colville district was richer in gold than the Fraser River country. Furthermore, American patriots wanted to bolster up the political frontiers against the rapidly-growing British Pacific Coast settlements, and to swamp out from the area west of the Rockies and south of the forty-ninth parallel all vestiges of the influence of the Hudson's Bay Company.

At all events several miners came into the region both from the south and from the west, but the remoteness of the area from established overland routes, the rivalry of new camps in Nevada and Colorado, and the attractions of both Union and Confederate armies for the adventurous, delayed the development of the Pacific Northwest to some extent until the late eighteen-sixties. 'Rushes' and 'excitements' were on a comparatively small scale except in western Montana, Owyhee, and the Colville district. All the loud boasting about greater Californias and the like could not explain away the fact that, in 1869 as in 1849, California was still the Golden State, producing 31·5 per cent of all the gold raised in the States, although together the north-western states and territories produced 36·2 per cent, of which rather more than a half was mined in Montana, rather less than a third in Idaho and the remainder in Washington and Oregon.[2] Excluding eastern Montana, within twenty years of the first Californian gold discoveries there were probably as many, if not more, miners scattered about the American Pacific Northwest as there were in California itself, although another two decades were to pass before three of these territories attained a population sufficient to warrant statehood, while Oregon had been admitted before mining pursuits within its boundaries were of much more than secondary importance. Throughout the Northwest region, however, the

1. *Daily Alta California*, 24 March 1858 (Editorial).
2. The total United States production of gold in 1869 was $63,500,000; California produced $20,000,000, and Montana (coming third with Nevada in second place with $14,000,000) raised $12,000,000; Idaho produced $7,000,000 and Washington and Oregon combined $4,000,000—the same value as the combined output of Colorado and Wyoming Territories. *Mining Journal* (London), 15 April 1871.

news and prospects of rich mineral finds after 1849 helped to attract immigrants, and many Cousin Jacks came into the region, especially into Montana where rich and extensive copper deposits saved Butte from the speedy death that was the lot of so many ephemeral gold-mining camps on the Pacific Coast.

With prospects of making a fortune quickly rather brighter in British Columbia than in Washington and Oregon, and with the Nanaimo coal-field on Vancouver Island affording fairly constant employment to miners content to work for wages, Cousin Jacks probably preferred the British to the American Pacific coastlands. Further inland, however, the Boise Basin attracted a fair number, many of whom, however, were 'birds of passage'. Although the first strike in the Boise region had been made in the early eighteen-sixties, it soon became apparent that this was, for the most part, not a 'poor man's country'. Capital was required to work the rich ledges of silver and gold-bearing ores, although in many localities the mountainous nature of the country facilitated mining development by enabling underground exploration by simply driving levels in from the hillsides and by making adit-drainage possible in a considerable zone of richly mineralised 'country'. A fair amount of placer gold was found in the districts around Idaho City, Placerville, Centerville, and Oro Grande in the Boise Basin, and some also in the Snake River Valley, and this yielded fairly good returns without the outlay of much capital, as did, in the earlier days, the quartz gold lodes of the Red Warren and Rocky Bar districts of the South Boise. Down in Owyhee County, Silver City's boom days began in 1863, and continued intermittently for a quarter of a century, gaining wide repute through its Poorman Mine which yielded gold and rubies with some 'pay dirt' running out as rich as five thousand dollars per ton.[3]

One of the first Cousin Jacks in Idaho was the Redruth mining captain William Nancarrow, who came out in 1870 to examine the Atlanta Star Mine on behalf of an English company. The report made by J. Ross Browne to the Federal Government in 1868 had spoken of some 'selected ores' from the Atlanta lode averaging as high as eleven thousand dollars per ton, and had blandly stated that in some places it was equally rich in gold. Mining operations were started, and just at the time the Comstock came into a boom spell it was reported that the average daily return per stamp head on the Atlanta was £15 sterling as against a mere £9 from the great Nevada lode. Another report spoke of silver ores returning as much as £6,000 per short ton, and there were

3. W.P.A. Guide, *Idaho*, p. 262.

the usual stories about the Atlanta's being a practically inexhaustible true fissure vein. Nancarrow's opinion was more sober, yet he reckoned that along a mere three-hundred-foot length of the lode which had already been laid open there were reserves worth over £600,000 sterling after deducting all the costs of raising, smelting, and reduction. The Atlanta was being worked by adit-driving, and the Redruth captain reckoned that it could be worked that way without any costly pumping machinery fully eight hundred feet deeper than it was then being worked.[4]

Mining prospects of this kind seemed to be almost too good to be true. Many English investors, already smarting from unhappy speculations in mines in California, Nevada, Colorado, and elsewhere, were liable to ask why the owners of such rich claims were seeking to sell them to foreigners. They were told, generally truthfully enough, that it was because

the pioneers of mining are for the most part poor men, who, although by their daily toil they may be able to discover these rich deposits, are unable to raise the capital required to develop and utilize them. In some instances they may find moneyed men to help them; but then when the discoveries are past doubt, the monied parties seek to take the whole to themselves, without benefiting the first discoverers. Hence it often happens that there is no other resource but to find a market outside their Territory, so as to sell as a whole what they would only have been too glad to continue to work if they could have done so on terms. Another point is that money is worth from thirty to forty per cent per annum in these parts, so that to raise assistance on mortgage is ruinous, and out of the question, especially with men who have become exhausted in securing and partially determining the value of their property.[5]

Capital was found in many instances, and with it came immigrant labour. A number of St Just men found their way to Silver City,[6] which also drew a number of working miners from Grass Valley[7] and other Pacific Coast camps. Their fortunes and misfortunes were similar to those of Cousin Jacks in other parts of the American West and elsewhere. Some did well, others badly, but the Idaho mining industry was unlucky in that its more extensive silver deposits were being developed when the Federal Government demonetised silver in 1873. Idaho, too, suffered from the way in which the financial crises of the late eighteen-sixties and early seventies checked railroad development into a Territory north of the first transcontinental railroad line, that of the

4. *Mining Journal* (London), 17 December 1870; 7 and 21 January 1871.
5. Ibid., 17 December 1870.
6. *Cornish Telegraph*, 11 April and 22 August 1876.
7. John Jewell who went from Grass Valley to manage some mines in Silver City in November, 1871, was probably a Cousin Jack (*Grass Valley Union*, 8 March 1872).

Union Pacific which had been completed in 1869. The troubles between prospectors and capitalists, too, were the prelude to an industrial history in which some of the most violent conflicts between capital and labour in America occurred. Forewarnings of the 'class war' of Coeur d'Alene in the eighteen-nineties were apparent in 1876 when two St Just men wrote home a bitter account of their treatment in the Silver City mining district.

Disillusionment among emigrants was common, and homesickness inspired many jaundiced accounts about conditions in the New World. Nevertheless the complaints the St Justers made of being swindled out of their wages, of dishonest mine management, and other similar criticisms were so frequent that they must be considered in any appraisal of the progress of mining settlements in Idaho and other western states and territories. The West Cornwall miners had gone out 'anticipating a successful career on the Pacific Coast of America'; after a far from comfortable Atlantic crossing and caring little for the hustle of New York, which, being strict Methodists, they regarded as a latter-day Sodom or Gomorrah, they decided to go to publicity-booming Silver City. Soon they were regretting they had ever heard the name of the Idaho gold camp, for

although it was a flourishing camp when we first struck it, very soon the cheating game was started here as in other parts of this great swindling country. One mine after another in succession postponed the pay until they were three or four months behind, and then, in order to have the benefit of the men's labour as long as possible, they gave (as we thought) all encouragement necessary to continue the workings, but at the same time took the proceeds to themselves. To-day we may venture to state that there are over twelve of the leading mines closed down without paying a cent for five months, or making any satisfaction to those who have ventured their lives in the depths of the earth in the most dreadful part of the globe. Although it has been reported that much gold has been raised in Idaho there is scarcely a mine which pays a dividend in the Territory, but they have been worked by assessments (or calls) until the stock-holders are wearied out. It is almost impossible to estimate the unjustifiable expenditure generally connected with these mines, especially as the reckless 'supers' can accumulate wealth to satisfy themselves in twelve months; for frequently the rock, valued at a hundred dollars per ton, will hardly realize twenty dollars when the 'supers' present the produce to the companies. So . . . the choice grain is the 'supers' portion while the gleanings fall to the companies, and the worthless chaff may be divided among the workmen.[8]

8. *Cornish Telegraph*, 11 April 1876, letter signed 'A.B.' and 'W.A.'

Such complaints revealed that as early as the mid seventies the antagonism of working miners and managers in Idaho was already acute and provided a prelude to the bitter labour wars two decades later. The mine superintendents were widely hated and distrusted, but they were by no means all scheming 'Yankees' as some Cousin Jacks maintained; not all the Cornish 'boss miners' in northern Michigan were regarded with much affection by their fellow countrymen, and in some mining camps the Cornish hands declared that 'Cousin Jack cap'ns' were the worst slave-drivers of all, although it could never be said that they were so ignorant of mining matters as one 'Yankee' foreman in Idaho who told the men under his charge to blast some rock in square pieces.[9]

Defrauded of their hard-earned pay, the St Just miners went on to voice further scathing criticism of Silver City, of Idaho, and of the United States in general. The mining camp was a sink of iniquity, the sabbath was not kept, the only Bibles were those brought out with them by some Cornish immigrants, but saloons were crowded, dancing was carried on to excess, and shooting scrapes and stabbing affrays were not uncommon. Furthermore

> four-fifths of those who were here in the flourishing times have not now a cent to help pay their fare down from these snow-clad mountains to some better climate. There are hundreds leaving Silver City weekly who have been here years and yet are obliged to walk to the railroad—a distance of two hundred and ten miles—without the means to purchase the necessities of life in this dreadful season of the year when the snow is higher than the telegraph poles. More than that the snow is forty feet deep in Silver City, and .. the climate is so disagreeable that often when it snows, with snow drifting, one can scarcely find his way to his next door neighbour's house. Some houses are entirely covered with snow; then a candle has to be kept alight all day, and a tunnel driven through the snow in order to find the doorway.

So they advised their friends back home in West Penwith, who rarely saw a snow-drift against a granite hedge more than five or six feet deep, to keep away from Idaho and the Pacific Coast. Even if a man made good wages, the cost of living was exorbitantly high. Those who quit Silver City and trudged to the distant railroad simply had not the savings to pay stage fare at twenty cents a mile. No honest miner could save money in a country of swindlers, cheats and robbers. One Idaho mine was reported to be paying off its debts, including wages, at five cents in the dollar, and the St Just miner wrote that

9. *Cornish Telegraph*, 22 August 1876, letter of a St Just miner, 'W.W.' dated Eureka, Nevada, 3 July 1876.

such cases are of frequent occurrence, and there is hardly a camp on the coast where working men have not been wronged out of their earnings in a similar way, the results of which are now plainly to be seen: families on the verge of starvation, men that have left their homes leaving their wives and families behind, in order to better their condition, families begging their way from camp to camp in search of work, travelling many hundreds of miles on foot, very often for days without the common necessities of life passing their lips except water.[10]

The same correspondent stated that in that summer of 1876 unemployment was rife throughout the Pacific Coast mining regions; those who found a job labouring from dawn to sunset for a dollar a day were fortunate, and

there are numbers of us (Cornishmen) who have had to travel, (some of us are still travelling) hundreds of miles, and blankets on their backs, sleeping in the open air, all though the rascality of the mining superintendents and trustees.[11]

Reports like these deterred emigration from Cornwall or, at least, made intending emigrants think twice before setting out for Idaho. Yet they demand some qualification. The disgruntled Cornish miners wrote during the trough of an economic depression; collapse had followed boom and disillusionment was aggravated tenfold by previous excessive hopes. Silver City, though it was to revive after this time, with people drifting away and those remaining giving way to feelings of apathy and despairing hopelessness, had had premonitory warning of the fatality which overtook so many Pacific Coast mining camps and transformed them into ghost towns. Still, the decline of one mining camp often coincided with and was, sometimes, caused by the rise of another, and in Idaho, if the Owyhee diggings and the Silver City region declined, it was not long before fresh feverish hopes were roused by the gold finds in the Coeur d'Alene district.

There was, too, generally a cautious and conservative streak in the Cornishmen's character. Forced to emigrate by economic necessity the majority of them were not looking for adventure but for financial security, hoping only to make a living and to save something for retirement in old age either in Cornwall or in America. Some places offered better chances for the realisation of these limited ambitions than others, while elementals such as climate and working conditions below and above ground certainly came to be weighty considerations. With

10. *Cornish Telegraph*, 22 August 1876; letter of 'W.W.'
11. Ibid., 22 August 1876.

possibly rather more than its due proportion of swindling or incompetent mining superintendents, its early adoption of dynamite blasting and single-hand drilling,[12] and more snow most winters than west Cornwall got in a century, Idaho was not very attractive to the Cousin Jack.

From the very outset of mining in Idaho, too, there was a fairly large Chinese population in the gold and silver diggings, which tended to cheapen labour. Some of the richest early diggings were in extremely wild, mountainous country; most of the rivers were swift-flowing and were dangerous to ford, but forded they had to be many times by immigrants coming from the west and the south-west; many men were drowned on the way and more became victims of rheumatism through wading or swimming in icy mountain torrents—especially those who would not wait for spring or early summer to travel to distant, newly discovered Idaho diggings. In the early eighteen-sixties the Bannack Indians were hostile and treacherous, and immigrants came into the Territory knowing that they were almost as likely to lose their scalps as they were to reach the gold-fields.[13] Despite all these disadvantages a number of Cornish people came to Idaho, although it was Butte City and not Boise that became the great Cornish mining centre of the inland Northwest.

As in the other Pacific Coast regions gold was the first lure that brought miners into Montana. The initial discoveries, near and east of the Continental Divide, were made in the early eighteen-sixties although half-breed fur-traders had found the first indications of gold near the Deer Lodge Valley in the late eighteen-fifties, while prospectors and miners going overland to or returning from California had 'fossicked' around although with little success at first. By the end of 1864 Bannack, Alder Gulch, Last Chance Gulch, Confederate Gulch, and Silver Bow Creek were mining camps. Effective vigilante action against the notorious Plummer Gang early in 1864 indicated that some sort of law and order had come to Montana, and it was the neighbouring Territory of Idaho that was to hold for a long time a most unenviable reputation for numbers of stage hold-ups. Montana, however, had its full share of shooting scrapes and the like, and hazards to life were just as great as they were in the western neighbour Territory, but the miner who saved a proportion of his earnings or, more rarely, really made a bonanza strike, was probably reckoned to have less chance of keeping them in Idaho.

12. *Grass Valley Union*, 8 March 1872, and see above, p. 120.
13. *Daily Alta California*, 24 and 25 June 1863.

One of the first reports of Montanan gold to appear in the Old Country was that of the discovery of the Dry Gulch diggings, about four miles from Helena, by a German or Norwegian miner called Brown, published in a Cornish newspaper late in 1865. The Cornish paper copied the account from the Toronto *Globe;* the *Globe* had taken it from the *Salt Lake Telegraph;* the Utah paper had acknowledged its indebtedness for the item to the Helena *Post*. Whether Cousin Jacks in their homeland were very sceptical about third-or fourth-hand Pacific Coast newspaper reports of rich gold strikes is doubtful, but the account of a gold deposit 'which, in richness and extent, has, perhaps, no parallel in the history of gold mining', could have raised visions of a new and greater California. The report said that when Brown struck into the lode

gold, in almost solid masses, glittered before his bewildered vision. For two weeks longer, unknown to others, he tunnelled into the golden wall. Secreted about him he had accumulated several gunny sacks, literally filled with the precious metal; when longer secrecy became impracticable, from the very extent of his unexpected wealth, Mr. Brown proceeded to record and secure his property, when the public were informed of the great discovery. It is said that he now keeps a strong guard night and day over his seemingly incomparable wealth, while he himself, unassisted, delves into the golden wall around him and continues multiplying his sacks of precious ore. One person who was admitted into Mr. Brown's drift states that it presented a scene of wealth more akin to a picture of imagination than actuality. Gold! gold! gold! met the view on every side—above, below, and all around—and reflected back its rich hues in the glare of the candle, as if this subterranean vault had been hewn out of a solid ledge of the yellow metal. This gold is found in a well-defined ledge, fully fifty feet in width. The gold vein is three feet wide, three-fourths of the entire substance being pure gold, the remainder mainly bismuth. On each side of the vein is a casing of one foot of quartz which will assay from five hundred to two thousand dollars to the ton. The very wall-rock is rich.[14]

Western newspapers could rarely be censured for understatements, yet the Dry Gulch diggings, located a year after the finds at Last Chance Gulch, really gave Helena the start which, in half a decade, boosted its population to around the ten thousand mark, while further discoveries of gold and silver in the district roundabout made it the city of millionaires, some accounts accrediting it with fifty at one time,

14. *W.B.*, 29 December 1865; the *Salt Lake Telegraph* which copied this item from the Helena *Post* was the issue of 11 November 1865.

others thirty.[15] At all events the region was rich enough to attract immigrant miners from 'all over', and doubtless Cousin Jacks came fairly early on the scene from Pacific Coast diggings, from Wisconsin and from the Old Country itself; they were hardly likely to be left far behind when it became known that an East Anglian Attleborough man, Reginald Stanley, had, with three American partners, located the Last Chance Gulch diggings, and made a fortune from his claim.[16]

Gold was found in the Butte district in 1866, then silver, while finally the great copper deposits were discovered which, all said and done, was what Cousin Jacks from both the Old Country and Michigan knew most about. For a few years, Butte's copper ores were sent to Swansea for smelting, involving such high transportation costs that only the richest ores could be worked at all profitably, but in 1884 local concentration into matte began; eight years later, with the construction of the great Anaconda smelter, western Montana threw off its thraldom to outside smelting firms. Silver had supplanted gold as Montana's main mineral source of income in 1875, but from 1885, although the more precious metals continued to be mined, 'copper was king'.

In and around Butte all three minerals—gold, silver, and copper—were successively mined. The first gold strikes in the vicinity were made in 1864, and there was to be a real small-scale boom or 'excitement' in 1867, during the summer of which year the population of the camp was not far short of five hundred; through shortage of water, however, scarcely two hundred were still living in the camp in 1870.[17] Four years later the first silver strikes in the district were made, and hardly a twelvemonth passed before there was talk of this locality being the second Comstock. Towards the end of 1875 it was estimated that there were three hundred miners in the camp, and more coming every day; lots and cabins which a year ago could have been bought for fifty dollars were now costing ten times that amount; the 'old' established Hotel de Mineral was crowded and was extending its buildings; the Girton, a boarding-house, was packed with seventy miners; new frame buildings were being quickly erected to shelter the immigrant population.[18] Winter started early in 1875, and hastily about eighty 'residential cabins' were built in a couple of months.[19] It was to the credit of Butte and its immigrant citizenry that, in these hectic days, building construction was not limited to mine-shaft houses and living cabins and

15. *W.B.*, 20 December 1894: W.P.A. Guide, *Montana*, p. 161.
16. *W.B.*, 20 December 1894.   17. W.P.A. Guide, *Montana*, p. 198.
18. *New North West*, 22 October 1875.   19. Ibid., 7 January 1876.

that a public school building went up quickly enough for eighty scholars to be on its registers the next March.[20] In April, 1876, it was reckoned that just over five hundred men were fully employed in the Butte mines, including fifty sets of leaseholders or tributers,[21] and work went on despite a heavy fall of snow in the district in mid May which damaged roads, destroyed bridges and even saw trees uprooted by the sheer weight of the snow.[22] By that time it is almost certain that some Cornish miners had come into the district, and they may have already been more interested in the copper lodes they located than they were in those carrying silver ores. At least fifty Cornish miners were listed in the Federal Census of 1880 for Butte, though hardly half a dozen had wives with them, but once copper mining really got under way the Cousin Jacks and Jennies began flocking into Butte on a scale similar to that in which they had earlier come to Upper Michigan and Grass Valley. Polglazes and Pascoes, Roddas and Rowes, Trewalters and Treglowns, Penroses and Trebilcocks were among the early Cornish settlers in Butte.

To Cornishmen Montana came to mean Butte, but a number of Cornish miners were also employed working the Drumlummon gold lode in the Marysville district in the eighteen-eighties and nineties. Located by an Irishman, Thomas Cruse, the claim was bought by an English company in 1883 for a million and a half dollars.[23] This Montana Company, as it was called, had a chequered history. First of all it had difficulties in raising the capital to pay the purchase price and to organise the working of the mine. Some of the early machinery installed proved defective. The hard winter of 1883–84 held up mining operations. The directorate decided that they should purchase the adjoining mineral property for £60,000, and to do so raised the share capital from £600,000 to £660,000 by issuing a further thirty thousand two pound shares. To keep the shareholders quiet a 'wretched' dividend of eightpence per share was paid late in 1884, while the local management of the mine was taken over by R. T. Bayliss from Superintendent Attwood. There was an immediate improvement, on paper at least, for whereas in four months' working during the summer and fall of 1884 Attwood had only crushed 12,000 tons of ore which had averaged but $11.46 a ton, Bayliss, in less than three months, starting in November 1884, crushed 5,538 tons which had averaged $23.38 per ton, and had stepped up daily earnings from just over $1,100 to nearly

20. *New North West*, 10 March 1876. 21. Ibid., 28 April 1876.
22. Ibid., 26 May 1876.
23. W.P.A. Guide, *Montana*, p. 305; *Mining Journal* (London), 8 March 1883.

$1,600, besides cutting the costs of wages by twenty per cent.[24]

After the first set-backs—faulty machinery, sub-zero winter temperatures, and friction among the personnel on the location—the Drumlummon began to pay good dividends to the shareholders of the British Montana Company, but its troubles were not over. Early in 1886 London Stock Exchange 'bears' attempted to force down the price of its shares, resorting to such unscrupulous practices as forging telegrams ordering the sales of shares by directors of the Company.[25] The Company, too, had to pay extremely dear for materials and wages; fuel cost three and a half times as much as it cost in Cornwall, dynamite was fifty per cent dearer, while the general run of wages for labourers and tradesmen were from four to six times as much as they were at the time in the Old Country.[26] Helena capitalists attempted a 'bear raid' on the Company's shares in July 1887, when there were the first mutterings of apex-claim litigation being organised against the Company, and at the end of that year the ores showed a fairly appreciable falling off in produce. Associated silver ores had helped to boost dividends paid till that time, but falling silver prices began to have a marked effect in 1888.[27] Dividends amounting to £537,000 were paid out by April 1891, and the remaining gains amounting to £136,000 spent on development work indicated a total return of £673,000 or £13,000 more than the total capital of the Company.[28] Damage by fire and flood did not improve the outlook in 1892, when the Company was reconstructed and had to face an apex claim for damages amounting to two and a half million dollars.[29] Thereafter litigation went on in a long and confused series of actions and appeals, which finally went against the British

24. *Mining Journal* (London), 14 March 1885.
25. Ibid., 27 March 1886.
26. Ibid., 25 September 1886. In a letter dated Helena, 20 August 1888, 'J.B.T.' stated that men working ten hours a day in the 'West' were paid from twelve to sixteen shillings a day, but that some west Cornwall mine-owners were grumbling about paying an average wage of £3 a month. (*W.B.*, 13 September 1888). The writer was probably Joseph B. Trevarthen of Troon, near Camborne, who after mining in Michigan and at Austin in Nevada for two and a half years, went home, but came back to Austin in 1882. In March 1884, he came north to Montana, and in 1887 left mining to start farming. Two years later, however, he was appointed deputy inspector of mines for Montana by the Territorial Governor. (*W.B.*, 18 July 1889.)
27. Ibid., 30 July and 17 September 1887; 17 March and 12 May 1888.
28. Ibid., 4 April and 3 October 1891.
29. Ibid., 1 October and 26 November 1892. Apex litigation was the consequence of the Congressional Act of 1872, usually called the 'apex law', which was designed to remedy conflicts arising from claim-owners possessing a metalliferous vein which dipped into the earth below the vertical boundaries of the claim. The apex of a vein is its surface outcropping; the 1872 Act defined a claim as a surface area of 1,500 by 600 feet, where the vein had its apex

Company. The complicated story leaves the impression that some investors did moderately well out of the Drumlummon lode, a few Stock Exchange manipulators in both London and Helena did better, while lawyers had certainly got good pickings from the lawsuits. Including a few small dividends paid in the late eighteen-nineties when it was mainly engaged in working over waste 'slickens', the aggregate dividends paid by the British Company amounted to about three million dollars, but this figure hardly squares with estimates that the Drumlummon yielded twenty million dollars worth of gold between 1885 and 1895, and that for the whole period when it was worked it yielded in the neighbourhood of fifty million dollars in gold and silver.[30]

Still, the Drumlummon and the other mines round Marysville, whose total gold production over the years came to about seventy-five million dollars, in their boom times supported a population of nearly three thousand souls in the eighteen-eighties and nineties; a handful of tributers or leases were working in the area as late as the nineteen-twenties, while still later dreams were entertained of new bonanza Drumlummon strikes being made. While harrassed shareholders in London, Manchester and elsewhere in the Old Country were sadly scanning the financial newspapers to see their shares dropping down to four or five shillings and nineteen shillings per share had been called up,[31] the working miners at Marysville were more interested in the fortunes of their cricket team. A report of a match played in August 1899, between the Marysville men and a team from Centerville, the Butte suburb, reads like an account of any match played between rival Cornish village teams in the Old Country, with some batsmen scoring 'ducks' and others having good 'knocks'.[32] Cornish names were prominent and probably half, if not more, of the players on both sides hailed from Cornwall. The fair coverage given in Butte and other Montanan newspapers to wrestling matches from time to time, especially in connection with Fourth of July celebrations, also shows the strength of

---

or outcropped. The claim owner could follow this vein an unlimited distance underground within its 1,500 foot length, but there was no limit on following its subterranean lateral extensions. Where metalliferous veins were extremely faulted, and even interlocked, as, for example, in the copper lodes of Butte this led to almost endless conflict and litigation. *Vide* J. K. Howard, *Montana, High, Wide and Handsome*, pp. 70-1; J. F. Marcosson, *Anaconda*, pp.112-13.

30. W.P.A. Guide, *Montana*, p. 305.

31. *Mining Journal* (London), 24 June and 30 December 1899; at the end of 1908 Montana Company shares were quoted at 6d. to 1s. (Ibid., 2 January 1909).

32. *Butte Tribune Review*, 2 September 1899. An earlier cricket match played between Marysville and the nearby mining camp of Gloster on the Fourth of July 1886, was reported in the Truro *West Briton* on 5 August 1886.

the immigrant Cornish element in the Montana mining towns and the way in which the Cousin Jacks took their sports and pastimes to the mining camps of the far West.

A far sadder account of the Cornish immigrants, however, is provided by the all too frequent accounts of mining fatalities in the newspapers of both the Old Country and the new. Even more shocking was the death rate caused by 'miner's con', or phthisis, attributable to the dusty mines in which the men worked, and by pneumonia caught by many men walking home in sub-zero temperatures after labouring in deep levels where the temperature sometimes reached 125°F. Butte got the reputation of having more citizens in its cemetery than anywhere else, but unlike the graveyards of Pioche and Tombstone comparatively few buried there 'died with their boots on', and if they did it was far more often by accident than mayhem. Thus back in Butte's early days, in the first three months of the year 1882, there were thirty-three burials in Mount Moriah cemetery; seventeen of these were children, of whom, it is true, one had been murdered, while three were still-born babes; seven male adults had died of pneumonia, two men had lost their lives in accidents, and one man had committed suicide; of the three women buried, one was also a pneumonia victim.[33]

'Miner's con' probably accounted for a high proportion of the miners who died in their early forties in Butte, after working for perhaps twenty or twenty-five years underground. Within a fortnight of each other in the fall of 1899, Joe Thomas of St Austell, aged 40, John J. Johns of Stokeclimsland, aged 41, and Samuel J. Rowe of Goldsithney, aged 43, all died of this occupational disease.[34] A few years later its continued ravages led to a demand that something be done to improve working conditions in the Butte mines, although this must be associated with similar movements both in Cornwall and in the Transvaal at the same time urging that mine-owners or political authorities do something to check the incidence of this particularly agonizing and dreaded occupational disease, a disease which generally incapacitated men for months, even years, before death came as a relief to their sufferings, while it also, on the other hand, tended to make a mild attack of influenza or a slight bout of pneumonia fatal. Early in 1908 it carried off John Williams, aged 49, Thomas Edwards, aged 31, Edward Pollard who had been totally incapacitated by rheumatism for the last eighteen months of his life, aged 42, shift-boss William Eddy, aged 43, who had also been ill for months and whose death had been 'expected' for some time, Thomas

33. *Butte Daily Miner*, 4 April 1822.   34. *Butte Tribune Review*, 7 and 21 October 1899.

Magor, aged 62, Richard Trenerry, aged 37, and James Waters, aged 44.[35]

These were only a mere fraction of the deaths caused by phthisis. The editor of the radical *Tribune Review*, urging the introduction of improved ventilation in the Butte mines, declared that in 'something like two weeks', in May, 1908, a dozen or more had died of 'con' in the mining city, and even avowed that men were dying of it almost daily.[36] This was the licensed exaggeration of a reformer-journalist, but since over the whole year, in fact over nearly the past decade at least, there was a funeral of a consumptive or pneumonia or bronchitis victim every second or third day on the average in Butte, there was justification for demanding that something be done about 'too much bad air and too many funerals'.[37] Attempts were made by the companies, not always without outside pressure and agitation, to improve matters, but as late as the beginning of the New Deal period the percentage of deaths in Butte from 'consumption', and other pulmonary and lung diseases, was 21 as against 16·2 for urban America.[38]

Mortality figures are dismal reading, but no survey of social conditions in Butte can ignore them.[39] In a four-year period ending 30 April 1909, Butte had had 77 fatalities from mining accidents; the toll of 'tuberculosis' rose from 49 deaths in the year ending 30 April 1906, to 73 in the year ending 30 April 1909, while out of 74 deaths from this cause in a year running from May 1910 to May 1911, the occupation of 47 was recorded as mining whereas only seven were designated as 'housewife'. Pneumonia and broncho-pneumonia in some years swept off a hundred or more victims, or nearly twenty per cent of all deaths in Butte. Organic heart disease, frequently caused by the strain of ascending and descending mine shafts, also killed many, no less than 44 of the 653 deaths reported for the year ending 30 April 1907, being ascribed to this cause. Infant mortality was also extremely high, 144 of the 582 deaths for the year ending April 30 1908, and 106 of the 560 for the following year being those of children under five years of age; no less than 90 still-births were returned in the year ending 30 April 1906, and 70 the following year. There were bad epidemics in some years, includ-

---

35. *Butte Tribune Review*, 4 and 18 January, 8 February, 18 and 25 April, 16 and 23 May 1908.
36. Ibid., 23 May 1908.   37. Ibid., 23 May 1908.
38. R. R. Renne, *Butte, Montana: Preliminary Report of an Economic Survey*, p. 20.
39. The following paragraph is based on the *Annual Reports of the City Officers of Butte*, which, while drawing a distinction between pneumonia and broncho-pneumonia, identified phthisis with tuberculosis and did not indicate how many miners who died of pneumonia, bronchial ailments, and so forth, had contracted phthisis.

ing measles in the summer of 1905, a persistent epidemic of diphtheria during the year which ended in April 1906, while in the course of the next year meningitis, probably caused by the Rocky Mountain tick, killed thirty-nine victims, nineteen of them in the months of March and April 1907. It was only in two months of the twelve, August and December 1906, that no deaths were reported as being caused by this malady. There were thirty-two fatal cases of scarlet fever in the year ending 30 April 1909, the toll being from one to six a month, but with not a single month passing without a death from this cause. Apart from the meningitis epidemic of 1906–7, however, comparable mortality returns could have been culled from any mining district in Cornwall during the same period, but Butte's phthisis record would have appeared comparatively worse, although so many migrant miners who had been to the Transvaal came back to die in Cornwall of what, in some mining parishes, came to be called the 'African' complaint.

The tragedies caused by the 'con' led some to go further than demand improved working conditions. There are Cousin Jacks to this day, both in the Old Country and on the Pacific Coast, who would not return to work in the mines at any cost, who regard proposals to re-open mines or to expand mining activities as tantamount to attempts to condemn labouring men to death. In the days when the mines were working to capacity there were too many men, not far advanced in years, walking about the streets, too feeble to work, haggard and sallow-looking, every now and then coughing hollowly and bringing a hectic flush to their sunken cheeks, men who hoped that they would be better when the weather improved a little but whom their friends knew would, before long, be in the graveyard. More often than not, too, they left widows and young children in impoverished circumstances to fend for themselves as best they could, which meant not only immediate under-nourishment and lack of common necessary comforts of life, but future debility and ill-health for themselves and even for generations yet unborn.

Phthisis should be regarded as one of the main causes of the unhappy relations of labouring miners and capitalists in Butte and elsewhere, and these relations were made far worse by accidents which, often with very considerable justice, could be attributed to managerial carelessness, parsimony and disregard for the lives and welfare of the men they employed. In the trade-union movement in Montana the Cornish were not very prominent, having brought with them to the mining camps of the Rocky Mountains and Pacific Coast the highly independent individualism which had hampered the effective development of unionism in Cornwall. The Cornish tribute and tutwork systems had made them averse to collective bargaining practices; they preferred, if possible, to make their own 'bargains' with mine-owners, and they were so naturally conservative in their ways that they were slow and reluctant to admit, or even realise, that the consolidation of mining adventures under the control of a few large corporations or 'trusts' made the ways they had known and to which they were used obsolete. Their conception of co-operative action, exaggerated by over-facile and frequent quoting of the county motto 'One and All', did not go much further than the fraternal organisation of mutual benefit and aid societies, although some Cornishmen realised the need for co-operative industrial action to maintain and improve wages and working conditions. It is possible that Cousin Jacks would have played a much greater part in trade-union development if so many of them had not come to regard the best way of dealing with an unsatisfactory employer was simply to leave at the end of a wage or tribute contract. In Cornwall such methods had been moderately effective on account of the large number of small, independent mining concerns, the possibilities of finding alternative employment at certain seasons of the year in fishing and farming, the fairly considerable proportion of miners who had their own smallholdings, and, as the nineteenth century wore away, the ease of migration to urban industrial centres and of emigration overseas. In America it was easy as one mineral strike followed another to move on when working conditions in a certain place became unsatisfactory, and particularly so for the high proportion of 'single' men who had no family ties to hold them to the places where they worked.

Even when they had families along with them, the tendency of the Cornishmen was to put most stress on the benevolent social roles of any union they joined. Sometimes an individual arose among them, like Richard F. Trevellick,[40] who argued the need of working-class solidarity against employers' blacklists and the like, but to the majority of Cousin Jacks any union was primarily a fraternal society. They joined a union in much the same spirit and purpose as they joined such organisations as the Sons of St George and the Independent Order of Good Templars. Some regarded the union they joined in much the same way as they had regarded the burial clubs of the Old Country; membership would ensure that they had a good crowd of friends and acquaintances at their funerals and their widows and children would receive some financial

40. *Weekly Tribune* (Butte), 20 July 1895.

help at the time of bereavement. Union funds sometimes helped in times of sickness or of accidental incapacitation. The first Miners' Union in Butte was organised as early as 1878, and among its officials were the Cornishmen Edward Trebilcock, John Eddy, and Joe Thomas. In its fourth year, by which time it had a membership of about five hundred, it spent $2,600, of which ten funerals of members cost $2,000, or $200 each; the calculation that 'time lost' and so forth meant that each funeral of a union man cost $1,500 indicates that nearly all the members attended the funerals of their fellows.[41] Originally formed to keep wages at a fair level, this union was strong enough to impose the 'closed shop' in the Butte area, but it was decidedly less militant than the later Western Federation of Miners and the Industrial Workers of the World, and the success of the Butte miners in maintaining high wage levels and gaining the eight-hour day in 1901, was due as much to the rivalry of the great copper-mining magnates, Clark, Daly, and Heinze, as it was to the strength of the union movement and its Montana leadership.

Occasionally a Cousin Jack demurred at paying union dues, but most of them joined just to be with the crowd. Working-class unity in Butte, however, as in other Pacific Coast mining regions, was strained and sometimes wrecked by feuds between different racial groups. The Cornish–Irish feud in the Montanan copper town was as bitter as it had been in Upper Michigan on the Comstock, and in Colorado, and later German, Italian, and Slav immigrants caused further troubles and friction. The weakness of the old Miners' Union became obvious in the years just before the First World War. The rival 'copper kings' had either died or composed their differences, but the union leaders, or would-be leaders, quarrelled among themselves, and failed to put up any effective resistance to the imposition of the 'rustling card system' by the mine-owners which was tantamount to the re-introduction of the 'black list'. The violence of a group of extremists in 1914 split the Union asunder, and lost the closed shop until the time of the New Deal.

Worse followed in 1917. The 'Wobblies' made a determined effort to get control of the working-class movement in the Montana copper-mines, and tried to embitter the relations of labour and capital by blaming the mine-owning class for the disastrous Granite Mountain Mine fire in June, which killed more than a hundred and fifty men.[42] All they

41. *Butte Daily Miner*, 14 June 1882.
42. Estimates of the toll of this disaster vary; 164 seems to be the most likely figure and is given in the W.P.A. Guide, *Montana*, p. 73.

succeeded in doing was to provoke a strike which led to the imposition of martial law on Butte for eighteen months—till hostilities in Europe were over. The 'Wobblies' brought on themselves charges of being pro-German and anti-American by the way in which, in Butte, they set about the organisation of a 'Committee on Grievances' in a stormy meeting at the beginning of the troubles. It was proposed that this committee of ten be selected from ten different nationalities, partly on account of the polyglot mining population of Butte, partly from the internationalist credo of the I.W.W.'s. The suggestion was acclaimed and adopted. Quickly a 'Republican' Irishman, a Finlander, a German, an Austrian, a Bulgarian and then a Russian were elected, followed by an Italian. No Serbian could be found to serve on the committees, and the Italian seemed extremely reluctant to be associated with it. Then the proposal to elect a 'Cousin Jack' was shouted down by a noisy group yelling 'We don't want any of them. We will let the Irish take care of them!' The Irish Republican 'Pearse and Connolly Club'[43] was believed to be the mainspring of the strike, while recent Finnish immigrants, good workers but mostly ignorant of the English language, were said to have been misled by agitators into believing that if they did not join the strike they would be conscripted into the army and sent back to Europe to fight.[44] The fact that the Federal Government maintained martial law in Butte until after the end of the European war indicated how seriously it regarded the racial antagonisms in the great Montana copper-mining district.

Although the majority of the Cornishmen in Montana were inclined to think that if a miner got a fair day's pay for a fair day's work he had little cause for complaint, the provision of reasonably healthy and safe working conditions in the mines depended upon managers realising that such meant increased efficiency and economy in working, and some did not learn this elementary lesson as quickly as others. When the fire in the Granite Mountain shaft, however, immediately cut the copper output of the Butte mines by over a quarter, the lesson was driven home that every possible precaution against outbreaks of fires in mines should be taken; all who had invested money in those mines were justly dismayed by the talk going around Butte, talk not only of the hundred and fifty or more men who had been suffocated to death, but the talk among miners generally in Butte that they had no wish to take any further chances of losing their lives by working underground. It was that fear,

43. Named after the leaders of the 1916 Easter Rising in Dublin.
44. *Mining Journal* (London), 14 July 1917.

not racial prejudice against America and her Allies that brought many Butte copper-miners out on strike in the summer and fall of 1917, but, on the other hand, one mass meeting of strikers urged the Federal Government to take over the mines, declaring that they would then gladly go back to work and prove themselves good patriots. So long as there were worth-while mineral veins in the depths of the earth, perhaps even a little longer than that, Cousin Jacks were prepared to face the risks involved in working them, whether in Butte or in the Old Country itself. Yet, had it not been for high war-time copper prices, the Granite Mountain disaster might have led to a permanent decline of mining at Butte, just as the Levant disaster, a few years later, coinciding as it did with slumping tin prices, practically killed the St Just mine.

Sudden disasters always distort perspectives; the immediate shock is paralysing. Yet even the Granite Mountain disaster killed less than one in a hundred of the miners employed in Montanan mines at that time, while nearly as many were dying every year in Butte from occupational diseases contracted in mining. The latter toll was continuous, and the fear of it drove many from the mines who would have laughed to scorn the idea that mining was an abnormally dangerous enterprise. The fatalistic Cousin Jack philosophy was that accidents could happen to anyone, and that meeting trouble half-way or worrying over what might never take place would not get a man anywhere; a man had to work to live, and the mines offered work, whatever its risks might be. Even in Cornwall about three in every thousand employed in the mines met with a fatal accident in the course of their employment every year.[45] By Old Country standards wages in Butte were high, and if the cost of living, too, was high, it was not excessively so, and the majority of working men in steady employment lived as comfortably, as long and with perhaps a few more luxuries as those who stayed in the Old Country or who had taken themselves off to try their luck on the Rand. Montana winters, compared to those of Cornwall, were hard, but Butte, by 1898, offered more in the way of creature comforts than the men of another questing generation found in the gold-fields of the Klondike, an El Dorado that seems to have drawn hardly any Cousin Jacks either from the Old Country or the new.

Montana, of course, was not quite the 'land of pilchards and cream', but there were times when pilchards, brought from Portleven, could be bought in Butte City for thirty cents a dozen.[46] Canned 'cream',

45. *W.B.* 25 September 1913.
46. *Butte Weekly Tribune*, 2 January 1897.

inferior to, but a passable substitute for, the clotted cream of Cornwall, cost ten cents a tin in 1893, and saffron was put up in small boxes selling at six for a dollar.[47] Tea was only twenty-five cents per pound, whereas in England at that time the cheapest tea was about thirty cents and the best nearly fifty. The treacle lovers from the Old Country might, in Butte, acquire a taste for maple syrup at a dollar a gallon, or for New Orleans molasses at seventy-five cents per gallon, while for a dollar and seventy-five cents the Montanan shopper could buy a five-gallon keg of 'Honey Drops Syrup'. Canned fruits and vegetables were far more common in Butte in the eighteen-nineties than they were in Cornwall, but whereas a dozen pounds of raisins cost a dollar in Butte, in Truro they were ninety cents; dried apples and apricots cost about twelve cents a pound in Cornwall while dried peaches were about fifteen to twenty cents per pound in Montana. Such prices, however, are mere samples, and it is more significant that in Butte, in 1893, the average daily wages of the working man represented half a hundredweight of sugar, or a hundredweight of the best quality flour, or fourteen pounds of tea; in Cornwall, many miners in 1893 made just enough in a day to buy fourteen pounds of sugar, less than a quarter hundredweight of flour, two pounds of the cheapest tea, and about seven pounds of corned beef, as against thirty pounds in Butte.

The gold strikes brought into Montana Territory a population of twenty thousand by the time of the 1870 census, and of these about a third were foreign-born. It is likely that a fair proportion of these were of Cornish birth; several of the Mineral Point 'boys' who were at Placerville in the Boise Valley, in 1862,[48] almost certainly drifted back east over the Montana line as soon as the news of the rich strikes there reached them; other Cousin Jacks were to come in from northern Michigan and the Mother Lode Country of California. In the fall of 1872, William Evans of Bodmin was publishing emigration propaganda on behalf of the Northern Pacific Railroad Company to attract settlers to Montana, claiming that it contained some of 'the richest mineral lands and the finest farming lands in the world,' calling it the potential 'paradise of miners', and claiming that it was a territory in which capitalists could invest with confident expectation of magnificent dividends. Evans went on to state that in Lewis and Clark County

47. This and the following prices in this paragraph are based on various advertisements in the Butte *Populist Tribune*, and in particular the issue of 23 December 1893, and the *West Briton*, and in particular its issue of 21 December 1893. The value of the dollar has been assumed to be 4s. 2d. at this time for rough comparative purposes.

48. *Mineral Point Tribune*, 25 March 1863.

there were six quartz mills in operation, and that miners there were receiving wages of six dollars per day with and seven dollars a day without board—wages which a Cornish miner at home did well to make in a full fortnight's work. There was the chance, too, for any individual to make a fortune.[49] While the majority of readers may have realised that a miner's prospects of striking a bonanza lode were small they could not deny that some men would be lucky, whereas for the ordinary working-man to make a quick fortune in Cornwall was practically unknown. Several years later, writing for no propagandist purpose, a Cornish miner[50] described how the poor Irish prospector, Thomas Cruse, found the Drumlummon Lode and sold it for a million and a half dollars, and while this writer, then in Helena, admitted that some men spent their lives prospecting and died poor, he asserted that 'hundreds of others' had made fortunes in a similar manner to Cruse in Montana.[51]

Propaganda like that which Evans put out on behalf of the Northern Pacific Railroad, despite the small headlines used by the local Cornish press in those days, struck the eye, but the casual reader could easily have missed a paragraph in July 1876, which baldly recorded one of the most sensational episodes in the history of this frontier. The *Cornish Telegraph*, under the heading 'Massacre by Indians' stated that

a scout, being the bearer of despatches, has arrived at Fort Ellis, Montana. He reports that two detachments of troops under General Custer and Major Rese attacked a body of two thousand five hundred Indians on the twenty-fifth of last month in a defile of the Little Horn. General Custer's command was overwhelmed and annihilated, himself being killed with sixteen of his officers and three hundred privates. Major Rese retreated with difficulty until he was joined by the reserves.[52]

It should, perhaps, be added that this brief paragraph, which overestimated the white casualties by seventy or eighty, and underestimated the number of hostile Sioux Indians by fully fifty per cent, was published in west Cornwall only four days after the *New North West*, published in Butte City, described the disaster which had befallen the flamboyant American cavalry general in a lengthy account of over three columns.

Prospectors and miners came or attempted to come in, while the Federal authorities made some ineffectual attempts either to keep them out or to expel them from the Indian territory if they crossed the boundary line. The task of the latter would have been impossible even if the Federal forces had been efficiently commanded by officers who were

49. *W.B.*, 21 October 1872. 50. Joseph B. Trevarthen, see above, p. 240, n26.
51. *W.B.*, 3 September 1888. 52. *Cornish Telegraph*, 11 July 1876.

conscientious in the performance of their duty. The consequence, aggravated by the destruction of the buffalo herds, was war with the redskins who, believing that the whites were bent on stealing their lands and their means of living and would never keep the promises they had made by the Fort Laramie or any other treaty, were now convinced that they must either fight or die. Gold-hunters and hide-hunters who had trespassed on Indian Territory and had been expelled by troops had no liking and even less respect for the Federal Government, while the troops themselves were but the ghost of the proud Union army of a decade before, and were unable to guard the boundary and keep the whites from encroaching on Indian lands, and equally incapable of protecting white lives from retaliatory Indian raids, and in some instances, as Custer's débâcle showed, of keeping their own scalps.

Custer's boastful announcement that his party had found gold in the Black Hills came just at the right—or wrong—time. Older mining regions had been hit by the financial crash of the previous fall. Miners making what they regarded as bare wages, and, in many cases, hardly enough to provide them with board and shelter, were ready to move to any new El Dorado or Potosi, whether it was in Indian country or not. Newspapermen seeking a story of hope and promise as a change from the dreary records of economic depression boosted the Black Hills all over the Pacific Coast and beyond. By the fall of 1874, within three months of Custer's 'strike', Californian newspapers were reporting that miners had run into trouble with the Indians, who, in the Black Hills, were 'not too gentle to shoot and scalp a miner'.[53] Before the battle of the Little Big Horn, on 25 June 1876, the usual contradictory stories of wealth and the reverse had come out of the Black Hills. Some made rich finds and, locating promising claims, wrote to friends to come and join them;[54] others reckoned that they had come to a god-forsaken, forlorn country and wished themselves elsewhere, while the accounts which they wrote home spoke of men not being able to earn a dollar a day and of able-bodied miners begging for bread.[55] Men from the Californian mining camps found it far harder to adapt themselves to wintry conditions in the Black Hills than those who had spent some time in the Montana and Colorado mines. The surge of immigrant miners into Dakota was perhaps greatest from Montana at first, and in

53. *Grass Valley Union*, 30 October 1874.
54. *New North West*, 10 March 1876 (Editorial).
55. *Grass Valley Union*, 28 April 1876.

that western bordering Territory far-sighted citizens considered that the more gold strikes were made in Dakota the quicker would the 'eastern door' to Montana be opened and safeguarded by 'a strong line of camps'.[56] Prospectors and miners who came a little late to the Black Hills and found the best claims taken up, were active in the Yellowstone and Big Horn countries before the Custer expedition came to grief in that district.[57] Early in 1876 parties from Utah and Colorado were coming up to Custer and Hill Cities through Cheyenne, the nearest Union Pacific Railroad depot to the newly discovered gold regions, and all the talk on the trains from New York to Omaha was said to be of 'the wonderful Black Hills'.[58] Hard on this, however, came accounts that for fifty men in the Hills making twenty dollars a day, a thousand could only scratch up a dollar each a day, and that the Indians were so hostile and well armed that a white man's life in the mining area was not worth five weeks' purchase.[59]

The formidable Indian force disintegrated after the battle of the Little Big Horn, although many groups of miners were attacked by hostile bands later in the year. Custer's fate taught caution to the soldiers in the region, while after the battle of Slim Buttes on 9 September 1876, the Federal authorities were able to force the Indians to cede the Black Hills to the whites, and more or less to confine the surviving Indians within narrowly circumscribed reservations as remote as possible from locations that attracted miners. It is possible that the fear of Indian attacks led the miners who came to the Black Hills to congregate in the rich strike areas in greater numbers for mutual protection than they otherwise would have done, and this was the cause of very early complaints of there being too many men looking for too little gold in such places, and accounted, too, for the reports of the best deposits being quickly exhausted. On 16 August 1876, W. P. Wheeler was writing from Deadwood City to his friend Alex Carmichael back in West Montana that there were then eight thousand men in a camp which might support two or three thousand, and he reckoned that half the people then in Deadwood would be forced to leave before winter began.[60] On the other hand, particularly around Custer and Hill City, the rugged nature of the country and the extremely hard rock led many prospectors to give up in despair after much hard hammering and drilling, and to seek elsewhere; every report of a new, more accessible

---

56. *New North West*, 10 March 1876.
58. Ibid., 17 March 1876.
60. Ibid., 15 September 1876.

57. Ibid., 10 and 17 March 1876.
59. Ibid., 9 and 16 June 1876.

strike provoked local 'stampedes', and the gold-seekers of the Black Hills in 1875 and 1876 were wandering about all over the place in a manner as wayward as that of the hide-hunters seeking the rapidly dwindling herds of buffalo, animals whose movements were as unpredictable as the wind.[61]

The great discovery in Dakota was that made in Deadwood Gulch in the fall of 1875, news of which was said to have almost depopulated Custer City overnight. Rich placer deposits were first found and were rapidly worked, but by the spring of 1877 mining experts were confident that the district contained some of the richest auriferous quartz ever discovered in the world. The Federal Government formally took over the region on 1 April 1877; till then, the only 'law' was the miners' meeting, and when Thomas H. White, a Cornishman from Pensilva, came in from Nevada early in 1877, he found posted up round shaft and adit entrances notices bearing such legends as—'Don't trespass, or you'll be shot.' or 'No admittance! A Warning—There is a shotgun set in the tunnel.' On the trails, no great distance from the diggings, hostile Indians still lurked, ready to kill a white man if they got the least chance.[62]

Thomas White, who had been run out of one Nevada mine for warning interested English capitalists that it was working a line rock deposit and not a true fissure vein as the men seeking to sell it claimed,[63] had come to Dakota prepared to find the new diggings a humbug. He quickly changed his mind, and advised English mine investors to come in and help develop the area. At the time of his arrival, however, the claims were practically all owned by poor men who lacked the necessary capital to develop them, and were using horse-whims to crush orestone, besides other simple, cheap and wasteful machinery to exploit their claims. These small operators, too, according to White, had run into difficulty with spotted ores, with too much water in the lower levels and shafts and not enough at the surface, and had suffered from long-lying snowfalls in winter. Provisions, through the difficulties of communications with the outside world, were dear. White warned Cornishmen not to come out to the Dakotas from the Old Country unless they could bring enough money to keep them for some time after their arrival. It was fairly easy to come by railway as far as Cheyenne or as

61. *New North West*, 14 April 1876, letter of Ben F. Workman, dated Spring Creek, Dakota Territory, 5 March 1876.
62. *Mining Journal* (London), 17 March 1877.
63. Ibid., 5 May 1877; letter of White dated Deadwood, 3 April 1877.

Sidney, in Nebraska, but from these places it was a long and costly stage journey to the Dakota diggings. The five-day stage journey from Cheyenne, which ran twice a week (that from Sidney being rather less regular though the distance was shorter)[64] cost fifty dollars; luggage in excess of twenty-five pounds weight had to be paid for at the rate of twenty cents per pound, while every meal along the route cost a dollar. When they arrived in Deadwood, they would find flour costing twelve cents a pound, butter from fifty to sixty cents, beef from fifteen to twenty-five cents, and other prices equally high. So, in Deadwood, in the early part of 1877, a man needed high wages to live—three or four dollars a day might just suffice, but it would only provide a man with pretty tough fare, while claimless men or those who could not work their claims when they lay buried deep in snow had undoubtedly had a very hard winter.[65]

Still, the miners came in. A Grass Valley Cousin Jack, James Gluyas, who came to superintend the Caledonia Quartz Mine in the late summer of 1879, and was not over-struck by his first impressions of the 'Hills', admitted that it was the 'biggest quartz mining country' he ever saw, but reckoned that most of the miners were 'eastern people'. Flour was still ten cents a pound, but wages were only $3.50 per day, while hundreds were out of work. And as for the climate, it was

peculiar. There seems to be no Summer. The wind blows as it blows at Virginia City, but the climate is not as good. Some days we have a burning sun, with a cold and chilling wind, and at the same time freezing. The weather will sometimes change four or five times a day—with rain, snow, freezing, hot, and sultry, all in one day. Of course this makes sickness, and every boarding house contains several sick men.[66]

Nevertheless, the lure of gold, even the hope of finding work, brought men to the Black Hills. By 1877 the Hidden Treasure, Father de Smet, and Homestake mines had gained renown, and it was not long before capitalists on the scene, like George Hearst, were on their way to making vast fortunes. The 'poor men's diggings' of 1875–76 had not lasted long, and it could not be claimed that the ordinary wage-earning miner was any better off in the Hills than he would have been anywhere else on the 'Coast'. A number of Cousin Jacks came, and it is interesting that it was in Dakota that a Cornish mine name was given

64. White reckoned that Sidney was 230 miles from Deadwood, and the cost of the stage journey $35. Cheyenne was 300 miles from Deadwood.
65. *Mining Journal* (London), 17 March 1877.
66. *Grass Valley Union*, 26 November 1879; letter of James Gluyas to William Henry Mitchell, dated Central City, Dakota, 12 November 1879.

to a mineral lode—the Caradon, perhaps named by White after his home district, a lode or ledge which he said was as wide as the Homestake, wider than the Golden Star, but not so wide as the Star.[67] Not all the money immigrant miners made in Dakota went in board and lodge, or was squandered in saloons, or lost in the stage-robberies for which Deadwood gained notoriety second only to the Boise Basin; some found its way back to Cornwall, and it is likely that a lucky Cousin Jack returning from the Black Hills was responsible for calling a moorland tenement on the road from Ding Dong to Morva, hard by the Men-an-tol, Dakota Farm.

A decade after the Black Hills gold discoveries, rumours began that tin had been discovered in Dakota, and this aroused more interest in Cornwall than the reports of the gold strikes in that region had done, although many were sceptical and recalled the fiasco of the Missouri 'tin mountain' of 1867.[68] Traces of what appeared to be stanniferous rock were found at Harney Peak in the early eighteen-eighties, and by 1886 American newspapers were hailing—prematurely—mining developments in Dakota which would make the States independent of foreign supplies of tin and 'astonish John Bull (and, presumably, Cousin Jack) in his own bailiwick'.[69] Samples of the Harney Peak orestuff was brought to the Chyandour smelting works near Penzance for analysis, and were reported to be inferior to the ores of Dolcoath, East Pool, Wheal Agar, and West Francis.[70] Since there had already been at least one case of an American 'mine' being 'salted' with Cornish tin, investors in Britain were warned to be wary of American tin-mining promotions, while Cornish mine adventurers were confident that they had, as yet, nothing to fear from Dakotan competition.[71] The Etta Mining Company, organised in 1883, had by the fall of 1886 spent three-quarters of a million dollars in attempts to work the Dakota tin deposits with apparently no returns worth the name,[72] but two or three Cornish mining experts went over to investigate the property of this Company in 1886, and one of them, Captain Gilbert of Mellanear, went out a second time in October 1886.[73] Early the following year some

67. The Star Ledge, in early 1877, was averaging 150 feet in width, the Caradon and Homestake about 100 feet each, and the Golden Star 50 feet. (Letter of T. H. White, dated Deadwood, 10 May 1877, *Mining Journal* (London), 9 June 1877.)
68. See above, pp. 161–63.
69. *Cornish Telegraph*, 15 July 1886, quoting *Iron*, n.d.; *W.B.*, 15 July 1886.
70. Ibid., 22 July 1886.   71. Ibid., 22 July 1886.
72. Ibid., 20 October and 27 November 1884 and 21 October 1886.
73. Ibid., 18 October 1886.

Cornish smelting firms sent out Captain John Curtis of Kennegie to investigate, and he came back in July declaring that 'Dakota is all bosh!' and that the Etta tin lode was only a hundred and sixty feet long, with the ore in it too poor to be worth working.[74] Despite the opinion of Cornish experts, who must, however, have been somewhat prejudiced, English capitalists were induced to buy the Etta properties for two million pounds sterling in the summer of 1887, after a Dolcoath captain, Davies, made assays of the Harney Peak ores which indicated that they might be payable but which also revealed that they contained impurities which were not associated with either Cornish or Australian tin ores.[75] Even the confirmation by Richard Pearce that cassiterite ores existed in the Black Hills did not imply that such ores could be economically worked, and Thomas White came home to the Old Country in 1889 declaring that the English capitalists had bought a worthless property.[76]

Conflicting 'expert' opinions, suspicions of financial shark company promotion, the reluctance of speculators to admit that they had made a mistake and had persuaded others to put their money in a losing venture, all helped to keep Dakota tin in the news for years. The English Harney Peak investors got Captain Josiah Thomas of Dolcoath, the top-ranking tin miner of the day, to go out and investigate the mining property in the summer of 1892. His report was honest and extremely cautious; tin ore of a sort existed, but most of it seemed to be, in his opinion, too poor to be profitably worked. He suspected that the alleged lodes would deteriorate and pinch out instead of improve in depth, a feature which in the experience of Cornish tin-mines was a symptom of imminent failure. Finally, he recommended the sinking of deep shafts to prove—or disprove—the value of the property.[77] The investors, however, felt that they had spent enough, and with the onset of the great depression a few months later the mine was suspended, although litigation and talk of attempts to rework it were to go on for many years. Perhaps a little prematurely but justly the Denver *Mining Industry and Tradesman* in February 1893, pronounced the epitaph on the Dakotan challenge to the Cornish tin-mining industry as 'one of the most gigantic pieces of mingled mining fake and foolishness this country has ever seen'.[78]

74. *Cornishman*, 7 July 1887; *W.B.*, 11 July 1887.
75. Ibid., 11 August and 20 December 1888, and 4 July 1889.
76. Ibid., 9 January 1890.
77. Ibid., 27 October 1892; *Cornish Telegraph*, 3 November 1892.
78. Quoted in *W.B.*, 2 March 1893.

Rumours of the tin finds in Dakota had affected the old Cornish industry only in so far as they disturbed the sensitive tin markets. The real danger to Cornwall lay in the East Indies whose soaring, cheap production of alluvial deposits was slowly crushing tin mining in the Old Country to death, although the end was not yet to be. As for the other metals of Dakota, gold and silver, they gave employment to many, including some Cousin Jacks, and their rapid exploitation in a few short years, speedily transformed what had been Indian hunting grounds into yet another State of the Union, to be admitted as South Dakota to the Federal sisterhood within a decade and a half of Custer's death. Cornish immigrants never came in such numbers to Dakota as they did to Michigan, California, Colorado, and Montana. The mining history of Dakota only began in 1875, and Dakota then was only another American region that offered opportunities to the adventurous, that had to compete for mining immigrants not only with Montana, Colorado, and Arizona, but with the diamond-fields of South Africa, and the tin workings of Australia. However, by that time there had been enough Cousin Jacks in the older mining camps of the States to develop and teach others the techniques of hard-rock mining that were needed to exploit the auriferous quartz deposits of Deadwood and Lead. The Cousin Jacks could hardly be blamed if their skills did not include the alchemy needed to transform tantalite into cassiterite and thereby accomplish the miracle of making the profitable working of Harney Peak 'tin ores' possible.

10 A home from home

Whether it was in the Black Hills of Dakota or in the Californian Sierras, in Bisbee or in Butte, in the lead region of the Upper Mississippi or in the copper and iron ranges of northern Michigan, Cornish immigrants, brought many of their own customs with them. Changes inevitably occurred and the Cousin Jacks, like all other immigrants, gradually became more and more American. Both the Old Country and the New, since the times of George III, have been subject to forces which have eroded provincial traits. Nearly six generations have passed since the first Cornish emigration to Wisconsin, and even in the later mining camp of Butte can now be met the third and fourth generation descended from immigrant stock. The American environment has inevitably influenced and modified ways of life. In Cornwall historical developments have wrought such changes that the old emigrants would never recognise it were they to come back, while their grandchildren and great-grandchildren who have only heard of the Old Country from oft-repeated and oft-times transmuted reminiscences of their forebears would find the reality even more unrecognisable. Even the last generation of emigrants, many of whom joined in the migration from the Upper Michigan Peninsula or who went directly to Detroit at the end of the First World War know and remember only a Cornwall of rough muddy or dusty roads, of squires and parsons dominating a caste-ridden society, and of woodland valleys and rough grazings thronged with rabbits and pheasants. If they return, they find, to their amazement, all the roads save farm lanes asphalted, the land-owning squires death-dutied to the point of extinction, Anglican parsons servantless and struggling to live on stipends that have lagged far behind the risen cost of living, and nearly all the game destroyed by disease or poachers. Farm mechanisation has transformed the rural landscape; hedgerows have been swept away and the numerous small paddocks of less than half a century ago merged into prairie-like fields for more economical mechanical husbandry; horses have practically disappeared from the farms and roads save for a few kept by 'dude-ranch' riding schools. Deforestation in the First World War has left many of the former woodlands of Cornwall to be covered by scrub and brush which is as rough as any second or third growth in the ravaged forest lands of northern Michigan. Some of the outbuildings and grounds around the Cornish manor-houses and parsonages might make a Southerner wonder whether Sherman had not gone along that way, too, and the gaunt minestacks and ruinous engine houses of long 'knacked bals' in the mining districts enhance such an impression.

If there have been numerous changes, there have also been many survivals both in the Old Country and the New. Any appraisal of social history must take account of three fundamental human needs—food, shelter and clothing. Changes have occurred in all three and they have all been profoundly affected by environment. To every place they went in America the Cousin Jennies brought the traditional Cornish pasty and saffron cake, and to this day they can be found in Grass Valley, Butte, Upper Michigan, Mineral Point, and elsewhere. Humorous tales of Cousin Jack's fondness for his pasty have served to make it known in districts far beyond those wherein they settled, just as in Britain the tourist traffic of recent decades has helped to popularise it far to the east and north of the Tamar boundary. The later Finn and Greek immigrants to northern Michigan, however, have proved more adept at 'crimping' over the pasty-crust than many English visitors to Cornwall, and over-repetition of the old story that the Cornish would put anything into a pasty, even the devil himself if he ventured across the Tamar, may be held responsible for 'up-country' English mistakes in mincing the meat and other ingredients instead of chopping and slicing them, in using inferior beef instead of steak, and even in attempting to cook the contents first and then wrapping the pastry around them, instead of baking crust and contents together.

Whether it is true or not that the pasty was 'invented' by the wives of Cornish miners as a good luncheon for their men to carry to work with them, it was, and is, certainly a handy meal in every sense of the word. It was quickly made, less than an hour being needed to prepare the ingredients and cook it; a man could easily carry it in his jacket pocket; well-wrapped in paper or cloth it would keep warm for a fair time, and, in any case, could easily be 'warmed up again' by anyone who liked his pasty that way. Whether eaten hot or cold, a pasty was an all-the-year-round meal, although some sorts were more common at certain season than others—leeks in the early spring, onion in the fall, swede turnips in the colder months of the year, apple from Michaelmas on perhaps to spring gooseberries in high summer, pork and rabbit when there was not 'an R in the month', and so forth. The all-season standby, however, was the steak and potato—beef an' tater—which really was a substantial meal, and it was this type that became best known in the mining regions of America. In Butte they went and still go by the name of 'a letter from home'.

Real letters from home brought to Cornish emigrant families little packages of saffron to make the other favourite Cornish delicacy—saffron cake, but wherever the Cornish settled in any numbers local stores soon started selling saffron. 'As dear as saffron' was a Cornish saying, and it was as dear in America as in Cornwall. In the Old Country it was customary to sell it by the drachm, and in Butte in the eighteen-nineties it was apparently being sold in half-ounce boxes, costing about eighteen cents each, which was about three to five cents dearer than in Cornwall. Still, eight drachms of saffron would flavour and colour several cakes, and wages in Butte were high enough for a miner to have a piece of cake of a saffron bun whenever he felt like a 'crib' or a snack. Another Cornish favourite was the so-called 'heavy cake', but this did not become so popular in America, nor, for that matter, did the Cornish 'seedy cake', perhaps because such cakes looked dull and colourless besides saffron and the traditional cakes of other immigrant races in America.

Despite some occasional imports of Cornish pilchards by Butte store-keepers marinated pilchards and mackerel, and still more star-gazy pie[1] rarely if ever made their appearance on the tables of the immigrants who, however, developed a taste for American freshwater river and lake fish. Since in Cornwall squab and rook pie had been regarded as luxuries, the Cornish immigrants, many of whom had regarded poaching on squirarchal preserves and warrens as no sin, helped exterminate the vast flocks of passenger pigeons which once darkened American skies.

Several factors account for the rarity of Cornish clotted or scalded cream in America. In Cornwall it had been too expensive for labouring miners, who were more likely to use 'sky-blue' skimmed milk than fresh milk and cream in their homes. Only in Wisconsin, in the American regions to which the Cornish came in considerable numbers, did dairy farming develop very quickly. Even as late as 1850, more than a

1. A fish pie, usually pilchard sometimes mackerel, made with the fishes' heads protruding up through the pie crust.

dozen years after a fair number of Cousin Jacks had come to Mineral Point, nine Cornish farmers,[2] owning or occupying between them 480 acres of 'improved' and 750 of 'unimproved' land, together owned only 25 milch cows, 53 other cattle, and 29 horses; some of them produced butter, and it is possible that a few of the 69 pigs were milk fed, although most of them probably roamed about the 'unimproved' lands rooting for a living. Thus the farmers seemed to be concentrating on raising corn, wheat, oats, hay and potatoes, and their aggregate butter production was returned as being only 1,400 pounds a year; this return, however, cannot be regarded as very trustworthy since the seven farmers listing butter among their produce used the 'round figures' 50, 200, 100, 200, 50, 600 and 100 respectively, suggesting that they either used themselves or sold to others what they roughly estimated to be between a single and a dozen pounds of butter a week. Furthermore, the farmer who produced 600 pounds of butter, William Rule, only had a solitary milch cow at the time of the census return, while the farmer with most cattle, Robert Rule who possessed four milch cows and six other head of cattle returned no butter at all. Even more strange are discrepancies in stock valuations in these early Wisconsin stock returns. James Treloar put down $75 as the value of his solitary milch cow and two horses. James Bennett returned a valuation of a mere $10 for six horses, two milch cows and sixteen other cattle, which suggests that the compiler of the returns slipped badly, since $170 was the aggregate valuation of John Rule's roughly equal stock holding of seven horses, two cows, four other cattle and ten swine. Perhaps Bennett was continuing the old Cornish practice of renting livestock, but since he was the only one of the nine farmers to return a wool clip—a small one of forty-five pounds weight—it is possible that sheep were being returned under the heading of 'other cattle'. All that is certain is that, by 1850, Cornish farmers in Wisconsin were pursuing mixed farming practices similar to those of the homeland. They were producing some butter although, apparently, no great amount, and the average production of rather over a pound a week from each milch cow indicates that they only kept enough milking animals to supply the needs of their own families and, sometimes, neighbours, while pigs and young beef-stock consumed the surplus. Only one of the nine farmers, Elizabeth Goldsworthy, was still using 'working oxen', which by 1850, too, were vanishing from the arable lands of the Old Country.

2. Viz. Sarah Jenkins, James Bennett, John Terrill, Elizabeth Goldsworthy, James Rule, John Rule, William Rule, Robert Rule and James Treloar (1850 Census).

In Upper Michigan there were similar indications of mixed farming in early State Census returns, but a higher proportion of horses and working oxen indicated the employment of draught animals directly or indirectly about the mines. In 1854 Houghton County had 98 horses, 11 mules and 59 working oxen, but only 20 milch cows, 37 swine and 12 sheep, the total population of the county then being 2,868; ten years later, with a total population of 8,225, there were still only 59 milch cows. Neighbouring Ontonagon County in 1854 had 46 milch cows and a population of 3,624, and in 1864 it had 105 milch cows and a population of 5,406. Such figures show that milk was scarce in the Upper Peninsula, and suggested that if any Cousin Jack wanted cream in those days he had better go home to Cornwall. In the high altitude mountain mining camps of the Pacific Coast milch cattle were probably even scarcer in the pioneering days, and cream rarely if ever seen.

Many Cornish immigrants to America brought with them a fair knowledge of gardening as well as of mining. Single men who predominated in the pioneer and in the less permanent mining camps are hardly likely to have troubled about gardens, but family men in the more enduring settlements often cultivated plots of land round their new homes. In Grass Valley, lying fairly low in the Sierra foothills and enjoying an equable climate, many kinds of vegetables, flowers and fruits could be and were grown. John Rodda of Penzance established a flourishing nursery business in Grass Valley in its early days; retiring in 1868 and finding the Old Country less to his liking than the new, he came back to Grass Valley in 1874.[3] Rodda specialised in shade trees and gooseberries, a fruit which must be reckoned as a prime favourite among Cornish people. Rodda's garden, covering about seven acres out on Pike Flat, was very like those around his native Mount's Bay, but in Grass Valley he was able to raise outdoor grapes in luscious profusion besides currants and gooseberries, cabbages and beets. When he wanted to extend his garden, Rodda must have thought that Grass Valley was just another West Penwith, for he had to blast away large rocks, which a local journalist remarked might be romantic and ornamental, but were 'impediments in the way of his crops'.[4]

Still there were many marked differences between the Old Country and the new. Cornish immigrants missed the gorse and heather of Cornwall, although around the city limits of Butte were grass and stunted shrub which looked something like the more desolate Cornish moor-

3. *Grass Valley Union*, 11 December 1866 and 5 September 1874.
4. Ibid., 12 February 1867.

lands after a droughty summer. Cornwall, however, had nothing that bore much resemblance to the sage and creosote brush of Nevada, the manzanita of the Californian Sierras, and certainly nothing like the fantastic cacti found around some Arizonan mining camps. The predominantly coniferous forests and woodlands of the mountain camps and of northern Michigan were different from the oak, beech, ash, elm, hazel and willow of Cornwall. In America the immigrant found many flowers which he had never known at home, but yet in Upper Michigan spring brought as luxuriant a crop of golden gleaming dandelions as ever was seen in Cornwall, while in the Mother Lode country of California the dock was as pernicious a weed as it ever was in any Cornish garden or field.

The natural differences which most struck and affected the immigrant were climatic ones. Cornish immigrants had grown up in a region which hardly knew extremes of summer heat or winter cold. During a summer heat-wave the temperatures might creep well up into the eighties for an hour or so in the early afternoon; in winter a 'cold snap' with freezing temperatures rarely lasted more than a few days, perhaps a fortnight at the longest. Rainfall was fairly evenly distributed through the year, with a tendency for the wettest months to be in the early winter and the driest in early summer. Yet a so-called mild and equable climate could have astonishing vagaries. Flurries of snow might fall any month from October to May; thunder might be accompanied by devastating hailstorms in the August harvest month; an almost summer-like and dry February in 1891 was followed in the second week of March by the 'great Blizzard' which, in a region entirely unprepared, caused heavy losses in the flocks of sheep. Gales reached near hurricane force at times, and only the hilly nature of the county saved it from flood devastation following torrential downpours of rain, though it is noteworthy that the worst mining disaster in Cornish history took place in July 1846, when a cloudburst over the badly drained valley lying between Mitchell and Newlyn East caused a torrent to pour down the shaft of East Wheal Rose lead-mine and drown thirty-one men.[5]

If Cornwall had no climate but simply 'weather', much the same could be said of many of the mining camps to which Cousin Jacks came in America. All that can be said without qualification is that west of the Rockies the length of winter increased with latitude, that in the area between the Rockies and the Sierras altitude and slope of the land were aggravating or ameliorating factors, and that in the higher Sierras

5. *W.B.*, 17 July 1846; the disaster occurred on July 9th.

of California winters could be extremely cold. Disastrously high winds and torrential downpours were more prevalent over the whole continental area than in the small English county, but any small mining district in America over the years suffered less from exceptional meteorological cataclysms than did Cornwall. In the arid American Southwest, men soon learnt the risk of summer flash-floods in valleys and gulches that were dry for the rest of the year. It took a year or two, but men developed a weather-sense up in Montana and Michigan which enabled them to detect the likelihood of a blizzard just as the weather-wise in Cornwall could with fair accuracy predict an approaching storm of rain or wind.

American winters, however, drove many immigrants home again or to more equable climes. Possibly climate should be reckoned as a potent second goad to those Wisconsin Cornish who trekked away to the Pacific Coast in 1849 and succeeding years. From mid November to mid March, fully a third of the year, temperatures averaged below freezing point in the district around Mineral Point. Further north, in the Upper Peninsula of Michigan, temperatures usually fell to freezing point by the end of October and rarely got above them till late in April while all navigation on the Lakes was impossible throughout January, February and March, and might be seriously limited in both December and April. Summers in both these regions were warmer than in Cornwall, though not excessively so; at Houghton the average July temperature is around 67°F., and in the Mineral Point district it is about 73°, whereas the Cornish average is around 60°. In both the lead-mining region of the Upper Mississippi and in northern Michigan springs and autumns are much shorter than those of Cornwall. First experiences of prolonged and intensely cold winters told hard on many immigrants who were thin-blooded and, in many cases, through their mining avocation, less physically robust than they might have been, although low humidity made freezing and even sub-zero temperatures in many parts of America far more endurable than high humidity made temperatures eight or ten degrees above freezing point in Cornwall. This, however, served to increase the risk of frost-bite which, unlike the chilblains that were so common in Cornwall in winter-time, could and often did maim and kill. In the high-altitude mining regions further west, winters as a rule were more extreme, while, in addition, low atmospheric pressure caused many severe cases of mountain sickness and, occasionally, worse ailments among those whose blood could not endure the rarified air.

These climatic differences, however, did not cause any marked difference between the clothing worn in Cornwall and that in American districts. Furs and buffalo robes were cheap and fairly ready to hand in many of the pioneer camps for really cold weather, and in both countries in those times the tendency in summer or when working in heat was to strip off outer layers of clothing rather than change into light-weight materials. Both the well-off Cornishwomen and the affluent in America followed Parisian fashions, and dress-conscious less well-to-do people sought to copy them. In America, however, the wage-labourer in fairly steady employment could allow from a sixth to a fifth of his income for clothing his family, which was far more than a working miner in Cornwall could afford. In about 1870 a working miner's family in Grass Valley could, without stinting the larder, spend five or six dollars a week on clothes;[6] in Cornwall they might have to 'tighten their belts' to spend the equivalent of one dollar, which even allowing for higher prices in California still would not buy more than twice the amount of clothing which it could in Grass Valley.

Shelter possibly took a larger part of a labourer's income in some American mining districts than rents did in Cornwall, where so many cottages were held on the three-life system.[7] In 1871 it was estimated that cottages costing from fifteen to twenty dollars a month rent could be built for five hundred dollars or less.[8] In some of the more permanent mining camps, especially in the copper regions of Northern Michigan and Arizona, housing was provided to their employees by the mining companies, and company houses were also to be found in some of the gold-mining towns of California.[9] At Mineral Point, however, the cottages of Shake-Rag-under-the-Hill were built by the miners themselves,[10] just as they were in so many instances in Cornwall, although in Wisconsin the building material was limestone with wooden shingles for roofing, while in the Old Country the cottages, in the main, were of cob or 'moorstone' granite with thatched roofs. Local building materials,

6. The budget of a family of four adults and two children in Grass Valley in 1870 showed a total expenditure of $755; groceries, including sugar, coffee, candles, flour, soap, potatoes, tea and butcher's bill came to $298·30; 'clothing and material' took $159·50; the next heaviest item was the purchase of a cow and calf for $50, and after that 'State, county, and town tax' $45; the separate items 'medicine' $13·75; 'nurse in sickness' $5; and 'Doctor's bill' $30, make up an aggregate of $38·75, which came next in amount to an item 'interest on money' $40·38, which can hardly be regarded as a usual family cost unless regarded as some sort of mortgage interest. The average wage-earner, working three hundred days a year earned about $900. (*Grass Valley Union*, 31 January 1871).

7. See above pp. 22–3.   8. *Grass Valley Union*, 4 February 1871.
9. See below pp. 287 ff.   10. See above pp. 46–7.

of course, influenced the pattern and design of American houses, and in many of the mining regions abundance of timber and the cheapness and labour-saving of building with that material led to most houses being constructed of wood. While the external appearance of such houses was quite different from that of miners' cottages in Cornwall, the interior furnishings and layout were very similar, with the exception of American-type stoves for the Cornish 'ranges' and 'clome ovens', and Yankee clocks, and, possibly, that symbol of well-earned American leisure, a rocking chair. In both countries there was much rough, home-made and make-shift furniture, with boxes and packing-cases serving as cupboards, tables and seats; such furnishings, in pre-mail-order, and pre-hire-purchase instalment buying days were not the mark of a physical frontier but only the frontier between economical living and moderate affluence. Mine carpenters, however, were quite adept at making serviceable furniture, but the woods available in America were of a different type from those most commonly used in making tables, chairs, mantels, cupboards, bureaux and the like in the Old Country.

Climatic conditions also influenced building design. Many of the American mining regions subject to hard winters with heavy snowfalls naturally adapted the roof-pitch best calculated to shed off snow. On the farms, to reduce the need of much walking about in sub-zero and blizzard conditions, was evolved the huge barn, built near the farm-house which it often both overshadowed and dwarfed. The smallish outbuildings and linhays of Cornwall might serve in favoured Californian climes, but not in Wisconsin.

Farm buildings, for the most part, were intended to be permanent structures and were solidly and substantially built. So were the frame dwelling houses, and some brick and stone houses that were, in the course of years, built in the more permanent mining locations. The ephemeral camps rarely got beyond the make-shift shelters that were hastily put up in hectic discovery and rush days, some of wood, some of canvas, many of anything that happened to be handy to give some protection from wind, rain, or snow. 'Coyote holes' grubbed in the earth sheltered some miners, others used brushwood to construct kraals an African Bushman would despise, while the adobe-using Indians of the Old Southwest would have scorned the first dwellings in Tombstone. Once the excitement subsided or the lodes petered out, the miners flocked elsewhere and such places soon became 'ghost towns'. Fires swept many of them away; winter storms and gales disposed of others in a few years. Some vanished without trace; in others a few buildings remained, and it was possible for some communal life to go on, though a Cousin Jack who knew them in their heyday would hardly recognise a Virginia City no bigger than Breage or Gunnislake, or a Rough and Ready not as populous as Pendeen or Commonmoor. And were he to return to the Californian camp of Red Dog all he would find would be a cemetery, surrounded by encroaching manzanita shrub, wherein many sleep well after their fitful feverish search for gold. It was not only men but towns on some mining frontiers whose lives were adventurous, often stormy, and in many instances tragically short. Nevada's Star City lasted barely five years; Red Dog, hardly existed a century ago and no longer exists today.

Through pulp literature, film, and television, western mining camps have come to be popularly associated with saloons, gambling hells and brothels, suggesting that they were poles apart from the Methodist chapel-dominated mining hamlets of Cornwall. In pioneer days there were much dissipation, sin, and lechery, but if the ore deposits lasted long enough and the camp was transformed into a town, respectability came in—rarely, be it added, without publicity as the haste of so many mining settlements to assume the name and the organisation of 'city' revealed. Even the vaunting of civic respectability often took the form of exaggerating the violence, lust, and crime prevalent in the pioneer days that had ended, presumably, a few weeks or a month or two before. Redemption, if it may be so called, came remarkably quickly. San Francisco's first great Committee of Vigilance started to clean up that city within a year of the beginning of its career as the gateway to the gold regions; the notorious Plummer gang were hanged in Bannack within eighteen months of the first gold discovery there; the 'law-enforcement' reign of the Earps in Tombstone was already past history four years after the first beginnings of that Arizonan mining camp. In the early days of any diggings the overwhelming predominance of men in the population inevitably caused social problems, but it could be questioned whether professional prostitutes were any more numerous than those 'camp followers' who accompanied armies on the march, and the pioneer miners might justly be compared with an army of occupation subjugating a hostile frontier country.

The purity of the Old Country and its reverse in the new have, however, been grossly overdrawn. The fathers of the Cousin Jacks who came to Wisconsin, Michigan, and California in the pioneer days, if not the immigrants themselves, had been smugglers and poachers in the Old Country, and long after the beginnings of the great migration to America, many living in Cornish coastal districts reckoned it no crime to salvage for their own use the cargoes of wrecked vessels. While there has been much written of the strength of Methodist and other dissenting sects in early Victorian Cornwall, it is unlikely that half the total population of the Old Country attended the services of any religious denomination at all, either regularly or very devoutly in the eighteen-forties.[11] Many an editor on the Pacific Coast would have welcomed such copy as any Cornwall County Assizes twice a year or Quarter Sessions every three months provided Old Country newspaper reporters, while even the minor cases dealt with summarily by the local magistrates often could be classed as 'news'. Homicide, it is true, was rare in Cornwall and rather more prevalent in the New World, but this was due to a number of factors. The so-called 'code of honour' disappeared earlier in England than it did in the American South whence there was a considerable exodus of defeated and impoverished younger adventurous men after 1865. In the pioneer mining camps men had to be prepared to defend themselves or go under since the institution of law and order lagged behind settlement—though not so long as is often supposed. Above all, the law of the land was different in the two countries—on the one side the English game laws restricting the keeping of firearms by the commonalty of men, on the other the Constitutional Amendment in 1791 that the right of the people to keep and bear arms should in no way be infringed. The sensational record of Idaho and the Black Hills in numbers and frequency of stage-holdups in the seventies and later, was due mainly to the coincidence of both temptation and opportunity to rob richly laden vehicles on lonely trails; all that is open to doubt is whether the bandits of the Deadwood and Boise districts should be regarded as latter-day exemplars of the south-west British exemplars of the 'horrid career' of the highwayman commemorated in the Devon ballad of *Widdecombe Fair*, or as the forerunners of the automobile-riding payroll and other smash-and-grab robbers of the sixth and seventh decades of the twentieth century.

In the early eighteen-seventies, Redruth was the scene of two violent crimes that had a distinctly American interest and flavour. In April 1870, there was a fatal stabbing affray in the Cornish mining town, and the fact that the killer had been in California evoked critical comment on Western ways. The Assizes found the culprit not guilty of murder but guilty of manslaughter, and he was promptly sentenced not to the three or four years imprisonment which he might have got in a

11. W. Francis, *Gwennap a Descriptive Poem*, (1845).

Californian court but to a life sentence.[12] Another fatal knifing brawl took place in Redruth six months later, which led one newspaper correspondent to comment that there was little excitement in Redruth, for 'we are getting used to such occurrences; I almost fancy myself in New York or San Francisco'.[13] The editor of the same local newspaper demanded that the Assize judge took action 'to check this infamous trans-Atlantic habit'.[14] The forces of law and order obliged, for although the coroner's inquest jury had found a verdict of manslaughter only, the killer was committed to the Assizes on a charge of murder and was sentenced to twenty years penal servitude.[15]

Cousin Jacks in Upper Michigan, Gilpin County, Virginia City, and Butte may have been reminded of their quarrels and brawls with the Irish when, in April 1882, Camborne was the scene of 'disturbances' which indicated that a Cornish mining town could erupt in a spasm of lawless violence that would have made news anywhere on the Pacific Coast—even in Pioche or Tombstone. The trouble grew out of a drunken affray in which a youth had been brutally handled by two young Irishmen. When the culprits were tried at the local sessions they were sentenced to hard labour for six and eight weeks respectively. The crowd which had thronged the courtroom thought these sentences far too lenient, surged around the prisoners as they were led away, and had it not been for the ten or dozen policemen protecting the two Irishmen they might well have been lynched. The men then turned their attention to one of the defence witnesses who had recklessly offered to fight anyone, and after beating him up pretty badly, raised the cry 'Drive the Irish out of Camborne'! Some Irish houses were wrecked, and the Hibernians took flight; the Roman Catholic church was badly damaged, and only the threat of using firearms by the owner saved from sacking the house of a mine purser who had given work to Irishmen in several of the mines with which he was associated. The cry 'Give them Camborne!' became notorious in consequence of this riot when it was raised against the Irish. With the arrival of nearly a hundred police in the town from outside, the rioting ceased. Undoubtedly religious prejudices and the fear of cheap Irish labour at a time when the crest of a moderate boom was on the point of collapsing were contributory factors in these troubles, but the whole incident, together with some earlier Camborne disturbances in 1873, was primarily a manifestation

12. *Cornish Telegraph*, 4 May and 10 August 1870.
13. Ibid., 9 November 1870.   14. Ibid., 9 November 1780.
15. Ibid., 16 November 1870 and 22 March 1871.

of that popular dissatisfaction with orthodox conventional processes of law and order which had led to the formation of Committees of Vigilance in so many pioneer American mining camps.[16]

Crime and sensations being news has obscured the opposite facets of law-abidingness and respectability in both the Old Country and the new. As early as 1866 a Grass Valley editor claimed that any stranger coming into that town on a Sunday would be struck by its religiousness, with church bells ringing and people soberly going to their chosen places of worship, while any drunken man causing a disturbance in the foothill mining town could be sure of being promptly arrested and fined $18.50.[17] Two days later he had to report two cases of assault and battery, but three weeks afterwards, the same editor, William H. Miller, was asserting on 28 June 1866, that

> Grass Valley is becoming much too moral to suit us; not that we are opposed to morality, but it interferes sadly with our business. For three weeks we haven't had an item from one of the Justice's Courts. To be sure, we have plenty of items about parties, festivals, balls, and all that sort of thing. But the readers want sensation items. Bloody murder, horrible suicide, elopement, *crim.con.*, these are what a goodly number want to have as headings to our items. We have searched in vain for such; we have wandered from one end of town to the other, and all to no purpose. Every section of the town wears that same moral aspect that is so trying to the man who is compelled to furnish so many columns of reading matter, or hear folks say 'there ain't much in the paper'. . . . If this sort of thing continues much longer, we shall have to turn in and publish a religious paper. We are so tired of politics that we hate that subject, and unless the times change speedily we shall have to get 'religion on the brain', and deal it out in doses to our readers.[18]

The quietness and respectability of Grass Valley appeared in a more favourable light to a Cornish woman who visited the Californian Cousin Jack community twenty-eight years later, and recorded her impressions that the

> difference between Grass Valley and many other mining camps is very discernable. On the Sabbath here the stores are closed, as also are the mines, and everybody dons their best and goes to Church. There is a very comfortable Methodist Church, a quite large Roman Catholic Church and nunnery, and a small Episcopal Church. It seems almost that Porthtowan people predominate; indeed, one might imagine that all Porthtowan had migrated

16. *Cornish Telegraph*, 21 and 28 April 1882.   17. *Grass Valley Union*, 5 June 1866.
18. Ibid., 28 June 1866.

hither, for it is a fact that there are many more Porthtowan folk here than there are at present home in the original little Porth.[19]

This correspondent, who reckoned that sixty per cent of Grass Valley's population, which then, in 1894, was about six thousand, was Cornish, also visited the New Almaden quicksilver-mines in California, where the majority of the three hundred employees were Cousin Jacks, Virginia City, Butte and Northern Michigan. She found Virginia City in decline, the population having dwindled to barely six thousand, of whom she reckoned half to be Irish, a fifth Cornish, and a fairly large number of Italians and Chinese; the days of Judge Lynch had gone for ever; churchgoing in the famous Nevada mining city, however, was 'not quite as popular ... as in some Cornish towns'.[20] When she got to Butte she did not hesitate to write that the activities of Vigilance Committees had 'long since brought to an end the terrible deeds of blood and rapine of the road agents in Montana'.[21] She made no mention of church life in Butte, but alluded to the feuds of Cousin Jacks and the Irish as being far too frequent. Moving on to Upper Michigan she found that

although the winters are long and severe, they seem to be the chief season for enjoyment. Snow shoeing, tobogganing, sleighing, skating, church socials, dances, and an occasional concert fill up the time during King Frost's reign. In the summer those who can afford to do so take a trip down the lakes, or go out into the woods in camping parties, when fishing or deer stalking can be indulged in.[22]

Recreational facilities were, apparently, far greater in Upper Michigan than they were in Virginia City, where

there is not a level space where to play a game of football or cricket; no drives, no walks, no race track, no rink, no parks; dancing as an amusement predominates; there are generally two or three dances a week, when the elite gather to enjoy themselves, while card parties occupy the other evenings.[23]

More social activities went on in the mining camps of Western America than this lady, perhaps the wife or daughter of an upper-class mining captain or engineer inspecting mines or mining properties was aware of, or thought fit to mention. Virginia City may have had no level sports field but it staged its share of wrestling matches and rock-drilling contests, while neighbouring Carson City was the venue of the

19. *W.B.*, 13 September 1894. 20. Ibid., 19 July 1894.
21. Ibid., 20 December 1894. 22. Ibid., 28 March 1895.
23. Ibid., 19 July 1894.

boxing match in which Helston-born Bob Fitzsimmons won the world heavyweight title from 'Gentleman Jim' Corbett, by his famous knockout solar plexus punch in the fifteenth round on 17 March 1897. Cousin Jacks in Butte after the Carson City fight sent the victor a telegram worded 'Accept congratulations from your admiring countrymen in Butte. Hurrah for Helston!'[24]

Boxing, however, was less popular among the Cornish in America than their traditional sport of wrestling. At one time hardly a Fourth of July passed in Grass Valley or Butte without a wrestling contest with a score or more contenders, the majority of them with very recognisable Cornish names. In 1868 the contests at Grass Valley, beginning on the Fourth were continued on the sixth and seventh of July; the total prize money amounting to $300.[25] Nearly as much prize money 'in gold coin', was given for the matches held in Grass Valley on the Fourth and seventh of July 1874, when Richard Andrews won the first prize of a hundred dollars, William Nankervis the runner-up's prize of seventy dollars, Joseph Coombs the third prize of fifty dollars, and consolation prizes of forty, twenty-five and ten dollars being awarded to Charles Temby, William Henry Mitchell and Samuel Nankervis. To this array of Cornish names could be added the names of sixteen other contestants in the matches of those two days; the four Roddas, Edward, James, Joseph and Benjamin; Henry Northey; Samuel Blight; Robert Quick; John Roberts; Richard Cluyas; William Rogers; John Hooper; John Sarah; Francis Argle; Nicholas Odgers; Matthew Tonkin; and Francis Tresize.[26] As early in Butte's history as 1882 the *Daily Miner* was advertising an 'Annual Wrestling Match in Cornish Style', extending over three days, from 24–26 May, with two hundred dollars in prizes, the Miners' Union Band in attendance, and an admission charge of one dollar.[27] A local journalist, reporting the last day's matches wrote:

The old Cornish sports, which have been handed down with comparatively little variation for hundreds of years in the beautiful sea-girt mountains of Cornwall are peculiar, and almost necessarily preserve their individuality. Young men and boys are trained up to them and look forward to participating in them from infancy. No one who has attended the three days' sport on the grounds south of East Park Street can fail to be particularly impressed with at least one unusual feature, and that is the general good humour and fairness of the contestants. It has indeed become a by-word and a reproach in almost

24. *Weekly Tribune* (Butte), 20 March 1867.
25. *Grass Valley Union*, 1, 7, 8 and 9 July 1868.   26. Ibid., 9 July 1874.
27. *Butte Daily Miner*, 16 May 1882.

all distinctly American sports that chicanery and manipulation destroy honest and hearty appreciation on the part of the public at large.[28]

Between the rounds the Miners' Union Band played selections, reminiscent of the brass band contests that became so popular in Cornwall around this time; intervals between the 'serious' contests were filled by a dog-fight, which only lasted a few moments and may not have been deliberately contrived, and by

a match between two fat men, W. W. Sager, weighing $316\frac{1}{2}$ pounds and Thomas Sodey, $296\frac{1}{4}$ pounds. After a long struggle to get within reach of one another, Sodey won a fall, and which both were so exhausted that they went to sleep under the judges' stand where they were at last account.[29]

Boxing and wrestling, still more dog-fighting, were regarded by many as brutal spectacles, catering only to the baser instincts of participant and spectator alike. The commercialisation of these sports, the gambling on the results of contests, and the suspicion that the outcome of matches was sometimes 'fixed up' beforehand brought such sports into still further disrepute. On the other hand many enjoyed taking part in such contests as well as watching them. Men who followed such a dangerous calling as did miners were not likely to be over-squeamish about the sight of bloodshed, of men being knocked unconscious, or of birds and animals being egged on to fight talon and claw to death. Even at that time Nevada was about the only state in the Union where prize-fighting was legal, while miners of Tombstone and Bisbee took—and still take—trips down into Mexico to attend bull-fights. In the Old Country cock-fighting and dog-fighting were outlawed before this time, but that was no indication that they had altogether ceased.

Some of these brutal old-time sports, it is fair to add, came under the ban of the law in the New World as well. In the Marquette Iron Range in northern Michigan, the Cousin Jacks had prepared to spend the 1875 Fourth of July at Negaunee with a three-day wrestling tournament, but called off a cock-fight which they had originally planned as part of the celebrations when it was pointed out to them that this sport was illegal in Michigan. The wrestling, although limited to entries from Marquette County, attracted over sixty contestants and a crowd of about five hundred spectators. At a later contest in Negaunee, in June 1882, there were again over sixty entries, and the Ishpeming Band was in attendance to play between matches.[30] Long before that time Tom

28. *Butte Daily Miner*, 27 May 1882.   29. Ibid., 27 May 1882.
30. I am indebted for this information about Marquette County wrestling matches to the late Kenyon Boyer.

Carkeek, an Upper Peninsula miner, had gained a reputation as a Cornish wrestler almost as great as that which Bob Fitzsimmons got as a boxer later, while John Rowett of Ironwood, a law-enforcement officer, was nearly as famous around the turn of the century, winning the title of world champion for Cornish wrestling in 1901 and retiring undefeated nine years later.[31] But the greatest of the Cornish wrestlers in America was probably Joe Williams who, when fifty years old, in 1876, defeated Alf Williams, the champion of the Gold Hill Camp in Nevada, who was barely half his age and who had rashly challenged all comers. Joe put Alf down on his back three times in as many minutes, though he was much the smaller man of the two, and then being fouled finished off the contest by hurling the younger man through the air and on to the ground with such force as to knock him unconscious. This match was staged in the Alhambra Theatre in Virginia City for a prize of two hundred and fifty dollars. Some in the audience may have agreed with a Grass Valley journalist who admitted that to an outsider the sport of wrestling might seem as 'rough as a grizzly's hug',[32] while the Gold Hill men in the audience, who saw their champion defeated and their bets lost, may, perhaps, have wondered if the jubilant Grass Valleyan supporter of Joe was lying after all when he told them 'Joe can throw a rhinoceros over his shoulder—I've seen him do it'.[33]

Rock-drilling contests, in which two- or three-man teams vied with each other to show their prowess, were far less spectacular than wrestling matches and had a limited professional appeal to hard-rock miners. A series of contests for the world rock-drilling championship and a seven hundred dollar purse was won by the Butte Cousin Jacks, James Davey and Peter Teague, in the summer of 1892, when they drilled $33\frac{3}{16}$ inches in a quarter of an hour.[34] On their return to Butte they were given a reception that would have gratified a candidate for the American Presidency. Butte's two most renowned bands, the Boston and Montana Band and the Alice Band, given a day off by their Cornish captains, 'Tom' Couch and W. E. Hall, to take part in the welcome, struck up 'See the Conquering Hero Comes,' as soon as Davey and Teague stepped off the train: hundreds of hands were stretched out to greet them amid cheers; they had almost to fight their way to a carriage which was waiting to take them through the streets of Butte thronged

---

31. *Marquette Mining Journal*, 2 April 1958.
32. *Grass Valley Union*, 4 June 1870.
33. *W.B.*, 11 September 1876, quoting *Virginia Evening Chronicle*, 8 August 1876.
34. *W.B.*, 18 August 1892.

with cheering spectators. Behind the carriage and the two bands, playing most of the time, went some five hundred miners marching on foot, and then a long line of carriages brought up the rear of the procession to the 'East Broadway resort', where the celebrations ended with speech-making, cheering, and the drinking of healths in champagne which was flowing like water. Not forgotten in the tumultuous welcome was John Pryor, the Redruth man who sharpened the drills and who was given a hundred dollars and a gold chain by the champions to commemorate their victory. Together Teague, Davey and Pryor had proved the truth of the Rocky Mountain 'Cousin Jack Song' which ran:

> They come from distant Tombstone and Virginia on the Hill
> You ne'er can beat a Cousin Jack for hammering on a drill.[35]

And even when it came to single-handed drilling, which had caused so much trouble in Grass Valley in the early eighteen-seventies,[36] the Cousin Jacks soon proved themselves as adept as others. Some twenty years after those bitter conflicts John Kitto won many prizes for his prowess, notably in the fall of 1894 when, at Howard's Hill in Amador County, he defeated an American rival who, in fifteen minutes, drilled a $25\tfrac{3}{4}$-inch hole by drilling $29\tfrac{1}{2}$ inches in the same time.[37]

The ringing of the hammers on drills may have been musical, but it was monotonous to those who spent their working lives among it. Other kinds of music appealed to these men on both sides of the Atlantic. Brass and silver bands and choirs were social institutions which Cornish emigrants brought to America with them. Occasional paragraphs in Cornish local newspapers reported the breaking up of village bands through emigration, and suggest that at band practices men discussed emigrating and that the set determination of one man to go often persuaded some of his fellow musicians to emigrate too. Outstanding among the Cornish bands were those in Butte, organised in friendly rivalry by the men or their mining 'bosses' like Tom Couch of the Boston and Montana and 'Cap'n' Hall of the Alice; the Butte Miners' Union also had its own band. In Grass Valley as early as 1868 the Cornish brass band turned out to serenade Alice Kingsbury, the star of a touring

---

35. B. A. Botkin (ed.), *A Treasury of Western Folklore* (New York, 1951).
36. See above, pp. 120 ff.
37. *W.B.*, 17 December 1894, quoting *Amador Record*, n.d. The match was probably held over the Thanksgiving week-end, and on the occasion over a thousand dollars in wagers changed hands. The achievement of Kitto in drilling roughly nine-tenths of the depth of the Butte two-man team at Helena in 1892 does not indicate the superiority of single-hand drilling, for the single-drill was much narrower, and it was possible that the rock in Helena was harder though, of course, the reverse may have been the case.

theatrical company then visiting the gold-mining town.³⁸ In Upper Michigan several brass bands were organised at different times, such as that at Ishpeming while much further east, in the mining regions of Pennsylvania, there was a strong Cornish element in many bands, notably in that of Carbuncle City, but in the fall of 1875 it was noteworthy that, of the Cousin Jacks in this band, two can hardly have been associated with mining in the Old Country, one of them, Williams, having been police constable at Trispen near Truro, the other, Edward Mitchell, coming from Padstow.³⁹

It was sometimes said that their work as miners developed the lungs and voices of bandsmen and vocalists, although many might assert that it had the opposite effect and weakened their lungs. Others have attributed the love of music and song of the Cornish immigrants to their Celtic heritage, and still others to the importance accorded to hymn-singing and religious music in the Methodist chapels of Cornwall. The contribution of the Anglican Church to the love of music should not be forgotten, but it does not seem that bell-ringing and bell-ringing festivals were carried from the Old Country to the New. Probably when churches in the homeland installed organs in place of instrumental music,⁴⁰ the 'fiddlers', serpent players, cymbalists and the rest drifted away into secular bands in which their talents seemed to be more appreciated. While the Cornish had a great reputation for singing almost everywhere they went, it was greatest in Grass Valley, where a Cornish choir was organised in the early eighteen-fifties and has remained in existence ever since, specialising in Christmas carol singing. Fourth of July wrestling contests at Grass Valley in 1874 started with glee singing by the Cousin Jacks, Thomas Hollow, Robert Quick, and Thomas Curnow.⁴¹ Just as earlier British immigrants preserved folk songs in the Appalachians long after they had been forgotten in Britain, so choirs in Grass Valley and chapels in Butte kept alive Christmas carols that were rarely heard in Cornwall after the end of the nineteenth century.

A number of these carols were to be found in the hymn books used by the various Methodist 'connexions', but after the reunion of the chief denominations in Britain were omitted from the revised British Methodist hymnal of 1933. In America the connection between the Cornish immigrants and Methodism was close, but there had been developments in the Old Country before many left Cornwall for America causing

38. *Grass Valley Union*, 3 January 1868.   39. *Cornish Telegraph*, 20 October 1875.
40. Vide Hardy's novel *Under the Greenwood Tree*.
41. *Grass Valley Union*, 9 July 1874.

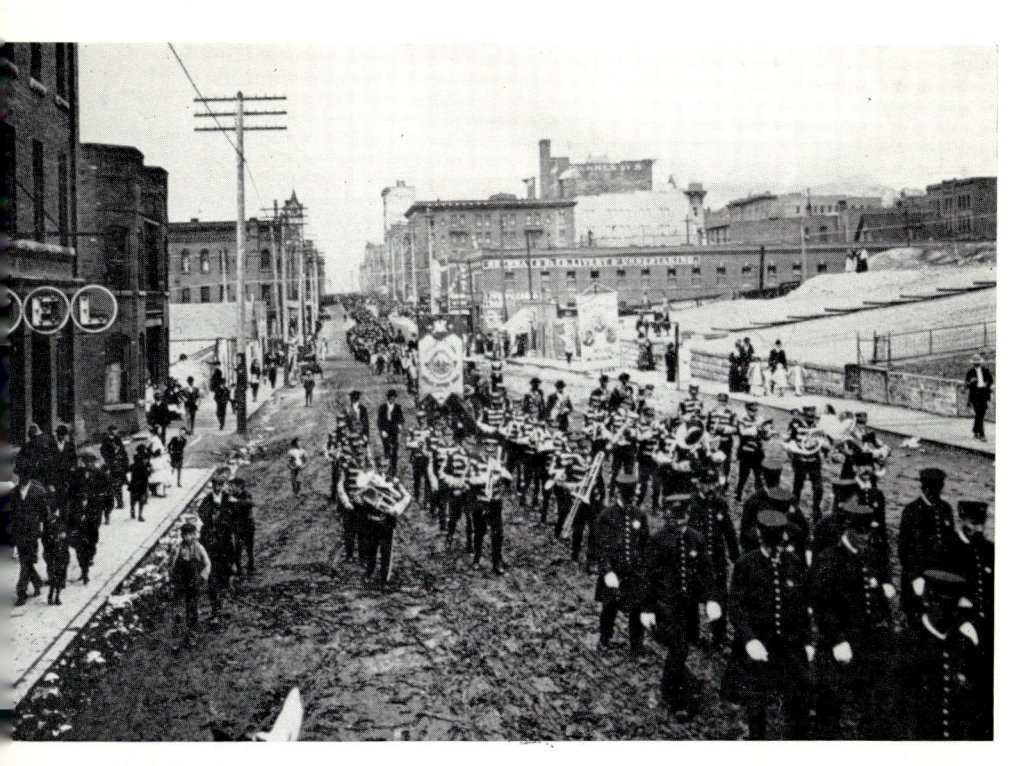

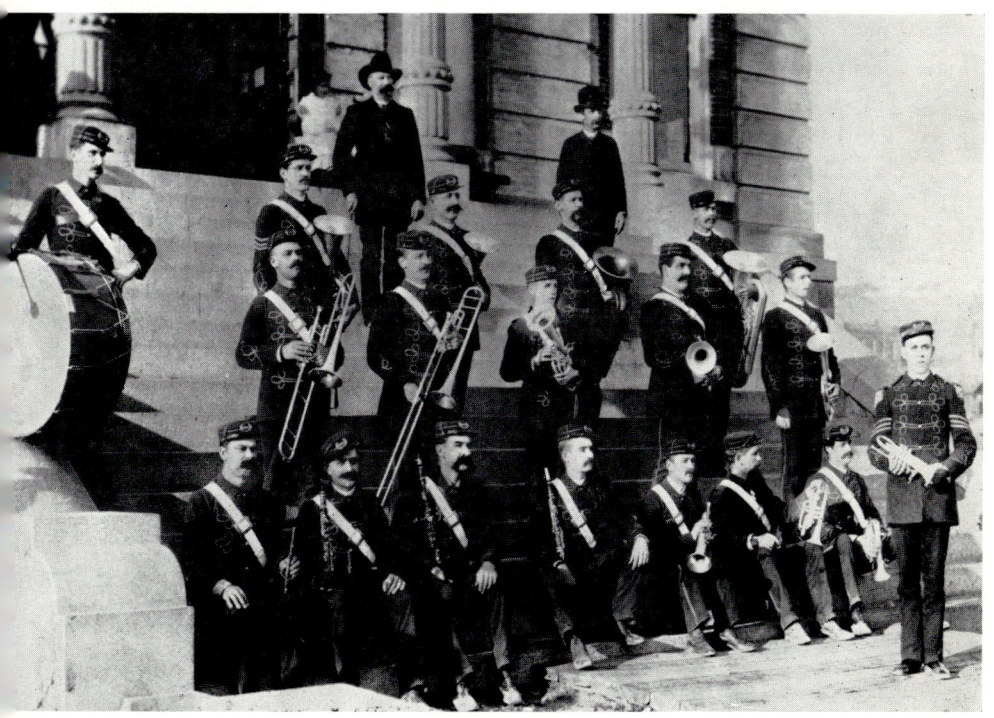

17 Miners' Union Day parade in Butte, Montana, and the Boston and Montana Band of Butte, 1888

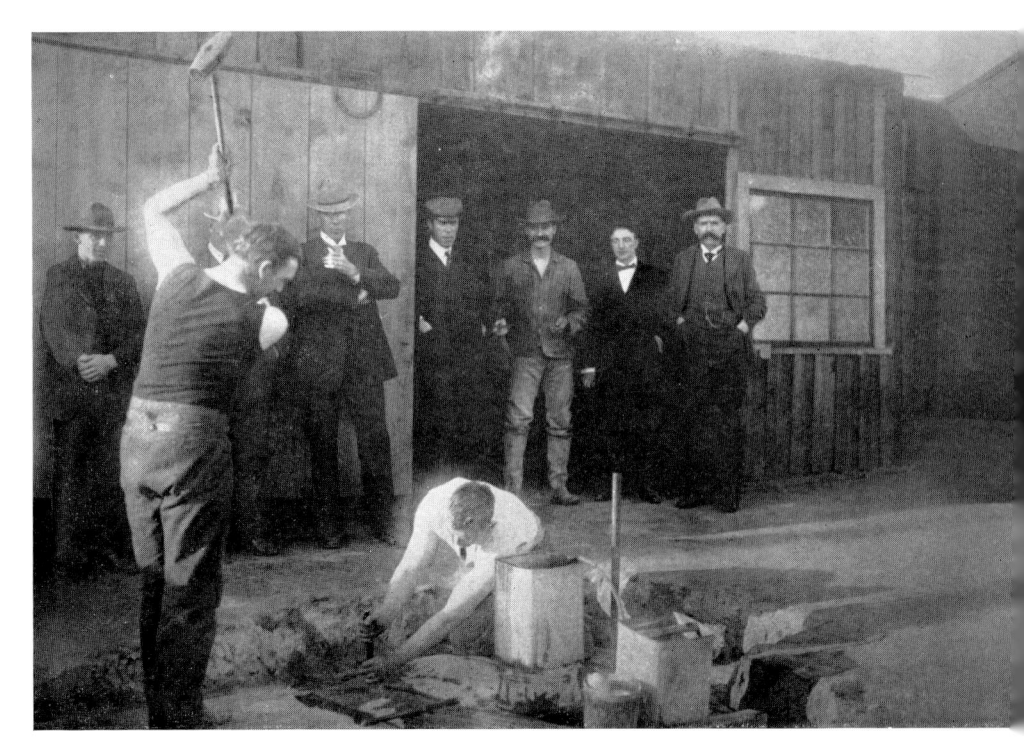

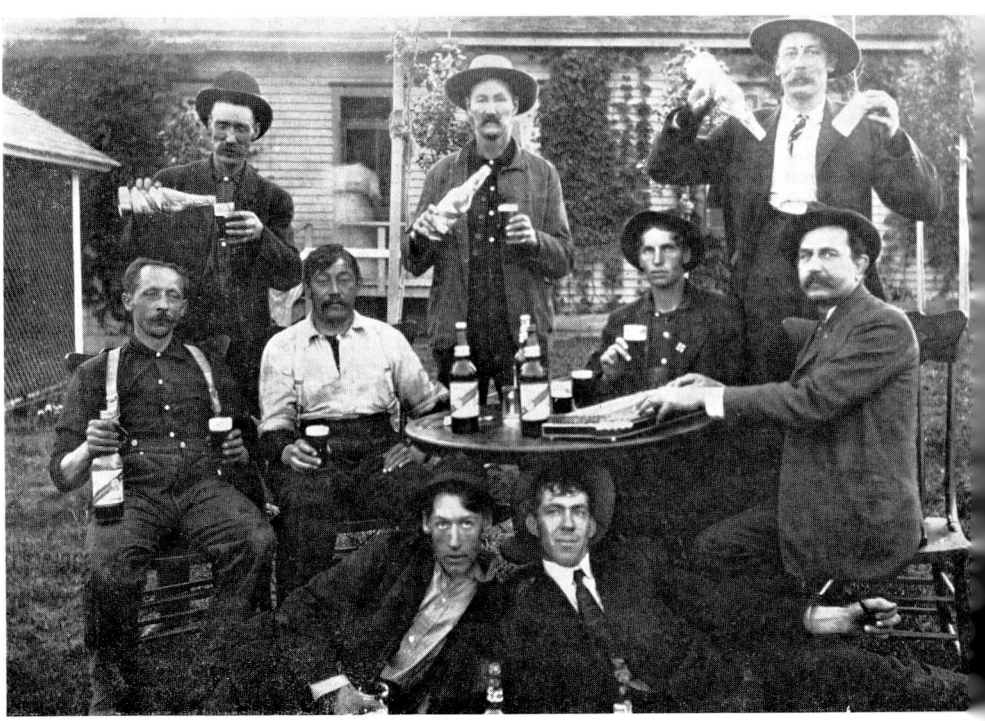

18 A rock-drilling contest and a drinking party at Butte, about 1890

divergencies from the American Methodist Episcopal church which, probably, conformed more closely with John Wesley's ideal ecclesiastical organisation than any set up in Britain by 'people called Methodists' in the half century after his death. By the time of the great migrations to Wisconsin, Northern Michigan and the Pacific Coast, Methodists in Cornwall had broken the bonds with the Anglican Church which Wesley had wished to retain, and had also split into a number of separate sects such as the Primitive Methodists and the Bible Christians. The latter had been founded by a Cornishman, William O'Brien of Luxulyan, and had become quite strong by 1830. O'Brien himself went to America on mission tours, but later left the denomination he himself had founded. Primitive Methodists obtained a considerable following in the Redruth–Camborne and St Austell mining districts, while the 'Bryanites', as they were often called, gained ground in the more purely agricultural regions of the county. Furthermore, one minor but voluble sect of Methodists, particularly strong in St Ives during the eighteen-thirties, advocated absolute teetotalism, and its example influenced other local Methodist congregations.

Inevitably these sectarian differences were carried by emigrants to their new homelands with them, but a sentiment of fundamental unity was shown by the way in which all the sects used the name 'Methodist' or 'Wesleyan'. Nevertheless Cornish provincialism exerted an influence on Methodist denominationalism which was not found elsewhere. Provincialism and 'clannishness' largely accounted for the popularity of local lay preachers in Cornish Methodist churches and for the difficult times 'foreign' ministers, born and bred east of the Tamar, had when stationed among them. Wesley's own episcopal learnings evoked no sympathetic response in a region remote from a diocesan centre until the see of Truro was established in 1876. The importance that John Wesley attached to singing was, perhaps, the feature that attracted many Cornish people to Methodism, although more may have joined because of Wesleyan individualistic egalitarianism which demanded that all participate in religious services, that the man—or woman—who could preach or exhort be allowed to do so, and that all who had voices have the opportunity to use them fully and freely to worship their God. There survived, however, some liking for older Anglican rituals, particularly those of baptism, marriage, and burial, and Cornish Methodists still took their children to the parish church to be christened, reckoned it the only proper place for a wedding to be solemnised and carried their dead to it for the last solemn rites.

Some of these provincial traits were so similar to characteristics developed by frontier conditions that features of religious life in American mining camps have been ascribed to the pioneer environment instead of to the pioneer himself. In comparatively early days in the Upper Peninsula of Michigan, American-born and trained ministers—and their wives—hardly knew where they stood with the Cousin Jacks and Jennies in their congregations. Some mine captains themselves had been local preachers in Cornwall and took the first steps in establishing meeting-houses, churches and Sunday schools; naturally enough they claimed a considerable say in the activities and policies of their churches, and ministers found themselves in subordinate roles. Mine-owners in northern Michigan often gave the ground on which churches were built, and possibly Methodist influences saved the Lake Superior copper region from much of the lawlessness of the 'tougher' gold and silver-mining camps of the Pacific Coast. Some of the most well-attended and famous of the earlier Methodist churches in Upper Michigan—Calumet, Hancock, Mohawk, Laurian in the copper region, Iron Mountain, Bessemer, and Wakefield in the iron ranges—had in their first years predominantly Cornish congregations, running even as high as ninety per cent.[42]

Many an 'up country' English minister and his wife stationed by Conference in Cornwall have found their congregations as difficult as Bartram and Elizabeth Gurney Taylor found the Cousin Jacks in Houghton in 1866. Taylor was surprised to find such predominantly male congregations, his wife to find the Cousin Jennies more fond of rather earthly gossip than holy conversation and that she was expected to run a choir rather than evangelize. While waiting for his wife to join him Taylor wrote her that one of his first services in Houghton had seemed rather constrained, but 'there was no loud shouting, there would have been much of it had I been of the shouting kind'.[43] Cornish immigrants brought with them their fondness for emotional revivalism, and this, in time, tended to alienate those who, having made a fair living, had aspirations to 'high-toned' social respectability and came to regard shouting preachers, vigorous singing, and emotional manifestations of conversion as undignified and vulgar. But the rank and file of Cousin Jacks and Jennies of those times liked their religion that way.

From the social point of view, singing was important. In the most

42. *History of Methodism in the Upper Peninsula of Michigan* (Historical Society of the Detroit Annual Conference, 1955), p. 15.
43. Taylor MS. letter of Taylor to Elizabeth Gurney Taylor, 3 August 1868.

unlikely places would be heard the hymns most frequently used in Methodist services. Patrons of saloons, inspired by strong liquors to song, were as likely to strike up *Rock of Ages* or *Lead Kindly Light* as to sing of *Sweet Betsy of Pike* or carol the Californian version of *O, Susannah*.[44] In Upper Michigan rivalry between choirs at annual singing contests held at the First Methodist Church Calumet evoked partisanship said to be similar to that aroused by Harvard–Yale football games.[45] Years afterwards people in the Copper Country recalled men who sang on their way to work and even as they went down the shafts. The Cornish may have had a superstition against whistling in a mine but it certainly did not extend to singing. At funerals, in Upper Michigan, when coffins were borne from church to graveside by pall-bearers, just as in the Old Country, the sombre hymn *Nearer My God to Thee* was sung, and from Cornwall, too, came the customs of long corteges and crowded attendances at funerals besides the wearing of deep mourning as tokens of respect for the deceased. In the Old Country labouring families struggled and often deprived themselves of the necessaries of life to pay a few pence each week into the 'Club' that would ensure them a seemly funeral when they died, and the same habits persisted after they had emigrated.

The same lugubrious temperamental trait is manifested in the liking for hell fire and damnation sermons and also in many of the superstitious beliefs which were held by Cornish people both at home and in the new lands to which they emigrated. One anecdote of the half-mythical brazen-lunged character, Dick Buller, in Upper Michigan, is that once he started singing the hymn *Deliverance* outside a church wherein the minister had just been threatening and describing the sufferings of the damned to his congregation. Dick's voice was so powerful that the terrified folk in the church thought that it was Gabriel sounding the Last Trump for the Day of Judgement, and several fainted right away in fear.[46] One fire and brimstone preacher, however, C. H. Northrop of Nevada City, got into trouble in 1868 when he threatened to expel members of his church who persisted in attending dances, which he denounced as immoral and licentious in such terms that a local newspaper said his remarks, even if tolerated in a pulpit, were too indecent to print.[47]

Stricter Methodists in Cornwall often condemned both dancing and

44. See above, p. 116.   45. A. Murdoch, *Boom Copper*, pp. 201 ff.
46. Ibid., pp. 201 ff.
47. *Nevada Gazette*, not dated, quoted in the *Grass Valley Union*, 22 January 1868.

card playing, but the most bitter controversies, perhaps, were about drink. In the new country by the mid nineteenth century the 'Maine Law' had become to some the ideal, whereby it was possible by legislation to achieve prohibition. The teetotal movement in the mining districts of California in the early eighteen-seventies should have warned social reformers who thought it possible to inaugurate the millenium by legislative fiat. Federal and State authorities did nothing, but any city was free to adopt 'local option' by majority vote of its inhabitants. Such an election was held in Rough and Ready township early in 1874; the prohibitionists carried the day, but the result was that while licensed saloons stopped selling 'drams' of spiritous liquors, somehow or other the law made it possible to purchase 'snake-bite liniment' in five-gallon kegs. It soon appeared that many of the citizens of Rough and Ready had taken excessive precautionary measures against a plague of serpents, got their five-gallon kegs into their homes, and

> Men drink who never drank before,
> And those who drank now drink the more.[48]

In Grass Valley the influence of the 'Ladies' Temperance Union' was strong enough to secure a local election on the issue on 6 July 1874, but certain 'Crusaders' injured their cause by denouncing all who refused to pledge themselves to vote for prohibition as drunkards and reprobates, which probably caused the decisive defeat of the total abstinence party by 648 votes to 298. While some Cornish 'Methodies' supported the movement, other Cousin Jacks in Grass Valley complained that the referendum on the issue interfered with the Fourth of July wrestling contests and reduced attendance at those sporting events.[49]

Teetotalism, whether of 'native' American or immigrant Cousin Jack 'Methody' origin, was also strong in northern Michigan. It was from the Copper Country that there came a Cornish 'plod' or yarn which, besides indicating how dialect survived emigration, revealed the paternalistic care some of the mining companies took of their hands and, furthermore, showed why Michigan people had their own explanation of the origin of the slang name for Cornishmen—that it was a corruption of 'cussing' Jacks. The story runs in a printed version:

'Well m'son,' quoth Captain Dick, 'I 'ear that al' bloody country is goin' dry. Dam-me, don't now but wot may be a good thing too. Good for some any 'ow, for I tell'e this 'ere drinkin' bizness is ruination o' moor than one fine lad

48. *Grass Valley Union*, 6 and 11 June 1874.
49. Ibid., 7 and 9 July 1874.

these days. 'Ow some poor devils do 'ate to let gow. Not sayin' don't like a bit o' today self naow an' then, but never 'as I made 'og over them. I remember back four years h'ago 'ow Jan Trevarthen used to 'it un. H'awful, m'son, h'awful. But dam-me, naow 'ow it 'appened. Well, I'll tell'e.

'Jan wuz sent down with me an' h'other boys to pull pumps to h'Algonquin mine. No more bloody h'ore left in the mine, mind you, but pumps h'as good h'as new. So dam-me, down we gows and w'en we gets there, 'ol bloody place wuz fixed h'up, nice h'as you please, for h'our comfort. Company h'even 'ad doctor there, to see that we didn't get 'urted. Fine fellow too, that there doctor. H'every time ussen come off shif, in we legs to that h'office, where we 'ad to change h'every bloody stitch we awned 'en put on dry clothin'. For, min' you m'son, it wuz some bloody wet, down in the bal. Afore we gets 'alf dressed, in comes that there doctor with gert jug o' whiskey, an', if I do say so self, a bit o' she gaws 'andsome when one's wet. But, min' you, m'son, it wuz only a taste 'e gives us. Well, one day, Jan 'e figgers as 'ow there mus' be some chance of havin' moor than that there doctor gives us, so, dam-me, down 'e brings a gurt large jug of 'is h'own an' 'ides 'im in be'ind 's locker, an' h'each time the doctor give 'im a drink from 'is bottle, why, dam-me, Jan'd sneak three or four drinks from 'is h'own. An' so it gows along till time came for us to win' up that job. The las' day, m'son, never shall I forget h'it. As 'ard work as men ever wuz put to an' h'almos' afore we 'ad the las' of the pumps h'out, that water wuz roarin' an' rumblin' like the crater of Vessuveyuss. So, h'into that bucket we piled an' h'up shaf we gows. Well, m'son, we h'all 'ad h'our bit an' Jan by way o' celebratin', 'ad 'is share an' moor too an' h'out 'e starts for that pub, afore the res wuz dressed. Dam-me, no sooner 'ad 'e started than back 'e come, white h's a ghost an' tremblin' from 'ead to foot. 'Boys,' says 'e, 'So 'elp me, God, I'll never take 'nother drink h's long as I live,' an' down 'e falls on the flure, cryin' like a baby. Believe it or no, m'son, h'owt we gaws, an' there wuz the 'ol bloody groun' strewn with snakes. Snakes, min' you. Well, dam-me, dam-me, we wuz all 'ard put to h'understan' that sort of 'appin' till the 'oistin' h'engeneer tol' us 'ow it wuz. Seems in h'early days, they 'ad found the 'oles comfortin' so daown they gaws, an' there they wuz, livin' for years. Well, dam-me, when the water in the mine started risin' h'out they come, for no bloody h'animal, but 'errin, can live in water any'ow. Finally we took Jan, cryin' all the way 'ome, he wuz, and from that day to this, never a drop 'as 'e touched. 'Eare, son, let's 'ave a bit o' that there Peerless. Wat's think on it?'[50]

50. Newspaper clipping, from the *Engineering and Mining Journal*, undated, preserved in the papers of Orrin W. Robinson, Michigan Historical Collections, Ann Arbor. Students of dialect might object that this tale, apparently written on the eve of the Eighteenth Amendment, is far from being an accurate written record of the Cornish brogue. While, in Cornwall, 'that' and 'that there' would certainly be 'thikky' or 'thiccy', and many vowels would be lengthened, like 'these' to 'thaise' or 'thaize', three categorical assertions must be made. First, the reproduction of Cornish or any other provincial dialect is almost impossible in English print.

The care of Upper Peninsula mine-owners for the welfare of their employees indicated in this 'plod', only came after bitter labour disputes and strikes in the early years of the present century, and had been rare, though not unknown, before the great industrial conflicts of 1913-14. In these conflicts Cousin Jacks were involved on both sides, and the troubles not only evoked strong and stormy manifestations of sympathy among the copper-miners in Butte but also, it can be asserted, encouraged the clay workers of the St Austell district to strike for better wages and working conditions in July 1913. Much of the trouble in Michigan in 1913 was caused by Calumet and Hecla and its associated companies abandoning the Cornish custom of paying miners who had worked a full five and a half day-shifts or five full night-shifts for wage for six shifts; the associated companies, along with a twenty-five cent increase in pay for a ten-hour shift attempted to pay the night-shift workers for only the five shifts which they actually worked in a week, but at the same time they continued to pay the day-workers for six shifts weekly. Wages on the Copper Range, at this time, were averaging around three dollars a shift, although the average rates paid by the Calumet and Hecla group of mines through 1912 had averaged $3.28 per shift as against $2.74 paid by the other companies in the region.[51]

On the whole, though wages were much higher in the generality of American mines than in those of Cornwall, hours of working were longer, and the underground captains, or boss miners, set a more gruelling pace. In some Cornish mines underground workers could 'touch pipe', or go off and sit down for five or ten minutes and enjoy a quiet smoke three or four times during a shift, while mine managers thought it adequate if men put in six or seven hours actual work in a shift. In any case, many Cornish miners worked on tribute, and if they were content with smaller earnings there was nothing to stop them 'going slow'. In most American mines, even sometimes when men were working on 'contract', this was not the case; a man was expected to work the whole shift, and in Upper Michigan in 1912 the shift was ten hours, with no time allowed for the transportation of workers from the

---

Second, instead of there being one Cornish dialect there are at least a dozen or more local variations in the Old Country itself, and two, three or four generations of sojourning in America had inevitably resulted in further variations. Finally, the writer himself, had to make modifications in order to be intelligible not only to non-Cornish American readers but also to the Cousin Jacks themselves who, whatever brogue they used in everyday conversation, could only read 'standard' English.

51. *U.S. Department of Labour Statistics: Bulletin No. 139; Michigan Copper District Strike: February 1914* (Senate Document No. 381, 63rd Congress, 2nd Series), pp. 12–13.

entrance of the mine and the ore face. Eight-hour maximum underground working shifts had already been adopted in Colorado, Idaho, Montana, Nevada, Utah, and Wyoming, but only since 1903, with the exceptions of Wyoming which had adopted such a law in 1871 and Utah which had done so in 1896.[52]

A ten-hour underground shift was long, and only the pressure of the Western Federation of Miners and other Unions secured 'shorter days' in American mines. On the other hand, long hours were normal in all employment at that time, and mining bosses reckoned that they asked from their hands no more than a fair day's work in return for a fair wage. A lot of the talk of slave-driving in American mines came from prejudiced sources—particularly from persons who feared that the attraction of higher wages overseas drew the best and most enterprising miners away from the Old Country. Decrying Nevada in 1880 was John Rule, who told the West Seton shareholders, that if men worked as hard in Cornwall as they did in Nevada they could save seven or eight pounds a month. Rule reckoned that if working miners in Nevada made five dollars a day, their living expenses came to thirty or thirty-five dollars a month. Water was scarce and expensive; in fact, in the seven years he had been in Nevada, it had, Rule said, rained only three times. He wound up his remarks by stating that men could be discharged without any notice as there were plenty of hands to take their place.[53] It was hardly to be expected that any of the shareholders at the meeting would have pointed out to Rule that his figures indicated that wages were still four times the cost of the common necessities of life, and that in Nevada, on Rule's showing, a miner could save up to ninety dollars a month, or easily twice as much as he could save in Cornwall assuming he could, by hard work, make as much as the returned Nevadan glibly asserted. A retort to Rule came from a Cousin Jack still in Nevada, William B. Johns who wrote home that the great drawback to mine employment in that state was Sunday working which had led a number of miners to go elsewhere; as for the 'aridity', while it cost a household in Eureka three dollars a month to get water, yet, in 1874 a torrential downpour had drowned fourteen men in Eureka, and if Rule wanted more water than there was at the Star mine in that district, he had better get a diving suit and go and live in Mount's Bay.[54]

52. Based on *U.S. Department of Labor Statistics: Bulletin No. 139: Michigan Copper District Strike: February 1914* (Senate Document No. 381, 63rd Congress, 2nd Series), p. 25.
53. W.B., 17 June 1880.
54. Ibid., 12 August 1880, letter dated Star Mine, Cherry Peak, White Pine Country, Nevada, 19 July 1880.

Rule had not mentioned the greatest drawback to working in some of the Nevadan, particularly the Comstock, mines—the excessive heat in the lower levels. On this point, the San Francisco *Mining and Scientific News*,[55] in a lengthy comment dismissing Rule's derogatory remarks as 'a harmless and extremely dry joke', admitted that 'during the last several years the men have had to contend with the hardship of excessively hot mines, but only for short intervals, for there cool water and ice are supplied without stint'. The outcry that these hot, dusty mines hastened the onset and ravages of phthisis only came later, but two years before John Rule had publicly criticised working conditions in Nevadan mines, a Virginia City newspaper had described what went on in the lower levels of the California and Consolidated Virginia Mines where the temperature was as high as 120 or 130°F. Exposed to an 'atmosphere' with a temperature thirty degrees above body heat, the human body began literally to cook; the first symptoms were the cessation of perspiration, followed by acute stomach pains, and then delirium. To reduce the risk of this happening to them, miners consumed ice and ice-water by the gallon, and poured it over each other, at the rate of ninety-five pounds weight per working miner a day. When the San Francisco periodical referred to men only being required to work for short intervals in such mines, a reader in the Old Country might have thought that what was meant were periods of days, weeks, or even months. The Virginia *Enterprise*, however, alleged that in the summer of 1878, in one level of the Con Virginia men could only work about ten minutes at a time, after which 'they fall back and let other men come to the front'. Under these conditions it cost sixteen dollars a day to get the work of a single man. When a miner stayed too long at the ore-face and began to show that the heat had begun to cook him by starting suddenly to run about or talk incoherently, according to the *Enterprise*

his companions 'doctor' him. It is rough treatment they give him, but it is found to be very effective. The man affected is seized and carried to the coolest place in the vicinity, where he is bound hand and foot and put through a process of rubbing.

The friction is applied to the stomach which is found to be the seat of trouble, and in which knots nearly the size of a man's fist are found to have formed. They must be rolled out, and as soon as they disappear perspiration again starts and the man regains his senses.

The rubbing is sometimes done with a gunny sack, but as this is liable to cause useless abrasions of the skin, a pick handle is preferred. To be rubbed

---

55. Issue of 7 August 1880, quoted in *W.B.*, 9 September 1880.

down with a pick handle in the hands of a muscular miner is not such treatment as any man in his senses would be likely to greatly desire, nor does the miner, even in his deliverance, desire it, therefore he is tied in such a way that he cannot resist.

The miners say that they bring a man out all right by their method of treatment in less than half the time it would be done by the physicians.

A day or two ago at the Con Virginia shaft the men took one of their companions who became deranged from the heat, tied him at the end of a rope, and lowered him about a hundred feet to a place where he could be conveniently 'doctored', then went at him with their pick handles, and soon brought him out all right. The shaft mentioned is so hot that beside it purgatory would be reckoned a cooling-off station.[56]

No mine in Cornwall was ever so hot as that, although some of those who worked in the lower levels at Levant in St Just had to endure temperatures as high as 117°F.

Later American mines were more mechanised than those in Cornwall, although pioneer workings made shift with 'machines' that had been used in European mines in the sixteenth century. Windlasses, horse-whims, rockers, 'cradles', arastras for crushing ores were all old devices which could be worked by the man or animal power of small pioneer groups, and similar crude implements continued to be used by small mine operators in the Old Country or by those who worked over the 'attles' or waste from older workings. Once a mineral discovery was proved, however, and sometimes even before it was or possibly ever could be proved, the American flair for publicity attracted capital which could be used to buy machinery to work it extensively. Pumps of Cornish design were installed by many mining companies in Upper Michigan and on the Pacific Coast, along with stamping mills, but Americans soon modified the designs of such machines, and by the end of the century Dolcoath Mine had installed 'Californian Stamps' which were more effective and economical than those of the Old Country, but the engineering firm of Harvey and Company at Hayle, who installed this plant in 1892 claimed that they had designed the original Californian stamps forty years before although the Pacific Coast engineers had 'considerably improved upon all their working parts'.[57]

From the outset there was one vital difference between the Old Country and America in regard to the employment of machinery. In Cornwall engines were devised and used to perform tasks beyond the

56. Quoted in the *Grass Valley Union*, 30 July 1878.
57. Address of Captain Josiah Thomas at the Royal Institution of Cornwall, on 27 July 1893 (*W.B.*, 3 August 1893).

capabilities of human or animal power; in the American mining regions machinery was regarded as a substitute for manual labour which was far dearer than it was in Cornwall. Furthermore, besides having the advantages of Cornish experience behind them from the very start of their mining ventures, the Americans were developing their mines at a time when many mining concerns in Cornwall had already reached and passed their peak of production, had installed machinery long before, and with declining profits were disinclined and even incapable of installing more modern equipment which was costly both to buy and to operate. The old Cornish engines, too, had been so strongly made that with comparatively little maintenance care they had continued in use for years; even if they were not so efficient as more modern American machines, their owners thought that they were too good and too valuable to be scrapped.[58] Cornish mining ventures, too, had been conducted with the primary aims of supplying moderately stable markets and of yielding regular dividends to shareholding 'adventurers'. In American gold- and silver-mining camps, however, speculative adventurers were avidly seeking quick fortunes, and were all the more ready to scrap efficient but slow machines for novelties which promised vaster and quicker returns. Only the successful mining machinery of the Americans survived; the failures—and there were an incredible number of them—rusted away into oblivion on the junk heaps. Machines, too, that were efficient in Cornwall were not nearly so effective in dealing with very different types of mineral ores in America, while the vaunted Californian stamps proved to be more efficient in crushing the ores of some Cornish mines than with those of others, and proved still more efficient in Tasmania than they ever did at Wheal Kitty or Dolcoath.[59] Milling machinery that could deal with soft ores was useless with some hard rock-stuff; atmospheric conditions of humidity or the reverse affected mechanical as well as human efficiency; varying qualities of fuel consumed by machines could mean all the differences between effective and ineffective and between economical and uneconomical working and so forth. Machines that were the best means of developing regular and wide lodes were costly and inefficient, perhaps even useless in country where ore-veins were narrow, irregular and faulted, and the Americans had far more of the former type than the miners in the Old Country, and wide, regular lodes also meant easier transportation of ores from the mine-face to the hoisting shaft. In hoisting machinery, underground 'tramming', and mechanical transportation of ores from shaft

58. *W.B.*, 22 June and 6 July 1899.   59. *W.B.*, 22 June and 6 July 1899.

to mill, the Americans were definitely far superior by the end of the nineteenth century. They had been readier to adapt mechanical drills in shaft-sinking and level-driving, more prepared to use new explosives like dynamite, and to employ machines to do work which either a machine or a man could do.[60] In American mines, too, the principle of the division of labour was earlier and more extensively applied.

In the mineral regions from Wisconsin and Michigan westwards, mining was the pioneer industry, and it did not begin before modern methods of financing and capitalist organisation were well developed. Almost any good American mining prospect after 1830 could call upon outside capital more easily and successfully than the Cornish mines which had been worked so long by local adventurers, who had contributed goods and services as much as if not more than money to their development. Many of these Old Country mines, too, were nearing exhaustion. Consolidation and amalgamation of mining ventures were facilitated in America by the fact that mineral titles were free of such encumbrances as lord's dues, Duchy of Cornwall rights, and so forth; true, some lawyers did their best to utilise ancient Mexican land grants in the Southwest to foment litigation, and some state legislatures passed laws about apex claims and restricting property-holding by foreigners which hampered mining enterprise although keeping lawyers busy. 'Wildcat' speculation was detrimental to mining progress in both countries, but its effects were often exaggerated by sensational contemporary publicity, and, furthermore, its results included careful appraisals of genuine mineral prospects and resources.

The capitalist nature of American mining from its early days contributed to the appearance of 'company' dominated communities. A rich strike in a hitherto unsettled and remote wilderness meant that any company securing several mineral locations or claims had to recruit labour to work them, and provide all the necessities of life besides tools and implements. Thus some of the early copper mining companies on Lake Superior 'imported' skilled Cornish miners and then, having got them to the mining location, had to provide them with food, shelter, clothing, and medical care besides materials needed for mining. Similar developments took place later in California, notably at Smartsville where, in 1877, a threatened wage cut provoked a strike against the Blue Point Gravel Mining Company which, during the dispute, was

60. *W.B.*, 4 February 1892, letter of 'Cousin Jack' dated Guston, Colorado, 8 January 1892; and Ibid., 30 January 1890, letter of Captain John Daniell, dated Opechie, Michigan, 2 January 1890.

bitterly condemned by the workers for trying to make them buy all their necessities of life from the company store. It was further alleged that this company had dismissed single men and replaced them by married simply to increase profits from store-keeping, had grossly favoured 'company-owned' boarding-house proprietors, had driven an independent butcher out of business by underselling him and by cutting wages of employees so that they could not pay him, and on all goods sold in its store had made profits ranging from twenty to a hundred per cent.[61] The wage dispute issue, however, soon came to predominate, with the workers demanding a minimum wage of three dollars a day and that the Company recognise the right of the miners to form a union to protect their interests. Since at that time the generally depressed state of gold mining had led to the North Bloomfield Gravel Mining Company in the same region reducing the general daily wage to $2.50,[62] without provoking any reaction by the men, it seems that the company store grievance was the trouble-breeding factor at Smartsville. There was some unrest at Moore's Flat, the same spring, where workers had tried to form a 'Protective Association' against a wage cut in 1865,[63] and miners there now copied the Declaration of Independence in a preamble to a series of union organising resolutions.

> We hold this truth to be self-evident, that the laborer is worthy of his hire, and ... it is only by unity of action on the part of the laboring miners, as against aggregated capital, that labor can hope, comparatively, to hold its own, and ... all experience teaches us, that without unification on the part of labor, all effort at the best is but spasmodic, ill-timed, and futile.[64]

These Californian labour troubles in 1877 were, apparently, limited to the gravel-mining or hydraulic companies, which worked on a wage rather than on a contract or 'tribute' labour system. As a rule, the company town was more associated with copper than with either silver- or gold-mining since the discoveries of the more precious metals usually led, in the regions where they had been found, to the proliferation of claims and companies, and the attraction was so great that no company had any need to attract labourers in by guaranteeing them living accommodation and board. Where companies owned a greater or smaller proportion of available dwelling houses, they were usually as fair in the terms of rents and tenures they demanded as were the owners of 'tied' cottages

---

61. *Grass Valley Union*, 18 February 1877, quoting the Marysville *Appeal* of 15 February 1877.
62. *Grass Valley Union*, 22 February 1877.
63. Ibid., 19 February and 29 March 1865.
64. Ibid., 15 March 1877.

in Cornwall, although the latter were more often associated with farmholdings than they were with mines or quarries. When the Great Depression practically ended the Copper Range of Upper Michigan in the early nineteen-thirties, many of the companies sold their houses to the 'tenants' for mere token payments.

Changing times rather than any essential difference between the Cornish and the American environment account for the variation in the types of amenities which mining companies afforded to their employees. In the mid nineteenth century more or less direct help to churches was given by mining magnates in both Cornwall and Michigan. Mine doctors were found in both countries, although the pioneer American camps were rather more likely than the Cornish mining hamlets to attract the less desirable members of the profession, men who through incompetence or other causes had found it personally expedient to leave older settled communities, or feckless adventurers whose personality unfitted them for success in conventional social surroundings. Many of these doctors were, by the standards of their day, quite competent to treat common ailments or injuries, and an outright incompetent might be driven out of a mining town, like 'Doctor' Blackwood from the South Pioneer Mine, in Arizona, in 1883.[65] Lack of adequate facilities, perhaps, more than a lack of professional skill, accounted for the high mortality consequent upon amputation operations in the eighteen-seventies and even later, but Nevadan and Montanan doctors seem to have done little in the late nineteenth century to arouse a public demand for preventative measures against phthisis. Butte's infant mortality returns in the first decade of the present century suggest that the effective treatment of pediatric complaints was less advanced than in the Old Country. After twenty years of quartz mining had incapacitated or maimed dozens of miners permanently in Grass Valley, a scheme to found a Miners' Home by an organised lottery evoked a feeble response.[66] In Upper Michigan the Calumet and Hecla Mining Hospital was opened at Calumet as early as 1867, and in 1872 the Cleveland Cliffs Iron Mining Company built a hospital at Ishpeming;[67] limited though such amenities were, things were not much more advanced in the Old Country until the present century. Before hospitals were provided, many operations had to be performed in the homes of the afflicted and injured, and the real wonder is not that mortality rates were high

65. See above, p. 226.
66. *Grass Valley Union*, 2 April, 17 September and 30 November 1871.
67. *Medical History of Michigan* (eds. C. A. Burr et al.), pp. 626–27.

but that they were not even higher. It was hardly surprising that the men in the Consolidated Virginia reckoned their treatment of occasional 'heat cases' preferable to calling on and waiting for professional medical attention.

Faith in 'quack' medicines, judging from the advertisements in contemporary newspapers, was common in both countries, but to America the Cornish immigrants brought belief in 'white magic', superstitious remedies, and the like, although these were not as prevalent in the new country as in the old. Treating a stye on an eye by passing the tail of a black cat over it nine times was known in Michigan as well as in Cornwall, but whereas in the Old Country black cats were regarded as lucky, in the States they were regarded as of ill omen. The belief that wearing ear-rings strengthened the eyesight survived emigration, as did the belief that a picture falling from the wall on which it hung forbode tidings of death. The 'apple a day, keep the doctor away' adage persists in both Michigan and Cornwall,[68] while the Tommy or Johnny Knockers, the mischievous sprites that haunted the mines of Cornwall, were believed by the Cousin Jacks to have accompanied them to Montana. Faith in the lucky horse-shoe nailed upside-down to the door of a house became quite common in both northern Michigan and California.[69] In the latter state, however, it was regarded by some not as a good luck charm so much as a safeguard against the entry of evil spirits and hobgoblins. A Marysville journalist seeing a horse-shoe nailed to a door in one of the main streets of that mining town in 1870 asked the owner why he had nailed it there; he was told that it was 'to keep the devil out', and that this was particularly necessary because, next door, was a Chinese wash-house and since the Chinese were continually burning paper to exterminate evil spirits their neighbour felt that if he did not 'keep that horse-shoe there as a defense, the tarnal imps would enter my store as fast as they come out of John's wash-house. As it is I am never troubled with them.'[70]

Belief in ghosts and in malign powers of darkness was by no means a monopoly of the Celtic immigrants. Many of the 'lost' mines of the Spanish Southwest were believed to be haunted; the famous tunnel that Sutro drove to drain the Comstock mines in the early eighteen-seventies gained such a notoriety before many years had passed. Some miners in the Caledonia Mine in the Gold Hill district got so badly scared by

68. R. M. Dorson, *Bloodstoppers and Bearwalkers*, pp. 116 ff.
69. *Grass Valley Union*, 30 July 1870: R. M. Dorson, op. cit., p. 116.
70. *Marysville Appeal*, 28 July 1870, quoted by *Grass Valley Union*, 30 July 1870.

inexplicable groaning sounds on a September night in 1878 that, after a vain search for an injured miner, they thought the noises were a supernatural warning of danger and left the mine, telling the shift-boss that, if he liked, he could fire off the dynamite charges for which they had drilled the holes, but nothing would persuade them to do so themselves. Commenting on the incident, the *Gold Hill News* remarked that if there was 'anything in ghosts at all, it does seem as though the mine in which so many have been killed would be the place to hunt for them'.[71] Nevertheless, in Cornwall ghosts frequented parsonages or manor-houses rather than mines and Cornish mines had few traditions of ghosts, although many firmly believed in Johnny or Tommy Knockers and in those mischievous piskeys whose jack-o'-lantern lights had led so many astray on moorland paths as they returned, on dark nights, from the bal, the chapel, or, more likely, kiddleywink.

What many Cornish miners and others did believe in was the divining or dowsing rod, and American as well as Cornish newspapermen now and again expressed opinions ranging from scoffing scepticism to out-and-out faith that any man 'with the gift' could sense the presence of subterranean water or mineral lodes by the twitching of a forked twig grasped in their hands.[72] Almost any forked stick would do, or even bent wire, but in Cornwall hazel rods were most favoured. Scientists in recent times have become more hesitant about expressing any dogmatic opinion on the practice, but in earlier times they were inclined to dismiss dowsing as worthless superstition, as did Nevada's *White Pine News*, in the spring of 1871, when it reported that

Several thousand dollars, in good gold coin, have been expended in sinking a shaft near the Treasure Hotel by authority of the divining rod. The shaft is now down one hundred and eighty feet through solid limestone. There are no indications yet of even the colour, but the work is still to go on—an example of faith not often witnessed.[73]

Still earlier scientists had dismissed the divining rod as necromantic fraud,[74] although in fact the weight of evidence may seem to vindicate dowsers or diviners. The divining rod was certainly more effective than some of the gold-finding devices flamboyantly advertised in the American newspapers in the days of 1849, like the 'goldometer' of Don

71. Quoted in *Grass Valley Union*, 27 September 1878.
72. Ibid., 3 and 9 February 1871.
73. Quoted by the *Grass Valley Union*, 28 April 1871.
74. Vannoccio Biringuccio, *Pirotechnia* (New York, 1943) p. 14. Biringuccio's work was first published in Venice in 1540, a year after his death. A recent account is that by Kevin Danahar, *In Ireland Long Ago* (Cork, 1962), pp. 128–33.

Jose d'Alvear, a 'Spanish geologist' who claimed to be the first real discoverer of gold in California and yet who was willing to sell a gadget which he declared would certainly indicate the presence of gold lodes for a paltry three dollars.[75]

The prevalence of old superstitions and outmoded beliefs among the Cornish and other immigrant groups in America has been given undue publicity by latter-day romanticists, anthropologists, and social psychologists. At least nine out of every ten immigrants were hard-headed realists who came believing that America afforded greater opportunities for achieving material welfare and comfort than the Old Country. The traditional free labour contract system of Cornish mining and the egalitarian tendencies in Methodist doctrine accounted for many of the Cornish immigrants affiliating themselves to the Republican party in politics. Caste antagonisms of the Old World were not to be found in the States, and hardly appeared even when it seemed that a moneyed plutocracy had gained inordinate power in the Republican party, for Cousin Jack had long been accustomed to making his own bargains about work with his bosses although he had been less at ease in dealings with aristocratic land-owning squires and parsons. When the champion rock-drillers, Davey and Teague returned to Butte from Helena in the summer of 1892, they were flaunting the colours of Benjamin Harrison.[76] Politics acquired in America and encouraged by Cousin Jacks still in the States contributed to the defeat of the 'Liberal' squire Pendarves Vivian by the 'Radical' lawyer Conybeare in the Camborne Mining Division election to Parliament in 1885.[77] Although seriously minded people were prone to deplore the hurly-burly of American elections, Cousin Jacks who had spent some time in America became active among those who, in the Old Country, demanded the extension of the franchise to all adult males, and they also became converts to the belief that, in politics, one man was as good as any other to represent the interests of all people in a legislative body and that government and administration should not be monopolised by a landowning caste.[78]

75. Advertisement in the *New York Advertiser*, n.d., quoted in the *Wisconsin Express*, 16 January and 6 February 1849.

76. *W.B.*, 18 August 1892.

77. *Cornish Telegraph*, 4 February 1886, letter of Anthony A. Johns, dated Pinal, Arizona, 5 January 1886; and Ibid., 11 February 1886, letter of George Lord, dated Alma, Colorado, 12 January 1886. The election had been held on 5 December 1885, and the result declared the next morning had shown Conybeare with 2,926 votes to Pendarves Vivian's 2,577.

78. *W.B.*, 21 May 1885, letter of Edward Skewes from Colorado, dated April 1885.

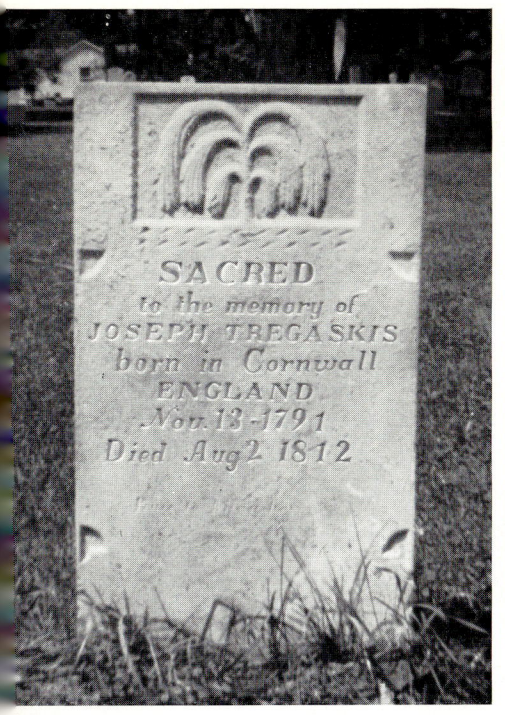

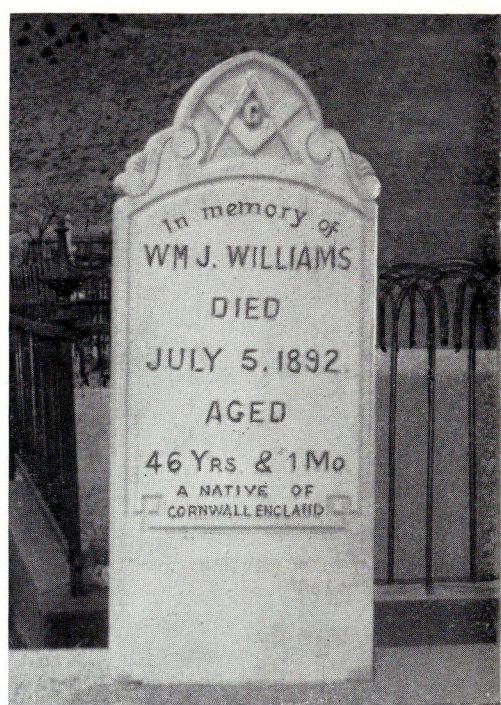

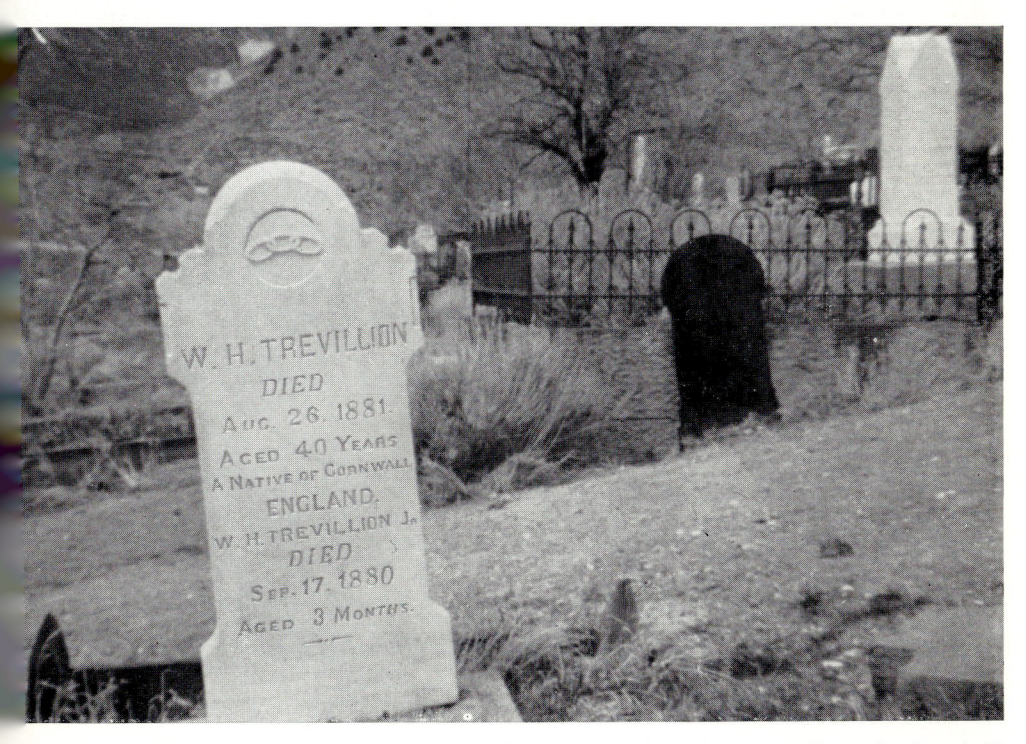

19 The graves of Cornish immigrants at Mineral Point, Wisconsin (top left) and at Virginia City, Nevada (top right and bottom)

1853.   In the
        Memory of my Dear Son, William Wearne.
born at Hayle March 20— 1822 Died Sunday Dec.r 19. 1852.
            aged 29 Years & 9 nine Months,
By Trade a Sailor, & a good workman well versd in History a good
Scholar very kind, and affectnate Brouth in Consoling & Simpithysing,
    He and his sister Biddy left this Place the 24 Nov 1852. for
California he dyed on the Isthmus at the 7 Mile House with the Pyanama feavor
75 others died on their Passage
            1.
William My Son for thee I murn   Your Chance of going you would give
My Harts ore whelmd with greef and Pain And be with us that youse you kind
Thou went away I then did say   5.
Never I shall see thee more again   tis if must so Biddy would go
            2.   With rashness she did spend her breath
Thou young in Prime and Best of time   your latest words is now my foe
My best advice to thee I gave   tis if you knew would be your death
To stay with me from dainger free   6.
That I with Peace go to my Grave   On Isthmus ground that obvious Beast
            3.   With no Respect that Monster Proud
But all in vain us for great gain   Seasd My Dear Sons affectnate Breast
My Sying Brest would glad Cesment   Interd without Coffin or a Shroud.
And save you from that dreadful Rain   7.
thats Subject to that Element   Three Months tardey glide away
         the   My Anchious Mind did sore an run
you would at last I do beleive   Then I did hear that sad dismay
give up all thoughts with settld mind   To God I die for thee My Son

20 'Trails for gold lead to the grave'—memorial verses on William Wearne, who died at Panama, 19 December 1852

If American politics, through popular suffrage, were livelier, more concerned with the needs and interests of the 'common' man, and evoked more general response than was usually the case in the Old Country, yet the vast majority of immigrant Cousin Jacks were content merely with voting for the candidates or party platforms which attracted them most or antagonised them least every second or fourth year. One of them, Edward Skewes, in Colorado, shrugged off Blaine's defeat in the 1884 Presidential Election with the remark that the first man anyone happened to meet in any American city or town was 'as well qualified for President as either Messrs. Blaine or Cleveland, and their moral lives would be just as blameless'.[79]

The majority of the Cornish immigrants were 'common men'. Forgotten men, too, in their homeland, for it was by emigration only that many of them obtained not only a better material livelihood but, for the first time, the full status and rights of citizenship. If several emigrated with dreams of fortune awaiting them, there were but few who gained or even desired fame. A conservative clannishness, resulting from provincial isolation in the homeland, tended to mark them apart from the generality of English-speaking immigrants, while the very fact they spoke English with such a distinct brogue further distinguished them from the English-speaking Americans among whom they had come to dwell. The distinctiveness was prolonged by the high proportion of marriages in the first generation of immigrants among themselves; with the second and third generations, however, marriages with other 'races' became general, and Cousin Jacks in the American 'melting pot' merged with other immigrants, new and old, to become 'brother Jonathans'. Many of them, however, never forgot the Old Country whence their forebears had come.

Their contributions to the development of America are hard to assess. The greater proportion of them were hard-rock miners who played an important role in the settlement of the mining frontier. In their highly skilled calling they contributed to the growth of that technological civilisation of which the United States became the outstanding pioneer example. Undoubtedly materialistic, this civilisation has facets which its many critics have ignored. It has lacked something in the way of traditions, but this has left it free from outmoded beliefs and customs, and this has enabled it dynamically to adapt itself to the ever-changing conditions of historically developing human environment. There has been glib talk of 'commonsense' from

[79] W.B., 21 May 1885, letter of Edward Skewes.

pamphleteers and politicians, but it was the working miners that were practical exponents of that virtue, men who knew how to develop the resources of the earth, not prospectors who could be lured away by any El Doradan will-o'-the-wisp. The aversion of the best Cornish miners towards the practice of 'picking the eyes out of a lode' contributed something towards the adoption of policies of conservation of natural resources in America.

That they, along with thousands from other countries, were immigrants, accounts in some part for the mobility which is so marked a characteristic of the American way of life. Having moved once, they were not, for the most part, unprepared or reluctant to move again, although it is difficult to make any generalisations on this matter. The miner went where mineral lodes existed; he moved on when they became exhausted or when he thought there were better prospects elsewhere. Other immigrant races, too, had the wanderlust, especially the English-speaking Scots and Ulster Irish who came out in Colonial times. The trait, however, was hardly shared by the 'peasant' folk who migrated in such numbers from Europe to America during the nineteenth century; by tradition virtually tied to the land they came, for the most part, to the New World seeking a place where they might settle for generations. Pennsylvania 'Dutch', 'Southern' Irish in Boston and New York, Germans in Wisconsin, Finlanders in Michigan were all far less likely to move a second, let alone a third or a fourth time than the Cousin Jack miner. That very mobility, too, tended to accelerate the dispersion of Cornish immigrants and their merging into a common new 'American' race.

Modern facilities of communication have helped to strengthen links with the Old Country that in many cases had become weak and tenuous, and whose links, too, have been re-forged by the genealogical penchant of the American people. The grandchildren and great-grandchildren of immigrants take as much and a similar type of pride in them as those whose forebears came over on the *Mayflower* or fought at Valley Forge. The search for and veneration of ancestors might be derided as a type of snobbery, but it is also a manifestation of American individualism against the excessive conformist tendencies of the modern age, and, in the case of the Cousin Jack, is a re-assertion of that provincialism which still persists in referring to the counties east of the Tamar as a half-foreign England, just as the Texans and the people of Maine refer to 'the other States'.

Individualism and provincialism, indeed, may explain why the Cornish immigrants have occupied a position out of all proportion to their numbers among the racial groups which have been settled in the United States. Many of them, on both sides of the Atlantic, prided themselves on being Cornish and 'English', a fact revealed now and then in Census Returns and on memorial stones to their dead. Despite their occupational and clannish concentration in certain areas, the Cornish have never been more than a small percentage of the population of the United States. All estimates of their numbers are pure conjectures, but it is hardly likely that first-generation Cousin Jacks and Jennies have exceeded a hundred thousand in any Federal Census, and probably scarcely half that number on the eve of the Civil War when the total population of the Union was approaching thirty-one and a half millions.[80] Second- and third-generation immigrants would swell the numbers, but hardly more than in proportion to the increasing total population of the United States. Furthermore, the second and third generations looked upon themselves as Americans, and although the children of the immigrants tended to marry among their own immigrant group, their grandchildren did not to such an extent. Nor should it be forgotten that while the majority of the immigrants looked on the Old Country as home, there were those who had left it without regrets, had shaken its very dust from their feet, and were resentful of even being reminded of Cornwall and of the hard lot they had endured there.

To many of the latter Americanization came easily, but several of them were environmental misfits who could never find a congenial home anywhere. Most Cornish immigrants, however, fell into two groups—those who settled in the United States and those who either returned home or continued their wanderings elsewhere. Even the latter 'birds of passage', when they were in America, contributed to its growth and development, while many of those who stayed in the United States moved from place to place, following mineral strikes on the vanishing mining frontier and eventually being caught up by the tide of urbanisation which has characterised the recent social development in both America and Britain. Detroit, San Francisco, Los Angeles, Seattle, New York, and Washington have all drawn Cornishmen from 'provincial' America just as in Britain, Plymouth, Birmingham, Cardiff, and, above all, London became the home cities of many Cousin Jacks, when the mines of the western county closed down and its farms became highly mechanised.

Provincial isolation made the Cornish peculiarly distinct from 'up-

80. The population of Cornwall itself in 1860 was about a third of a million.

country' England, and some provincial characteristics do not quickly pass away but can even be transplanted to another continent. While in some respects, on the western side of the Atlantic, the Cornishman quickly became Americanised, in others he retained his distinctiveness. The pressures towards conformity have been great, but that has been true of the entire western world. In any case, Cornish immigrants hastened the development and settlement of the American mining frontiers, particularly in Upper Michigan, the Upper Mississippi Valley and the vast region that came to be known as the 'Pacific Coast' in mining circles. To those frontiers they brought qualities of individual initiative and of neighbourliness which had been—and still are—fundamental characteristics of that which is most progressive and desirable in both the Old Country and the new, qualities by no means limited to a passing frontier, whether that frontier be forest, prairie. or mine. Before they came to the mining frontiers of the United States, the Cornish had had experiences of mining frontiers in their own homeland —indeed, it could be claimed that Cornwall could well have been called the mining frontier of England at the time the Thirteen Colonies were winning their independence. Fortunately, political divisions did not end Anglo-American economic interdependence, nor check the migration of common people to places where great and more assured opportunities of material welfare seemed to exist. The hosts of Cornishmen who emigrated to the United States, who found a living and settled there, helped to weld once more and more strongly the ties of common interests and affection that bind together the English-speaking people on both sides of the North Atlantic.

APPENDIX I

Public sales of copper ores in Cornwall and Devon 1845–75

Year	Amount of Copper Ore (1,000 tons)	Value of Copper Ores (£1,000)	Yield in Metallic Copper (per cent)	Year	Amount of Copper Ore (1,000 tons)	Value of Copper Ore (£1,000)	Yield of Metallic Ore (per cent)
1845	162·6	920	12·9	1861	176·1	1,013	11·4
1846	150·4	796	11·9	1862	186·7	977	11·7
1847	148·7	889	12·0	1863	176·2	872	11·2
1848	155·6	720	12·9	1864	166·7	859	10·4
1849	145·0	764	12·1	1865	164·9	807	10·1
1850	150·9	840	11·8	1866	148·8	679	9·1
1851	154·3	783	12·2	1867	125·7	548	8·3
1852	152·8	976	11·7	1868	121·8	554	8·1
1853	180·1	1,155	11·8	1869	103·2	431	6·9
1854	184·9	1,193	12·0	1870	90·2	375	6·3
1855	195·2	1,264	12·6	1871	74·3	292	4·9
1856	206·2	1,242	13·5	1872	67·5	316	4·4
1857	192·1	1,201	12·2	1873	61·7	271	4·0
1858	183·4	1,058	12·1	1874	51·3	218	3·6
1859	183·5	1,096	12·2	1875	47·8	239	3·3
1860	180·0	1,079	11·7				

(Based on R. Hunt, *British Mining*, p. 892, and *Mining Journal*, London, 7 August 1875.)

APPENDIX II

Shipments of rough copper from Lake Superior 1845–61

Year	Tons (2,000 lb.)	(Value $)	Chronological notes
1845	—	390	
1846	29	2,610	
1847	239	107,550	
1848	516	206,400	
1849	753	301,200 ⎫	Migration of miners
1850	640	264,000 ⎭	to California
1851	872	348,800	
1852	887	300,450	
1853	1,452	508,200	
1854	2,300	805,000	
1855	3,196	1,437,000	Completion of 'Soo' canal
1856	5,726	2,400,100	
1857	5,759	2,015,650	Financial panic
1858	5,896	1,610,000	
1859	6,041	1,932,000	
1860	8,614	2,520,000	
1861	10,337	3,180,000	Outbreak of Civil War

(Based on *Mining Journal*, London, 28 February 1863, quoting *Lake Superior News*, n.d.)

APPENDIX III

Production of fine copper, Michigan and Cornwall 1862–72

	MICHIGAN		CORNWALL	
	Ingot Copper (tons)	*Value ($)*	*Refined Copper* (tons)	*Value ($)*
1862	8,000	3,402,000	13,163	4,689,600
1863	6,500	4,420,000	12,870	4,185,600
1864	6,500	6,110,000	13,151	4,123,200
1865	7,000	5,145,000	12,623	3,873,600
1866	7,000	4,760,000	11,745	3,259,200
1867	8,200	4,140,000	11,373	2,630,400
1868	9,935	4,592,000	10,260	2,659,200
1869	12,200	5,368,000	9,348	2,068,800
1870	12,946	5,696,240	9,078	1,800,000
1871	12,857	6,171,360	7,818	1,401,600
1872	12,132	7,774,720	7,103	1,516,000

For comparative purposes the Cornish output has been calculated in short (2,000 lb.) tons; the dollar is based on the par value of 4s. 2d., but this does not take into account the 'greenback' inflation caused by the Civil War, and it is doubtful if throughout this period the dollar averaged half its par value. At best these statistics provide only a rough indication of the rise of the Michigan and the decline of the Cornish copper-mining industries. Further discrepancies in different sets of statistics are caused by Michigan not 'shipping' its entire 'raised' output some years and even more in others; by the Cornish-mine accounting year ending on 30 June; by the Cornish use of a 21-cwt. ton in ticketing sales of copper ores; and by private sales of Cornish ores, probably up to a fifth of the entire output in some years. (Based on Hunt, op. cit., and *Mining Journal*, London, 28 August 1875.)

Bibliography

The following is by no means an exhaustive list of works on Cornish miners and American mining history. It does, however, record the main sources and authorities used in the present work from the American side. I have refrained from an extensive relisting of works on Cornish history found in the bibliography of my *Cornwall in the Age of the Industrial Revolution* (Liverpool University Press, 1953).

ORIGINAL MANUSCRIPT SOURCES

Boston and Lake Superior Mining Association Correspondence: Michigan Historical Collections, Ann Arbor.
Brockway Papers: Michigan Historical Collections, Ann Arbor.
Census, Federal, returns of
  1. Arizona, 1880.
  2. California, returns for Bloomfield, Eureka, and Grass Valley, 1870
  3. Colorado, returns for Boulder and Gilpin Counties, 1880.
  4. Illinois, returns for Jo Daviess County, 1850 and 1860.
  5. Michigan, returns for Houghton and Ontonagon Counties, 1860.
  6. Nevada. 1870.
  7. North Carolina, returns for Rowan County, 1860 and 1870.
  8. Oregon, 1860.
  9. Wisconsin, returns for Iowa County, 1850.
Census, State, returns of
  1. Michigan, Upper Peninsula returns for 1854 and 1864.
  2. Wisconsin, Iowa County returns for 1836 and 1842.
Marquette Historical Society: Northern State Normal School Collection of Local Biographies.
Robinson, Orrin W. Papers: Michigan Historical Collections, Ann Arbor.
Taylor Papers: Michigan Historical Collections, Ann Arbor.
Wearne, Richard: MSS Journal.
Wilson MSS Boulton and Watt Correspondence: Royal Cornwall Polytechnic Society.

BIBLIOGRAPHY

OFFICIAL PUBLICATIONS

Butte, *Annual Reports of the City Officers*
Fuller, G. N. (ed.) *Messages of the Governors of Michigan*, Lansing, 1936
Michigan, *Vital Statistics of: Annual Reports ... of Births, Marriages and Deaths*, 1868–
United States Department of Labor Statistics: Bulletin No. 139: Michigan Copper District Strike; February 1914 (Senate Doc, No. 381, 63rd. Congress, 2nd. Series).
W.P.A. Guides–*Colorado, Idaho, New Mexico, Montana, Nevada, South Dakota, Utah* and *Wisconsin*.

JOURNALS, NEWSPAPERS, AND PERIODICALS

Alta California San Francisco
Bath and West of England Society Journal
Boulder Daily Camera, Colorado
British Columbia Historical Quarterly
Butte Daily Miner
Butte Tribune Review
California Folklore Quarterly
Colorado Chieftain, Pueblo
Colorado Magazine,
Colorado Quarterly
Cornish Telegraphy, Penzance
Cornishman, Penzance
Detroit Evening News
Grass Valley Union
Hunt's Merchant's Magazine and Commercial Review
Madison City Express, Wisconsin
Madison Express, Wisconsin
Milwaukee Journal
Mineral Point Tribune
Miner's Free Press, Wisconsin
Mining Journal, London
Mining Journal, Marquette, Michigan
New North West, Butte
Populist Tribune, Butte
Reese River Reveille, Nevada
Royal Cornwall Gazette, Truro
Silver Belt, Globe
Tombstone Epitaph, Arizona
Tombstone Prospector, Arizona
Tribune Review, Butte
West Briton, Truro
Wisconsin Express
Wisconsin Historical Collections
Wisconsin Magazine of History
Wisconsin Society of Science, Arts and Letters, Transactions

BOOKS, ARTICLES, AND PAMPHLETS

BANCROFT, CAROLINE, 'Folklore of the Central City District, Colorado', *California Folklore Quarterly*, **4,** 1945.
BENEDICT, C. H., *Red Metal: The Calumet and Hecla Story*, Ann Arbor, 1942.
BIRINGUCCIO, VANNÓCCIO, *Pirotechnia*, New York, 1943.
BOTKIN, B. A., *A Treasury of Western Folklore*, New York, 1951.
BURR, C. A., et al. (ed.), *Medical History of Michigan*, Minneapolis, 1930.
CHICANOT, E. L. (ed.), *Rhymes of the Miner*, Gardenvale, Quebec, 1937.

BIBLIOGRAPHY

COPELAND, L. A., 'The Cornish in South West Wisconsin', *Wisconsin Historical Collections*, **14,** 1898.
DARLETH, A., *The Wisconsin*, New York, 1942.
Detroit Annual (Methodist) Conference, *History of Methodism in the Upper Peninsula of Michigan*, 1955.
DICKINSON, H. W., and TITLEY, A., *Richard Trevithick, the Engineer and the Man*, Cambridge, England, 1934.
DORSON, R. M., *Bloodstoppers and Bearwalkers*, Harvard, 1952.
FORSTER, J. H., 'War Times in the Copper Mines', *Michigan Pioneer and Historical Collections*, **18,** 1891.
FRANCIS, W., *Gwennap—A Descriptive Poem*, Redruth, 1845.
HALFORD, C. N., *History of Grant County, Wisconsin*, 1900.
HARRIS, N. R., *Cornish and Welsh Mining Settlements in California*, unpublished thesis, Berkeley, California.
HAVIGHURST, W., *The Long Ships Passing*, New York, 1942.
HOWAY, E. W., *Early History of the Fraser River Mines*, Victoria, B.C., 1926.
KOHL, J. G., *Reisen Im Nordwesten Der Vereinigten Staaten*, New York, 1857.
LE BOURDAIS, L., 'Billy Barker of Barkerville', *British Columbia Historical Quarterly*, July 1937.
LEWIS, O., *Sea Routes to the Gold Fields*, New York, 1949.
LILLARD, R. G., *Desert Challenge—An Interpretation of Nevada*, New York, 1949.
LORRE, L. P., 'The First Steam Engine in America', *Transactions of the Newcomen Society*, **10,** 1903.
MARTIN, J. B., *Call it North County*, New York, 1945.
MERK, F., *Economic History of Wisconsin during the Civil War Decade*, Wisconsin State Historical Society, 1916.
Michigan, Biographical Record of Houghton, Baraga, and Marquette Counties, Chicago, 1930.
Michigan, History of the Upper Peninsula of, Chicago, 1892.
Michigan, Memorial Record of the Northern Peninsula of, Chicago, 1895.
MORGAN, D. L., *The Humboldt—River of the West*, New York, 1943.
MURDOCH, A., *Boom Copper*, New York, 1943.
NEAL, ROBERT M., 'Pendarves, Trelawney and Polperro—Shake Rag's Cornish Houses', *Wisconsin Magazine of History*, 1944.
NELSON, J. R., *Lady Unafraid*, Caldwell, Idaho, 1951.
NUTE, GRACE LEE, *Lake Superior*, New York, 1944.
PERRIGO, L. L., 'The Cornish Miners in Early Gilpin County', *Colorado Magazine*, **14,** 1937.
PIKE, Z. M., *Journal of a Voyage to the Sources of the Mississippi*, Philadelphia, 1808.
PITEZEL, J. H., *Lights and Shades of Missionary Life*, Cincinnati, 1860.
PRYCE, W., *Mineralogia Cornubiensis*, London, 1778.

BIBLIOGRAPHY

RICKARD, T. A., *History of American Mining*, New York, 1932.

ROBINSON, O. W., *Early Days of the Lake Superior Copper Country*, unpublished typescript in Michigan Historical Collections, Ann Arbor.

ROWE, MARY E., 'Little Bit of Cornwall Lives on Shake Rag Street', *Milwaukee Journal*, 9 October 1948.

ROWSE, A. L., *The Cornish in America*, London 1969.

SAWYER, A. L., *A History of the Northern Peninsula of Michigan*, Chicago, 1911

SCHAEFER, J., *The Wisconsin Lead Region*, Madison, 1932.

SMITH, ALLARD, J., 'The Lead Region of Dubuque County, Iowa', *Transactions Wisconsin Society of Science Arts and Letters*, **12,** 1900.

SPRAGUE, M., *Money Mountain*, Boston, 1953.

STRATTON, D. H., 'The Cousin Jacks of Caribou', *Colorado Quarterly*, **1,** 1952–3.

Wisconsin, *Commemorative Biographical Record of Rock, Green, Grant, Iowa and Lafayette Counties*, Chicago, 1901.

Wisconsin, *History of Iowa County*, Chicago, 1891.

Index

Abrams, Edward T., 86
Accidents, mining, 180, 184, 198, 211, 213, 221, 226, 242, 248, 263
Adventure Mine, Mich., 79, 80
Agate Harbor, Mich., 71
Agriculture, conditions of in Canada and Cornwall in 1830 compared, 28–30; depression in Cornwall, after 1815, 20–1; methods in the U.S., 27; in Illinois, 49–50; in Michigan, 260; in Washington, 130ff; in Wisconsin, 48, 260–1
Alaska, 227
Albion Mining Company, 71
Alder Gulch, Mont., 236
Alford, William, 62
Algonquin Mine, Mich., 281
Alice Band, 274, 275
Alleghenies, 27
Allen, A., 223
Allen, Philip, 45, 57
Allison Ranch Mine, Calif., 223
Amador County, Calif., 125, 275
American Flag Mine, Ariz., 219
American Institution of Mining Engineers, 205
American River, 97, 103
Americus Mine, Colo., 198
Anaconda, Mont., 238
Andersonville, 150
Andrews, Richard, 272
Anglesey, 3, 18–19, 23, 141
Apache Indians, 113, 138, 146, 150, 170, 177, 209, 215, 216, 224
Apex Claim Law and Litigation, 240ff
Appalachians, 276
Argle, Francis, 272
Arizona, 1, 113, 125, 147, 148, 150, 170, 177, 185, 186, 198, 207–8, 210, 213, **215-27**, 257, 265, 267, 289

Arizonan Emigration and Colonization Societies, 216
Arkansas, 179
Arndt, Charles, 41
Arnold, William, 217–18
Aspinwall, 103
Aspinwall, William, 74ff
Astoria, 229
Atlanta Star Mine, Idaho, 231–2
Atlantic (ship), 69
Attleborough, Norfolk 238
Attwood, Supt, 239
Austin, Nev., 184, 193, 194, 240n
Australia, 25, 32, 35, 61, 113, 125, 138, 171, 183, 189n, 206, 220, 256, 257
Austrian immigrants in America, 200

Bad Ax, battle of, 43
Baffin Land, 3, 97
Baltic, 26
Baltic Mine, Mich., 174
Baltimore 85
Banca, Isle of, 157, 171 *see also* East Indies
Bands brass and silver, 116, 272, 273, 274–6
Bannack, Mont., 228, 236, 267
Bannack Indians, 236
Barberton, South Africa, 220
Baripper, 205
Barker, Billy, 141–2
Barkerville, B.C., 142
Battle Mountain, Nev., 190
Bayliss, R. T., 239–40
Bear Flag Revolt, 138
Beard, William, 227
Bell-ringing contests, 276
Bennets' Temperance Tavern, Buffalo, 74
Bennett, James, 262

305

INDEX

Bennett, John, 118
Bennett, Joseph, 58
Bennett, William, 43
Bennett, —, of Globe, Ariz., 225
Bennetts, John, 221
Bessemer, Mich., 278
Bickford, William, 4
Biddick Diggings, Wisc., 58
Big Bug Mine, Ariz., 216, 219
Bi-metallism, 206, 210, see also Silver
Bingham, N.M., 209
Bingham, Utah, 186
Bingham Canyon, Utah, 212, 213, 214
Birmingham, 295
Bisbee, Ariz., 186, 211, 220, 223, 224, 225, 258, 273
Black Country, 24
Black Hawk, 63
Black Hawk War, 43, 44, 133
Black Hills, S.D., 70, 125, 176, 229, **251-6,** 258, 268
Blackhawk, Colo., 205
Blackwood, Dr, 226, 289
Blaine, J. G., 293
Blasting methods, 86-7, 120, see also Dynamite
Blight, Samuel, 272
Blizzard, 'great' of 1891 in Cornwall, 263
Blue Gravel Company, 111
Blue Jay Mine, Ariz., 223
Blue Point Gravel Mining Company, 287-8
Blue Tent Consolidated Mining Company, 111
Bluett, J., 223
Boarding houses, 91-2, 118, 238
Bodie, Nev., 186, 192
Bodmin, 249
Boer War, 150, 206, 227
Bohemian Company, 71
Boise, Idaho, 176, 186, 229, 231, 249, 255, 268
Boise River, 152, 170
Bolivia, 25
Bonaparte, Napoleon, 178
Boot Hill Cemetery, Tombstone, 220
Borlase, William Copeland, 214
Boston, Mass., 70, 71, 72, 74, 88, 98, 100, 112, 165, 201, 294
Boston Company, 71
Boston and Colorado Smelting Company, 205

Boston and Lake Superior Mining Association, 73-6, 77
Boston and Montana Band, 274, 275
Botany Bay, 25
Boulder, Colo., 204
Boulton, Matthew, 18, 19
Boulton and Watt, 18, 20
Bowen, John, 184
Bowman Lake, Calif., 111
Boxing 271-2, 273
Brazil, 25
Breage, 101, 159, 181, 221, 267
Breeden, 'Charley', 41
Brew, Chartres, 140
Briggs Mine, Colo., 198
Brilliant (steamer) 34
British Columbia, 61, 132, 140, 231, see also Fraser River
British Hollow, Wisc., 58
British Museum, 133
Broad, F., 223
Brown, John, 143
Brown, —, prospector, 237
Browne, J. Ross, 231
Bruce Iron Mines, Ontario, 67, 175
Brunson, Alfred, 67-8
Bryan, William Jennings, 225
Buchanan, James, 212
Buckskin Joe, Colo., 195
Buffalo, N.Y., 74
Buffalo Commercial Advertiser, 70
Bull fighting, 273
Bull Run, 144, 145n
Buller, Dick, 279
Butler, Henry, 102
Butte City, Mont., 58, 82, 114, 160, 176, 184n, 186, 198, 211, 238-9, 241, 242-4, 258, 259, 260, 262-3, 269, 271, 274-5, 276, 282, 289; martial law in, 247-8; polyglot mining population of, 247, 250
Buzzo, Capt. Henry, 81n
Buzzo, 'Rev.' J., 81

Caledonia Mine, Nev., 280
Caledonia Quartz Mine, S.D., 254
California, 23, 39, 48, 52, 53, 55, 56, 57, 59, 60, 61, 70, 85, 87, 92, 95, **96-126,** 127, 133, 134, 136, 137, 138, 139, 145, 150, 153, 159, 169, 175, 177, 179, 181, 182, 185, 186, 187, 196, 197, 210, 215, 229, 230, 232,

INDEX

California—*continued*
  236, 249, 251, 257, 258, 263, 267, 287, 290; agricultural prospects of, 107–8; 109; lecture on at Truro, 101; routes to, 100–1, 102–7
California, Bank of, 217–18
California Gulch, Colo., 195
California Mine, Nev., 284
California Stamps, 285, 286
Calumet, Mich., 58, 154, 278, 289; conglomerate lode, 65, 79, 87, 175
Calumet Mine, Mich., 173
Calumet and Hecla Mine, Mich., 78, 84, 88, 89, 166, 167, 169, 170, 172, 223, 282, 289
Camborne, 1, 3, 7, 19, 25, 43, 44, 45, 46, 48, 58, 62, 108, 114, 161, 180, 198, 205, 223, 277, 292; anti-Irish riots in 1882, 11, 269–70
Cambrian Iron Works, Penn., 173
Canada, 25, 28–30, 50, 148, 183, *see also* British Columbia, Fraser River
Cañon de Oro, Creek, Ariz., 219
Cape Cornwall, 8, 17, 188
Cape Horn, 100, 132
Captains, mining, positions of, in Cornwall and America, 76
Caradon, 19, 62
Caradon Lode, S.D., 255
Carbuncle City, Penn., 276
Cardiff, 295
Cariboo, B.C., 140–2, 152, 159
Caribou, Colo., 116, 199, 204, 205
Carkeek, Tom, 273–4
Carlyon, G., 223
Carmichael, Alex, 252
Carn Menellis, 62, 205
Carnon Stream, 134
Carols, Cornish, 276–7
Carson City, Nev., 271–2
Carson River, 106, 186
Carter, Henry, 11
Carter, 'King' John, 10
Carter, John Henry, 180
Cedar Valley, Ariz., 219
Celtic language of Cornwall, 10
Celts, 2
Cerillos, N.M., 209
Cerro de Pasco, 27
Census returns, federal of 1850, 48; of Gilpin County, Colo., 196; of Rowan County, North Carolina, 143n; of Grass Valley, Calif., 114ff; state census returns of Michigan, 96
Centerville, Idaho, 231,
Centerville, Mont., 241
Central City, Colo., 198
Central Mine, Mich., 79, 84
Chacewater, 116, 160, 172, 180, 195
Chancellorsville, battle of, 148
Chaplin iron deposits, Mich., 175
Chartism, 11, 63
Cherry Creek, Colo., 196
Cheyenne, Wyo., 252, 253
Chicago Tribune, 210
Chile, 25, 116, 164, 166, 187
China, 116
China clay, 78
Chinese immigrants, 111, 118, 123, 136, 137, 194, 222, 223, 225, 236, 290
Chippewa Indians, 64, 67, 70
Chippewa Mining Company, 71
Chivington, Major, 146, 147
Choirs, 116, 276ff, 279
Cholera, 31, 55, 102
Church and chapel in Cornwall, 1
Chyandour Smelting Works, 255
Circle City, 227
Cimarron, N.M., 209
Civil War, American, 78, 90, 94, 120, 138, **146–54**, 158, 161, 162, 163, 165, 196, 207, 230
Class divisions in Cornwall, 13–14, *see also* Landlords
Clark, William, 37
Clark, William A., 246
Clay, Henry, 60, 64
Clay working in Cornwall, 11, 78, 282
Clear Creek County, Colo., 205
Clemens, Christopher, 61, 139
Clemo, Capt. Henry, 73–6, 79
Cleveland, Grover, 293
Cleveland, Ohio, 69
Cleveland Cliffs Iron Mining Company, 289
Cliff Mine, Mich., 78, 80–1, 84, 85, 90, 172, 176, 223
Clift, William, 118
Clifton, Ariz., 223
Clifton Copper Mine, Ariz., 219
Climate, of Cornwall, 263; of Michigan, 90, 92; of Washington, 129–30
Clothing, 265
Cluyas, Richard, 272

307

INDEX

Clymo, Francis, 39, 42
Coal mining, 18, 24, 35
Coburg, Ontario, 28
Cochise, 207, 210
Cochita, N.M., 209
Cock fighting, 57, 273
Cocking, W., 223
Coeur d'Alene, Idaho, 233
Coleman, Henry, 118
Coleman, Walter R., 220
Collins, Colonel, 103-4
Coloma, Calif., 95
Colorado, 39, 73, 125, 146, 170, 177, 179, 185, 186, **195-206,** 209, 210, 211, 220, 225, 227, 230, 232, 246, 251, 252, 257, 283, 293
Colorado River, 216
Columbia River, 229
Colville, 230
Comanche Indians, 177, 209, 215
Combe Martin, 19n
Commonmoor, 267
Company houses, 265, 287-8
Comstock Lode, 1, 113, 170, 176, 179, 182, 184, 185, 186, 195, 209, 231, 238, 246, 284, 290
Comstock, Supt, 226
Confederate Gulch, Mont., 236
Confederate States, 138, 143, 145, 146-8, 163
Connecticut, 38, 63, 85, 197
Connolly, Governor of N.M., 146
Connor, Colonel P. E., 211-12, 214
Consolidated Virginia Mine, Nev., 284-5, 290
Contention Mine, Ariz., 219, 220
Conybeare, C. V., 292
Cooke, Jay, 218
Coolgardie, 189n, 206
Coombs, Joseph, 272
Copper mining in Connecticut, 63; in Cornwall, 18, 157-8, 160-1, 171; in New Jersey, 63; in Wisconsin, 62; by Indians, 63, 65-6; mining companies in Michigan, 71; prices, 157-8, 163, 170, 171; prices in the Civil War, 152-4,; smelting, 4; U.S. imports in 1841, 64, *see also* Bisbee, Butte City, Calumet and Hecla, Chile, Globe
Copper Falls Mine, Mich., 79, 172
Copper Harbor, Mich., 71

Copper Queen Mine, Ariz., 223, 224
Copper Rock, of Ontonagon, 66-7
Copperheads, 149, 150-1, 153
Corbett, 'Gentleman Jim', 272
Corin, John, 219
Cornish, character and characteristics of the, 1, 193; class divisions among, 13-14, 258; conservatism and traditionalism of, 83, 86, 235; cooking, 259-61; dialect, 58, 83-84, 280-1; farming of, 6-9; feud with Irish, 11, 82, 117, 269-70; illiteracy among, 83; immigrants, numbers of, 112ff, 295; mining conditions among, in 1866, 155-6, 158; mining 'bosses', 169; mining terms, 81; pasties, 75-6, 259-60; provincialism, 169; pumps, 119; stamps, 120; surnames, 173, 239; tin-mining, 2-3
Cornish Bar, B.C., 139
Cornish Telegraph, 250
Cornwall, 1-2; carols of, 276-7; Celticism of, 2, 10, 12; climate of, 6, 8, 9; Duchy of, 4, 13, 42, 132, 187; Royal Institution of, 208; saints of, 2
Cortez (ship), 135n
Cotton famine, 155
Couch, Thomas, 178-8
Couch, Capt. Tom, 274, 275
'Cousin Jacks', an alleged origin of name, 280
Cows, renting of in Cornwall, 7, 14
Coyle, Mrs, 221
'Cradle' in mining, 110
Crase, William H., 118
Cream, clotted or scalded, 260-1, 262
Creede, Colo., 195
Creegbraws, 198
Cricket, matches in Montana, 241
Crinnis Mine, 223
Cripple Creek, Colo., 195-6, 206, 209, 220, 227
Crisis of 1850, 59, 137
Crowan, 10, 187
Cruse, Thomas, 239, 250
Cuba, 61, 103
Cumberland, 24
Cumberland Gap, 37
Cumberland Mine, Ariz., 223
Curnow, Capt. John U., 85, **174**
Curnow, Thomas, 276
Currency inflation, 152, 165-6

INDEX

Curtis, Capt. John, 256
Custer, General G. A., 170, 250, 251, 257
Custer, S. D., 252, 253

Daily Miner (Butte), 272
Dakota (Territory), 67, 71, 170, 198, 228, **251–7**, 258
Dakota Farm, West Cornwall, 255
Dale, Edward, 104–7, 109
d'Alvear, Don Jose, 291–2
Daly, Marcus, 246
Daniell, Capt. John, 87–9, 172, 174
Davey, James, 274–5, 292
Davey, Peter, 28–30
Davey, Stephen, 212–13
Davidson, Mount, 191
Davies, Capt., 256
Daw's Saloon, Grass Valley, 117
Deadwood, S.D., 71, 252, 253, 254, 255, 257, 268
Death Valley, Calif., 178, 186
Deer Lodge Valley, Mont., 236
Delabole, 31
Delaware and Pennsylvania Mine, Mich., 172
Denbighshire, 19
Denver, Colo., 148, 186, 201, 205, 206, 227, 256
Denver Republican, 206
Deseret, *see* Utah
Desmond, Tom, 226
Detroit, Mich., 3, 68, 74, 76, 77, 86, 258, 295
Devon Great Consols, Mine, 79
Diamond Grove Diggings, Wisc., 58
Diamonds, 'discovery' of in Arizona, 215, 217–18
Ding Dong Mine, 255
Disasters, mining, 246–8, 263, *see also* Accidents
Divining rod, 43, 291
Dixie, Nev., 190
Dodgeville, Wisc., 51, 55, 58, 150, 163
Dog fighting, 273
Dolcoath, 88, 205, 206, 223, 255, 256, 285, 286
Donner Party, 177, 179
Dorson, Richard M., 84
Douglas, James, 136n, 137
Dowsing rod, 43, 291

Drilling rock, 88, 274–5; single-hand, 120, 236
Drumlummon Gold Lode, Mont., 239ff, 250
Dry Gulch Diggins, Mont., 236
Dublin, 74
Dubuque County, Iowa, 44
Dubuque, Julian, 37, 39
Duloe, 1
Dunn, Charles, 61
Dunstan, R. W., 162–3
Dynamite, 120ff, 213, 236

Eagle Harbor Company, 71
Eagle River, Mich., 71, 73
Eagle River Mine, Mich., 73
Earp brothers, 219, 267
East Caradon Mine, 223
East India Company, 128
East Indies, 157, 161, 171, 206, 257
East Pool Mine, 255
East Vulcan Mine, Mich., 85, 174
East Wheal Rose, 263
Eclipse Mine, Calif., 145n
Economist (ship), 30, 31
Eddy, John, 246
Eddy, William, 242
Edwards, Thomas, 242
El Dorado dreams and myths, 97, 127, 135, 177, 251, *see also* Gold
Elizabethtown, Ky, 218
Emerald Mine, Ariz., 220
Emigrant ships and voyages, 30, 31, 32, 33, 51, 102
Emigration, costs of, 33; reasons for from Cornwall, 155
Emma Mine, Utah, 203, 212–13, 214
Empire Mine, Calif., 119, 120, 121, 122, 123, 223
Enterprise (Virginia City), 284–5
Erie Canal, 68
Erie, Lake, 69
Essex, 24
Etta Mining Company, 225–6
Eudie, James, 142, 143n, 145
Eudy, Joseph, 145
Eureka, Nev., 191, 193, 283
Eureka, Mine, Calif., 124, 223
Eureka Stockade, 138
Evans, William, 249, 250

Fairplay, Colo., 195

INDEX

Falmouth, 73
Family life in mining camps, 199, 219, 222, *see also* Women
Farming, in Illinois, 49-50; in Michigan, 90-1, 260, 262; in Wisconsin, 48, 261, *see also* Agriculture
Father de Smet Mine, S.D., 254
Feather River, 126
Felch, Governor Alphonse, of Michigan, 131
Fever River, 53
Financial panics, 20, 25, 167-8, 182
Finnish immigrants, 259, 294
Fires, 192-3
Fishing, pilchard in Cornwall, 6, 9
Fitzsimmons, Bob, 272, 274
Flagstaff Mine, Utah, 213
Floods, 193, 263
Florida, 72
Fluming companies, 111-12
Food riots in Cornwall, 11, 19
Fort Ellis, Mont., 250
Fort Laramie, Wyo., 105, 251
Fort Yuba, Calif., 179
Fort Yuma, Ariz., 215, 216
Fourth of July celebrations and sports, 241-2, 272, 273, 276
Fox, G. C. and R. W., and Company, 73
Francis, Rebecca Jewell, 92-3
Franco-Prussian War, 171
Franklin Mine, Mich., 172
Fraser River, 23, 49n, 61, 70, 113, 115, 126, 127, **133-42**, 179, 185, 228, 229, 230
Fraternal societies, 245-6
Frémont, J. C., 178
Frobisher, Martin, 3, 97
Fugitive Slave Law, 59
Funerals, 279
Fur trade and traders, 26, 42, 127, 128, 178, 229
Furness, 175
Furniture, 266
Fuse, safety, 4

Galena, Ill., 53, 55, 103
Galveston, Texas, 100
Game Laws in Cornwall, 54
Gardening, in Grass Valley, 262; in Washington, 130
Gardner, John, 41

George III, 10, 258
George, Richard, 118
Georgetown, Colo., 198, 205
Georgia, 97, 99n, 178
Geronimo, 207
German immigrants, 246, 294
Gettysburg, 151
Ghent, Peace of, 27
'Giant Powder', *see* Dynamite
Gila River, 176, 179, 215, 216
Gilbert, Capt., 255
Gill, James, 142-3
Gill, W. E., 109-10
Gillies, William, 219
Gilpin County, Colo., 195, 196, 215, 269
Glen Woody, N.M., 209
Globe, Ariz., 186, 211, 219, 223, 224, 225, 226, 227
Gluyas, James, 254
Gogebic Iron Range, Mich., 90, 173
Gold, discoveries in California, 96ff; in Colorado, 195; in Dakota Territory, 252ff; on Fraser River, 126, 127, 133ff; on the Gila, 179; in Montana, 236ff; in Nevada, 179; in North Carolina, 142ff; mining methods, 110-12; production of the U.S. in 1869, 230; quartz and hydraulic mining, 112; rushes, 96ff, 133, 134, 135, 178, 185, 186-7, 194-5, 215, 227-8, 231, 236-8, 251-2, 253-4, *see also* Australia, Witwatersrand
Gold Bluff, Calif., 185
Gold Hill, Calif., 112, 123, 125
Gold Hill, Nev., 179, 180, 183, 184, 187, 274, 290
Gold Hill, N.C., 143, 144
Gold Hill Mine, Calif., 119
Gold Hill News (Nevada), 291
Golden Star Mine, S.D., 255
Goldfield, Nev., 179, 183, 193
'Goldometer', 291-2
Goldsithney, 242
Goldsworthy, Elizabeth, 261
Goldsworthy, Thomas M., 150
Goldsworthy, William, 51
Gonamena Mine, 223
Grand Canyon, 178
Grand Central Mining Company, Ariz., 219-20

310

INDEX

Grand Rapids, Mich., 90
Granite Mountain Mine, Mont., 246, 247, 248
Grant, President U. S., 202, 203
Grant County, Wisc., 41, 42, 61, 139
Grass Valley, Calif., 58, 95, 112, 113, 114–15, 116, 117–18, 119, 120, **121-4,** 125, 126, 135-6, 145n, 151, 152, 169, 184, 187, 193, 197–8, 211, 223, 232, 239, 254, 259, 262, 265, 270, 272, 274, 275–6, 280, 289; Cornish Carol Choir, 276
Grass Valley Telegraph, 135
Grass Valley Union, 121
'Great American Desert', 177, 180
Great Lakes, 33, 37; navigation of, 68–9, *see also* Erie, Huron
Great Plains, 102
Great Salt Lake, 177, 178, 213
Great Wheal Busy Mine, 176
Greek immigrants, 259
'Greenbacks', 165
Grenfell, John T., 107–8, 109
Grenville, George, 10
Gribble, Mary, 145
Grose, —, strike leader, 167
Guatemala, 49n
Gundry, Joseph, 45
Gunnislake, 116, 188, 267
Gurnard's Head, 8, 188
Gurney, Richard, 90–1
Guston, Colo., 203
Gwennap, 173

Hall, Capt. Josiah, 84
Hall, Capt. W. E., 274, 275
Hallamannin Mine, 176
Halsetown, 221
Hamilton, Nev., 192–3
Hammond, James, 137n
Hancock, Charles, 223
Hancock, R., 223
Hancock, Mich., 93, 143n, 278
Hankin brothers, 141
Hardscrabble, Wisc., 58
Hargreaves, James, 133
Harney, Peak, S. D., 255–7
Harper's Ferry, 143
Harris, D., 223
Harris, Capt. S. B., 174
Harris, Capt. William, 87

Harrison, Benjamin, 292
Harvard, Mass., 279
Harvey and Company, 285
Hayle 34, 50, 52, 167, 168, 285
Hazel Green, Wisc., 58
Health of miners, 56, 193, *see also* Occupational diseases, Phthisis
Hearst, George, 254
Heather, William, 211
Heavy cake, 260
Hecla Mine, Colo., 204
Heinze, Frederick A., 246
Helena, Mont., 186, 237–8, 240, *see also* Last Chance Gulch
Helena Post, 237
Helston, 48, 159, 272
Henningsen, —, company promoter, 216
Hermosa Mine, Ariz., 219
Hidden Treasure Mine, S.D., 254
'High grading', 73, 200
Hill, Alfred, 196
Hill, George, 197
Hill City, S.D., 252
Hillsboro, N.M., 211
Hoar, Richard, 85
Hodge, Harry, 184
Hollow, Thomas, 276
Holton, C. P., 145–8, 196
Homestake Mine, S.D., 254, 255
Hooker, General J., 148
Hooper, John, 272
Hope, B.C., 139
Hornblower, Josiah, 3
Hoskins, John, 90
Hospitals, 289
Houghton, Douglas, 62, 64–7, 70, 72
Houghton, Mich., 81, 85, 93, 94, 167, 262, 264, 278
Houses and housing, 46–7, 51, 265–7
Howard's Hill, Calif., 275
Huachuca Mountains, Ariz., 219
Hudson's Bay Company, 127–8, 133, 134, 137, 230
Hulbert, Edwin James, 66, 87, 175
Humboldt River, 106, 177, 186
'Hungry Forties', 101
Hunt, James, 219
Hunting, 53–4
Huron Lake, 51, 57, 58
Huron Mine, Mich., 173
Hydraulic mining, 111, 112

311

INDEX

Idaho, 39, 150, 174, 186, 227, 228, 230, **231-6**, 268, 283
Idaho City, Idaho, 231
Idaho-Maryland Mine, Calif., 119, 223
Illinois, 6, 37, 39, 42, 44, 48, 49-50, 53, 197
Illinois River, 38
Illiteracy, alleged of Cornish immigrants, 83-4, 143n, 183
Illogan, 43, 44
Immigrants, ages of, 197; numbers of Cornish in California, 112ff, in Michigan, 94; in U.S., 295, *see also* Irish, Italian, Scandinavian
Independence, Mo., 179
India, 25
Indians, Red, 26, 37, 42, 43, 64, 65-6, 101, 104, 128, 133, 135, 140, 170, 177, 178, 198, 207-8, 209, 210, 211, 215, 216, 224, 229, 230, 236, 250-1, 252, 253
Indian Territory, 146
Infant mortality, 55, 289; in Butte City, 243
Iowa, 37, 39, 42, 44, 48, 65, 197
Iowa County, Wisc., 43, 48n, 61, 150
Ireland, 24, 143
Irish immigrants, 184, 200, 246, 247, 250, 271, 294; in Michigan, 167; potato famine, 155; riots against in Camborne, 269-70
Iron mines and mining, 170, 172-5, *see also* Marquette, Mesabi
Iron County, Mo., 162
Iron Mountain, Mich., 175, 278
Ironwood, Mich., 85, 274
Ishpeming, Mich., 172, 174, 273, 276, 289
Isle of Man, 24
Italian immigrants, 198, 200, 246
I.W.W. (Industrial Workers of the World), 246-7

Jackling, Daniel, 213
Jackson, Dr C. T., 72-3
James II, 11
James, Edward, 43
James, Elizabeth, 45
James, Joseph, 43
James, Samuel, 129-33
Janin, Henry, 217-18
Jefferson, Thomas, 26, 178

Jenkins family, 48n
Jenkins, John, 226
Jenkins, Nicholas, 140
Jenkins, Sarah, 48
Jennings, Thomas, 204
Jerome, Ariz., 211, 223
Jewell, John, 232n
Jo Daviess County, Ill., 44, 53, 143n
Johannesburg, 58, 114, *see also* Transvaal
John (ship), 33
John, H., 223,
Johns, John J., 242
Johns, William B., 283
Jolly, Capt. John, 174

Kamchatka, 129
Kansas, 195
Karkeek, George, 91
Kaskaskia, 37
Keam, Thomas V., 148, 206-8
Kennegie, 256
Kentucky, 38, 39, 218
Kern River, Calif., 134, 185
Keweenaw, Mich., 67, 68, 71, 72, 77, 78, 79, 80, 85, 93, 94, 152, 170, 172, 175, 176
Keystone Mine, Ariz., 223
Kimberley, 217, 218
King, Clarence, 218
Kingsbury, Alice, 275-6
Kingston, N.M., 209
Kingston, Ontario, 69
Kinsman, John, 184
Kitto, John, 275
Klamath River, 229
Klondike, 206, 248
'Knacked bals', 259
Knights of Labor, 200, 225
Knockers, Johnny or Tommy, 290, 291
Kohl, Johann Georg, 83, 86-7, 89
Kruger, Paul, 214

Lafayette County, Wisc., 43
Laity, John, 226
Lake Superior mines, *see* Mines, Michigan
Lake Superior Mining Company, 71, 73
Lanarkshire, 24, 25

312

INDEX

Lancashire, 24, 35, 155, 158, 175
Land and mining titles, 102
Landlords and landowners, 21, 30, 129, 292; mining rights of in Cornwall, 4, 13
L'Anse, Mich., 92
Lanyon, 2
Lanyon family, 45
Last Chance Gulch, Mont., 23, 176, 236, 237, *see also* Helena
Latter-Day Saints, *see* Mormons
Launceston, 114n
Laurian, Mich., 278
Lawlessness, 136, 180–1, 199, 208, 220–1, 224, 236, 255, 267–70, 271; in California, 101, 108–9; in Wisconsin, 40–1, *see also* Plummer Gang, Smugglers and Wreckers
Lawrence, Joseph, 117
Lead, S.D., 257
Lead mines and mining 18, 26, 37, 39, 41ff, 47, 53, 63, 70, 72, 110, 149, 169
Lead weed, *see* Masonic plant
Leadville, Colo., 53, 186, 195, 203, 207, 227
Leasehold system 21–2, *see also* Three life system
Le Conte, Joseph, geologist, 217
Lee, John D., 177
Leggett, James M., 221
'Letters from Home' *see* Cornish Pasty
Levant Mine, 198, 222, 285
Lewis, Meriwether, 37
Lewis and Clark County, Mont., 249
Lightning Creek, B.C., 141
Lincoln, Abraham, 143, 149, 150, 154
Lincoln County, Nev., 180, 184
Lincoln, Mount, Colo., 204
Linden, Wisc., 34, 45, 51, 53, 60, 105, 139, 150
Liskeard, 28, 58, 174
Little Big Horn, Mont., 170, 251, 252
Little Cottonwood Canyon, Utah, 213
Little Turtle, 64
Liverpool, 74, 100
Lizard Point, 129
Loam, Michael, 89–90
'Local option', 280
Lockridge, —, speculator, 216
Lockport, Ill., 49
London, 35, 71, 97, 100, 112, 224, 240, 295

London School of Mines, 205
'Long Tom', 110–11
Looe, 1, 223
Lord, Capt. Edward J., 174
Lordsburg, N.M., 224
Los Angeles, 295
Lost Grove Diggings, Wisc., 58
Lousiana, 26, 37
Lucky Lass Mine, Ariz., 220
Lumbering, 91
Luxulyan, 277
Lydford Law, 13
Lyon County, Nev., 190
Lyon, Fort, 146

McGowan, Ned, 136n, 137–8
Mackinac Straits, 64, 67
McKinney, Patrick, 40
McSorley, Edward, 193–4
Madison, Wisc., 96
Madison County, Mo., 148, 162
Magor, Thomas, 242–3
Maine, 74, 294; Law, 280
Man engine, 89–90
Manchester, 241
'Manifest Destiny'
Manacles, 33
Maricopa County, Ariz., 219
Marquette, Mich., 170, 172, 173, 174, 175, 273; iron range, 78, 90
Marshall, James, 95, 97, 98, 99n, 178
Martin, B., 223
Martin, Benjamin, 145
Martin, G., 223
Martin, J., 223
Martin, John, 221,
Martin, Thomas, 221
Martin, W., 223
Mary Ann (ship), 74
Maryland, 38, 39, 144, 183
Marysville, Calif., 188, 290
Marysville, Mont., 241
Mason, Colonel, 96, 97
Mason–Dixon Line, 196
'Masonic plant', 43
Massachusetts, 197
Matthews, Matthew, 142
Maxwell Grant, 209
Mayflower, 294
Meadow Lake, 125
Meadow Valley, Nev., 184

313

INDEX

Mellanear Mine, 255
Men-an-tol, 255
Monominee Iron Range, Mich., 90, 173
Merrifield, Charles W., 156–8
Mesabi Iron Range, Minn., 90, 173
Meteor (ship), 69
Methodism, 11, 12, 15, 16, 24, 25, 45, 46, 55, 57, 63, 116, 132, 149, 233, 267, 268, 270, 276, *see also* O'Brien, Wesley
Metropolitan Iron and Land Company, Mich., 174
Mevagissey, 55
Mexican War, 178
Mexico, 25, 96, 97, 103, 187, 208, 216, 224, 226; Gulf of, 100
Michigan, 6, 39, 48, 57, 60, 62, **63–95**, 96, 100, 113, **163–70**, 172, 178, 183, 184n, 197, 198, 204, 211, 221, 223, 224, 234, 238, 249, 257, 258, 259, 262, 263, 264, 265, 267, 269, 271, 277, 278, 279, 280–1, 282, 287, 288, 289, 290, 294; in Civil War, 152–4
Michigan, Lake, 38, 53, 68, 129
Michilimackinac, 3
Middle West, 129
Mifflin, Wisc., 58
Miller, William H., 270
Millerites, 96
Milwaukee, Wisc., 51, 53, 72
Mine, captains, 173–5, 234; doctors, 289; management, criticism of, 80, 164; supply problems in early Michigan, 76–7; upkeep and maintenance of in Cornwall, 20
Mineral Park, Ariz., 219
Mineral Point, Wisc., 41, 45, 46, 48, 57, 58, 62, 66, 72, 102, 134, 149, 211, 249, 259, 261, 264, 265
Miner's Con, *see* Phthisis
Miners' Leagues and Unions, 121ff, 246–7
Miner's licence fee, in California, 151–2
Mines, in Cornwall: Crinnis, 223; Dolcoath, 88, 205, 206, 223, 255, 256, 285, 286; East Caradon, 223; East Pool, 255; East Wheal Rose, 263; Gonamena, 223; Great Wheal Busy, 176; Hallamannin, 176; Levant, 198, 222, 285; Mellanear, 255; South Tolgus, 176; Tresavean, 90; Wendron Consols, 176; West Francis, 255; West Seton, 283; Wheal Agar, 255; Wheal Kitty, 286; Wheal Providence, 85; Wheal Reeth, 176; Wheal Vor, 223

In Arizona: American Flag, 219; Big Bug, 216, 219; Blue Jay, 223; Clifton Copper, 219; Contention, 219, 220; Copper Queen, 223, 224; Cumberland, 223; Emerald, 220; Hermosa, 219; Keystone, 223; Lucky Lass, 220; Old Dominion, 223; Rustler of the West, 223; Silver King, 223, 226; Silver Thread, 221; South Pioneer, 289; Tough Nut, 220, 221; Vulture, 216

In California: Allison Ranch, 223; Eclipse, 145n; Empire, 119, 120; 121, 122, 123, 223; Eureka, 124, 223; Gold Hill, 119; Idaho-Maryland, 119, 223; North Star, 121, 122; Wisconsin, 60

In Colorado: Americus, 198; Briggs, 198; Hecla, 204; Sleepy Hollow, 198; Terrible, 202

In Idaho: Atlanta Star, 231–2; Poorman, 231

In Michigan: Adventure, 79, 80; Algonquin, 281; Baltic, 174; Calumet and Hecla, 78, 84, 88, 89, 166, 169, 170, 172, 223, 282, 289; Central, 79, 84; Cliff, 78, 80–1, 84, 85, 90, 172, 176, 223; Copper Falls, 79, 172; Delaware and Pennsylvania, 172; East Vulcan, 85; 174; Franklin, 79, 172; Huron, 173; Minesota, 67, 68, 79, 80, 86–7, 164, 166, 176, 223; National, 79; New Port, 85; 174; North West, 79; Norwich, 79–80; Osceola, 85, 198, 223; Pabst, 174; Pewabic, 79, 84, 167, 172, 174; Pittsburgh, 164; Quincy, 79, 167, 172, 174, 223; Republic, 173; Rockland, 78–9, 82; Schoolcraft, 167; Tamarack, 174, 223; Toltec, 80; Trap Rock, 81; Washington Iron, 173

In Montana: Granite Mountain, 198, 246, 247, 248; Speculator, 198

In Nevada: Caledonia, 290; Consolidated Virginia, 284–5, 290;

314

INDEX

Mines—*continued*
  Star, 283, *see also* Comstock Lode
    In New Mexico: Ortez, 208
    In South Dakota: Caledonia
  Quartz, 254; Father de Smet, 254;
  Golden Star, 255; Hidden Treasure,
  254; Homestake, 254, 255; Star,
  255
    In Utah: Emma, 203, 212-13,
  214; Flagstaff, 213; Prince of Wales,
  213; Utah Copper, 213
Miners' Free Press (Mineral Point), 41, 57
Minesota Mine, Mich., 67, 78, 79, 80, 86-7, 164, 166, 176, 223
Mining, in Cornwall, 4ff, 155ff, *see also* Copper, Tin; accidents, 180, 184, 198, 211, 213, 221, 226, 242, 248, 263; disasters, 246, 247, 263; dues of the Duchy of Cornwall, 42; engineers, American and Cornish compared, 88-9, 109, *see also* Daniell, Hornblower, Pearce, Thomas; frauds, 233-5, 255, 287, 291-2, *see also* Emma Mine; machinery, 3, 18; 202, 203, 285-7; *see also* Cradle, Rocker, Stamps; pioneers, poverty of, 232; population, in Cornwall, 156; rights and land claims, 40, 41, 42; speculation, 240-1, 286, 287, in Colorado, 201-3, in Michigan, 77-8; techniques, Cornish, 119, terms, Cornish, 81
Mining Industry and Tradesman (Denver), 256
Mining Journal (London), 224
Minnesota, 90, 173
Missionaries, 178
Mississippi, river and valley, 1, 27, 37, 39, 70, 102, 128, 129, 133, 151, 177, 229, 296; state, 146, 148
Missouri, river, 37, 38, 39, 201; state, 57, 59, 161-3, 171, 179, 197
Mitchell, 263
Mitchell, Charles, 118
Mitchell, Edward, 276
Mitchell, William Henry, 272
Mogollan Mountains, 210
Mohave County, Ariz., 219
Mohawk, Mich., 278
Montana, 39, 57, 125, 150, 170, 186, 199, 206, 210, 211, 213, 224, 228, 230, 231, **236-50,** 251, 252, 257, 264, 271, 283, 289
Montana Company, 239-41
Monterey, Calif., 99
Montreal, 34, 51
Moore's Flat, Calif., 11-12, 117, 288
Mormons, 96, 106, 177, 203, 211, 212, 213-14, 215
Morris, W. H., 227
Morro de Velho, 25
Morva, 255
Mother Lode, Calif., 1, 113, 117, 126, 127, 134, 135, 249, 263
Mountain Meadows Massacre, 177, 211, 214
Mount's Bay, 262, 283
Moyle family, 211
Moyle, Capt. J. H., 172
Moyle, John, 147
Mystery (ship), 102

Nanaimo, B.C., 231
Nancarrow, James, 57
Nancarrow, W. S., 190
Nancarrow, Capt. William, 231-2
Nankervis, Samuel, 272
Nankervis, William 272
Nash, Martin, 41
National Mine, Mich., 79
Navajo Indians, 207, 208
Nebraska, 197, 254
Nebraska Mine Company, 80
Negaunee, Mich., 172, 273
Negroes, in British Columbia, 137
Neil, David M., 227
Nevada, 39, 70, 73, 125, 150, 177, **179-95,** 199, 209, 210, 230, 232, 253, 263, 273, 274, 283, 289, 291
Nevada City, Calif., 119, 126, 188, 279
Nevada County, Calif., 111, 113, 121, 122, 151-2
New Almaden, Calif., 113, 124, 271
New Baltimore, Wisc., 62
New Diggings, Wisc., 58
New England, 38, 116, 127, 128, 148, 175, 188
New Jersey, 38, 42, 94, 175, 184, 197; copper mining in, 3
New Mexico, 113, 146, 147, 148, 177, 207, **208-11,** 224
New North West, 250
New Orleans, 103

315

INDEX

New Port Mine, Mich., 85, 174
New South Wales, 133
New York, 70, 71, 72, 85, 96, 97, 100, 112, 132, 144, 165, 169, 175, 201, 203, 216, 233, 252, 269, 294, 295
New York Tribune, 70
New Zealand, 25, 125
Newfoundland, 33
Newlyn, 1, 102
Newlyn East, 263
Nicaragua, 103, 138
Nolan, Pat, 82
North American Mining Company, 71
North Bloomfield Gravel Mining Company, 111, 288
North Carolina, 97, 99n, **142-5**, 178
North Pacific Railroad Company, 249, 250
North–South Crisis, 59–60, 137, see also Civil War
North Star Mine, Calif., 121, 122
North Wales, 24
North West Mine, Mich., 79
Northamptonshire, 24
Northey, Henry, 272
Northrop, Rev. C. H., 279
Norwich Mine, Mich., 79–80

Oakland, Calif., 217
O'Brien, William 277
Occupational diseases of miners, 160, 180, 199, 213, 221, 242, 248, see also Phthisis
Odgers, Nicholas, 272
Ohio, 37–8, 64; valley, 37, 38
Ojibway Indians, 92–3
Old Dominion Copper Mine, Ariz., 223
Old Dominion Mining Company, 224, 225
Olympia, Wash, 132
Omaha, Neb., 252
Omega, Calif., 123
Ontario, 28–9
Ontonagon, Mich., 63, 64, 65, 66, 67, 77, 78, 80n, 81, 82, 91, 93, 94, 152, 176, 262
Ontonagon Mining Company (French), 94n
Ontonagon River, 3
Ophir Creek, New South Wales, 133
Opie, James, 91–2

Oregon, 103, 128, 129, 134, 135, 137, 174, 229, 230, 231; crisis, 50n
Oro Grande, Idaho, 231
Ortez Mine, N.M., 208
Osceola Mine, Mich., 85, 198, 223
Osler, Edward, 148
Ouray County, Colo., 203
Owen, David D., 62.
Owyhee, Idaho, 229, 231, 235
Oxen, draught in Wisconsin, 261
Oxfordshire, 24
Ozarks, 162

Pabst Mine, Mich., 174
Pacific, Coast, 95, 296; War of 1880, 25
Paddock, Oscar, 102
Padstow, 30, 276
Pall Mall Gazette, 156
Panama, 52, 102, 105, 108, 132
Parnell, Supt., 225
Parys Mine, Anglesey, 141
Parys Mountain, Anglesey, 3
Pascoe, Peter, 173-4
Passenger pigeons, 68
Pearce, James A., 169n
Pearce, Richard, 205-6, 256
Pearce, Stephen, 199
Peck, George W., 74-5
Pelynt, 1
Penaluny, Thomas, 142
Penberthy, James, 219
Penberthy, John, 219
Pendeen, 198, 213, 267
Pennsylvania, 6, 38, 42, 60, 70, 85, 86, 94, 173, 175, 196, 203, 276, 294
Penpons, 211
Penrose, 19n
Penrose, Charles W., Mormon hymn-writer, 211
Pensilva, 253
Penton, Wisc., 58
Penzance, 1, 8, 35, 156, 255, 262
Perranzabuloe, 39, 119, 160
Perry, Sarah, 45
Peru, 3, 97, 187
Peters, Thomas J., 134-5
Pewabic (ship), 69
Pewabic Mine, Mich., 79, 84, 167, 172, 174
Philadelphia, Penn., 144, 165, 201
Phillips, B., 223
Phillips, William, 57

INDEX

Phoenicians, 1, 65
Phoenix, Ariz., 219
Phthisis, 160, 180, 199, **242-4,** 289
Pike Flat, Calif., 262
Pike, Zebulon M., 37, 178
Pike's Peak, 1, 176, 195, 199, 215, 227
Pilchards, 248, 260
Pima County, Ariz., 219
Pinal City, Ariz., 219
Pinal County, Ariz., 217
Pinos Altos, N.M., 210
Pioche, Nev., 180-2, 183-4, 186, 191, 193, 214
Pitezel, Rev. J. H., 81
Pittsburgh, Penn., 3, 71, 79, 165
Pittsburgh Mining Company, 71
Placerville, Calif., 186
Placerville, Idaho, 231, 249
Platte River, 105, 106, 177
Plattville, Wisc., 58, 140
Plummer Gang, 236, 267
Plymouth, 33, 295
Poaching, 123-4
Politics, Cornish immigrants and American, 58, 59-60, 85, 87n, 94-5, 292, see also Copperheads, Civil War
Polk, James K., 96, 97, 138
Pollard, Edward, 242
Polperro, 1, 223
Pontiac, 64
Poorman Mine, Idaho, 231
Pope, H., 223
Pope, William, 198
Population, growth in Michigan, 71
Porkellis, 45
Portage, Mich., 93
Portage Lake, Mich., 79, 82
Porthtowan, 270-1
Port Isaac, 31
Portland, Ore., 134, 229
Potato blight, 6
Potomac River, 144
Potosi, 251
Potosi, Wisc., 58
Potteries, 24
Pottsville, Penn., 60
Prairie du Chien, Wisc., 26, 67
Prescott, Ariz., 216, 223
Prideaux, Capt., 184
Prideaux, James, 43
Prideaux, William, 43

Pridman, Rev. E., 159
Prince, Capt. Vivian, 172
Prince of Wales Mine, Utah, 213
Prisk, William, 219
Probus, 145, 146, 196
Prohibition, 60, 280-1
Prouse, —, miner, 179
Prouse, John, 197
Prout, James, 180
Provision prices, Cornish and Butte compared, 260; in Deadwood, 254; in Grass Valley, 265, see also Wages
Prussia Cove, 10
Pryor, John, 275
'Puddingstone', 87, see also Calumet conglomerate
Puget Sound, 135, 137, 138
Pumps, Cornish, 2, 285, see also Boulton and Watt, Mining machinery

Quack medicines, 290
Quebec, 30
Quick, Robert, 272, 276
Quicksilver, 113, 121, 124-5, 271
Quincy Mine, Mich., 79, 167, 172, 174, 223

Rablin, Lavinia, 45-6
Rablin, William 45-6
Raby, Martin, 141
Racine County, Wisc., 129
Ragtown, Nev., 189
Railroads, 137, see also Union Pacific
Ralegh, Sir Walter, 3
Ralston, A. J., Bank of California director, 179
Ralston, N.M., 210
Ralston Party, 179
'Rand', see Witwatersrand
Rathbone brothers, 74
Real del Monte, 25
Recreations, 271ff, see also Cricket, Wrestling
Red Dog, Calif., 116, 126, 267
Redruth, 1, 7, 32, 48, 58, 62, 114, 116, 154, 161, 168, 181, 198, 199, 231, 232, 268-9, 277
Redruth Hollow, Wisc., 58
Red Warren, Idaho, 231
Reese River, Nev., 179, 184, 186, 191
Reese River Reveille, 194
Relief of distress, 158-60

INDEX

Religion in Cornwall, 15–16, *see also* Methodism
Republic Mine, Mich., 173
Republican Party, 292, *see also* Politics
Retallick, William, 49–50
Reynolds, Seymour, 197
Richards, Edward, 221
Richard, Va., 144
Ridgway, Wisc., 58
Rio de Janiero, 102
Rio Grande, 209
Rivot, M., 94
Robartses family, 11
Roberts, Charles C., 136n
Roberts, James, 198
Roberts, John, of Camborne, 108–9
Roberts, John, of Grass Valley, 272
Robinson, Orrin W., 84, 281n
Roche, 101, 198
Rock drilling, 120; contests, 274–5
Rocker, in gold mining, 110
Rockland Mine, Mich., 78–9, 82
Rocky Bar, Idaho, 231
Rocky Mountains, 102, 129, 178, 204, 227, 263
Rocky Mountain News, 227–8
Rocky Mountain Tick, cause of meningitis epidemic, 244
Rodda, Benjamin, 272
Rodda, Edward, 272
Rodda, James, 272
Rodda, John, 262
Rodda, Joseph, 272
Rogers, Henry, 10
Rogers, William, 272
Rogue River, 229
Rook Pie, 260
Rosevear, John, 183
Rough and Ready, Calif., 116, 126, 267, 280
Rowan County, N.C., 142, 143n, 144, 145
Rowe, Samuel J., 242
Rowett, John, 274
Royal Cornwall Gazette, 99, 148
Rugby, 101
Rule, Alfred, 180
Rule, John, of Mineral Point, 48
Rule, John, of West Seton and Nevada, 283–4
Rule, Robert, 261
Rule, William, 261

Rumphrey, Andrew, 57
Rustler of the West Mine, Ariz., 223
Ryan, Thomas, 41

Sabbatarianism, 270–1
Sacramento, Calif., 135, 179, 227
Sacramento River, 97, 138, 229
Saffron cake, 260
Sager, W. W., 273
St Agnes, 4, 104, 180, 212, 213
St Allen, 45, 62
St Austell, 19, 58, 85, 161, 242, 277, 282
St Blazey, 214, 222
St George, Sons of, 245
St Ives, 199, 277
St Just in Penwith, 17, 19, 35, 48, 58, 114, 119, 148, 160, 199, 232, 233–5, 248, 285
St Just in Roseland, 148
St Keverne, 33, 129, 131n
St Lawrence River, 33, 51, 137
St Louis, Mo., 37, 163
St Mary, Falls of, 39, 67, 68, 72
St Neot, 28, 29, 30
Salmon River, Idaho, 227
Salt Lake City, 106, 214
Salt Lake Telegraph, 237
Sampson, W., 223
Sancreed, 107
Sanders, George N., 148
San Francisco, 98, 103, 109, 113, 122, 127, 132, 135, 137, 215, 216, 217, 218, 227, 228, 267, 269, 284, 295; Vigilance Committee, 137
San Joaquin River, 138
Santa Fe, N.M., 146, 208, 209, 211
Santa Rita, N.M., 209, 210
Sarah, John, 272
Sault Ste Marie, 68, 75, 78, 90, 172, *see also* St Mary, Falls of
Scadden, Henry, 118
Scandinavian immigrants, 153, 167
Schoolcraft Mine, Mich., 167
Schools, in Butte, 239; in Tombstone, 222
Scobel, Edward, 117
Scotland, 158
Scots-Irish, 294
Seattle, Wash., 295
'Seedy cake', 260
Sennen, 17

INDEX

Servants, scarcity, of in California, 118
Seymour, Ariz., 219
'Shake-Rag-Under-the-Hill', Wisc., 46, 58, 265, see also Mineral Point
Sharon Springs, Kans., 195
Sheldon, John F., 41
Sherman, General W. T., 259
Shift working, 282–3
Shipwrecks, 33–4, 69
Shone Flat, Calif., 108
'Shooting scrapes', see Lawlessness
Shullsburg, Wisc., 58
Sidney, Neb., 254
Sierra de Oro, N.M., 209
Sierra Nevada, 1, 102, 106, 107, 112, 177, 179, 185, 186
Silver, 178, 182, 203; demonetisation of, 182, 232, see also Comstock Lode, Leadville
Silver Bow Creek, Mont., 236
Silver City, Idaho, 231, 232, **233–5**
Silver City, Nev., 184
Silver City, N.M., 209, 210, 211
Silver King Mine, Ariz., 223, 226
Silver Thread Mine, Ariz., 221
Simmons, John, 197
Sinaloa, 216
Sioux Indians, 250
Skewes, Edward, 293
Skewis, John, 180
Slack, Isaac, 217–18
Slavery, in Upper Mississippi Valley, 44
Sleepy Hollow Mine, Colo., 198
Slim Buttes, Mont., 252
Slough, Colonel, 147
Slump of 1873 in U.S., 167–8
Smartsville, Calif., 125, 287, 288
Smelting tin, 4
Smithsonian Institution, 66
Smoky Hill Trail, 195
Smugglers and wreckers in Cornwall, 10–11, 267–8
Snake Hollow, Wisc., 41, 58
Snake River, 231
Snell, Nicholas, 142
Socorro, N.M., 209, 211; mountains, 209
Sodey, Thomas, 273
Sonora, 108, 113, 216, 227; Company, 216; desert, 142
'Soo Canal', see Sault Ste Marie
Soulsbyville, Calif., 113, 116

South Africa, 25, 184, 206, 214, 217, 218, 220, 257, see also Transvaal, Witwatersrand
South America, 25, 98
South Carolina, 59, 152
South Dakota, 125, see also Dakota
South Park, Colo., 195
South Pioneer Mine, Ariz., 289
South Pioneer Mining Company, 226
South Tolgus Mine, 176
South Wales, 24
Spargo, John, 183
Sparnon, Henry, 226
Speculator Mine, Mont., 198
Springflower (ship), 30, 31
Squab pie, 260
Staffordshire, 24, 63
Stamp Act, 10
Stamps, Californian, 285, 286; Cornish, 120
Stanley, Reginald, 238
Stannaries, 12–13; court, 156, see also Cornwall, Duchy of
Star City, Nev., 183, 186, 188, 189, 192, 267
Star Mine, Nev., 283
Star Mine, S.D., 255
'Stargazy pie', 260
'Starvation Trail', 195
Stephens, Capt. W. W., 84–5, 174
Stevens, James H., 221
Stockton, Calif., 135
Stokesclimsland, 242
Story County, Nev., 183
Stoves, 266
Strikes and lock-outs, 11, 225, 226, 247, 248; in Almaden County, Calif., 125; in Butte City, 247; at Globe, Ariz., 226; in Grass Valley, 120ff, 135; in Michigan, 167, 282ff; at Moore's Flat, Calif., 111–12; at New Almaden, Calif., 124–5; at Smartsville, Calif., 125; at St Austell, 282
Superior, Lake, 39, 63, 67, 70, 71, 72, 88, 98, 153, 163–7
Superstitions, 290–1
Surnames, Cornish, 114–15, 143n, 223
Sussex, 24
Sutro Tunnel, Nev., 290–1
Sutter, Johann A., 98, 99n
Swansea, 205–38

319

INDEX

Taos, N.M., 209
Tamar River, 10, 17, 24, 79, 175, 259, 277, 294; valley 161, 188
Tamarack Mine, Mich., 174, 223
Tamarack Mining Company, 88, 89
Tamblyn, Nicholas, 55
Tarryall, Colo., 195
Tasmania, 286
Tavistock, 19
Taylor, Elizabeth Gurney, 82, 93, 278
Taylor, Rev. Bartram S., 81-2, 278
Teague, Peter, 274-5, 292
Tecumseh, 26, 38, 64
Teetotalism, see Prohibition
Temby, Charles, 272
Temperance, see Prohibition
Temperatures in mines, 284-5
Templars, Independent Order of Good, 245
Tennessee, 197
Terrible Mine, Colo., 202
Terrill, —, miner, 199
Terrill, Stephen, 43
Terrill Range, Wisc., 58
Texas, 59-60, 179, 294
Thomas, Joe, of St Austell, 242
Thomas, Joe, of Butte, 246
Thomas, Capt. Josiah, 88, 256
Thomas, Philo W., 62
Thomas, Mrs S., 57
Three life system of leases, 22-3, 265
Threshing machines, 13 1n
Thunder Bay Island, 69
Thurston County, Wash., 129ff
Ticketings of copper ores in Cornwall, 79
Timber trade, 26
Times, The, 98
Tin, mines and mining in Cornwall, 19, 78, 156-7, 185; prices, 157, 161, 171, 182n; alleged discovery in Dakota, 255-7; alleged discovery in Missouri, 161-3
Tippett, Edward, 197
Tippecanoe Creek, 38
Tip Top, Ariz., 219
Tithes, 9, 29, 30
Titus, Colonel, 216
Toay, John, 57-8
Tocapilla, 25
Toledo, Ohio, 64

Tol-pedn-penwith, 17
Toltec Mine, Mich., 80
Tombstone, Ariz., 176, 208, 214, 218, 219-23, 225, 227, 266, 267, 269, 273
Tombstone Epitaph, 220, 222
Tombstone Mill and Mining Company, 220
Tonkin, Matthew, 272,
Tonapah, Nev., 179, 183
Toronto Globe, 237
Tough Nut Mine, Ariz., 220, 221
Toy, Joe, 223
Trade unions, 125, 200-1, 225, 244-5, 283, 288, see also Strikes
Transportation problems of mines, 224-5
Transval, 220, 242, 244, see also South Africa, Witwatersrand
Trap Rock Mine, Mich., 81
Travelling hazards in the high Sierras, 187ff
Treasure City, Nev., 183, 187, 193, 194
Trebilcock, Edward, 246
Trebilcock, James, 174
Trebilcock, William, 174
Tregaskis, James, 175
Tregaskis, Richard, 66
Tregeagle, —, unjust steward, 11
Tregellas, Martin, 180
Treglone, John, 190
Tregumbo, Capt. John, 85, 174
Trelan, 129
Trelawney, Bishop Jonathan, 11
Treloar, James, 261
Tremayne, Capt. J. L., 219
Trench Company, 216
Trenerry, Richard, 243
Trenton limestones, 43
Tresavean Mine, 90
Tresize, Francis, 272
Trethevy, 2
Trevarrow, Richard, 92
Trevarthen, Jan, 280-1
Trevarthen, Joseph B., 240n, 250
Trevellick, Richard F., 245
Treverbyn, 28
Trevithick, Richard, 3, 25, 131n, 205, 206
Tribune Review (Butte), 243
Tribute and tribute workers, 5, 7, 8, 20, 166, 200, 204, 245, 282, 288
Trispen, 276

INDEX

Troon, 240n
Truckee River, 186
Truro, 101, 109, 114, 160, 174, 181, 205, 207, 210, 276, 277
Tuolumne County, Calif., 113
Tutwork and tutworkers, 5, 20, 245
Tyneside, 24, 35
Tywardreath, 159

Union Army, 146ff
Union Pacific Railway, 212, 233, 252
Unionville, Nev., 190
Upper Canada, 25, 28 ff
Upper Peninsula of Michigan, see Michigan
Utah, 177, 181, 182, 186, 203, 206, **211-15,** 252, 283
Utah Copper Mine, 213

Valley Forge, 294
Valparaiso, 102
Vance, —, seaman, 102-3
Vancouver Island, 132, 133, 137, 231
Van Diemen's Land, 54
Vermont, 183, 197
V Fluming Company, 111
Vicksburg, 153
Victoria, Vancouver Island, 137, 138, 140, 141, 142
Vineyard, James R., 41
Virginia, 39, 59, 97n, 187, 197
Virginia City, Nev., 23, 71, 85, 176, 179, 183, 184, 186, 187, 188, 193, 227, 267, 269, 271, 274
Vivian, A. P. Pendarves, 292
Vivian, Francis, 43
Vivian, John, 85
Vivian, —, strike leader, 167
Vivians, engineers of Pittsburgh, 3, 79
Voyages, conditions on emigrant, 31-3
Vulture Mine, Ariz., 216

Wabash Valley, 38
Wadebridge, 18n
Wages, 159, 165-6, 167-8, 191-2, 200, 226; in California, 288; in Dakota, 251, 254; in Michigan, 170, 282; in Montana, 240, 249, 250; in Nevada, 283
Wakefield, Mich., 278
Wales, 158; links with Cornwall, 2; copper smelting in, 4, see also Swansea
Walker, Capt. William, 138
Walls, William, 86
Walsh, Henry, B., 41
War of Independence, 26, 27
War of 1812, 25, 26, 27, 38
Warwickshire, 24
Washington, 129ff, 135, 230, 231
Washington, D.C., 96, 144, 151, 295
Washington Iron Mine, Mich., 173
Washington, Treaty of (1846), 133, 138
'Washoe', 115, 179, 182, 185, 186, 215, 227, see also Nevada
Waters, James, 243
Watt, James, 18
Wearne family, 34-5, 50, 51, 52, 54, 55, 105; Richard, Snr, 34, 51, 55, 105; Richard, Jnr, 61, 139; William, 52
Weber Creek, Calif., 104
Webster, Daniel, mine manager, 74-6
Weetman Court House, 144
Welsh immigrants, 219
Wendron, 45, 173, 187
Wendron Consols Mine, 176
Wesley, John, 11, 12, 15, 24, 277
West Briton, 149, 181, 210
West Francis Mine, 255
West Indies, 25, 129
West Penwith, 262
West Seton Mine, 283
West Virginia, 173
Western Federation of Miners, 11, 201, 225, 246, 283
Wheal Agar Mine, 255
Wheal Kitty Mine, 286
Wheal Providence Mine, 85
Wheal Reeth Mine, 176
Wheal Vor Mine, 223
Wheeler, W. P., 252
Whinner, Mrs, 222
White, John, 190
White, Thomas H., 253, 256
White Oaks, N.M., 209
White Oak Springs, Wisc., 58, 103
White Pine, Nev., 122, 125, 179, 182, 183, 186, 188, 190-1, 194, 195
White Pine News, 291
Whittle, Capt. Thomas, 174
Wicks, John, 175
Widdecombe Fair, song, 11, 268

INDEX

Wight, Isle of, 6
Willcox, Ariz., 224
Williams, Ariz., 223
Williams, Alf, 274
Williams, Joe, 274
Williams, John, of Rowan County, N.C., 145
Williams, John, of Butte, 242
Williams, Major John, of St Just, 148
Williams, John Carne, 140
Williams, —, policeman of Trispen, 276
Williams Creek, B.C., 141
Williams and Foster Company, 205
Winnemucca, Nev., 183
Wisconsin, 6, 23, 27, 34, 26, **37–61,** 62, 67, 92, 94, 95, 96, 100, 102, 103, 104, 105, 107, 108, 113, 116, 127, 129, 130, 139, 140, 149, 150, 152, 169, 175, 196, 197, 238, 258, 260–1, 264, 265, 267, 277, 287, 294
Wisconsin Mine, Calif., 60
Witwatersrand, 114, 150, 160, 206, 227, 248
'Wobblies', see I.W.W.
Wolf baiting, 57

Women, on mining frontiers, of Michigan, 90ff; of Nevada, 191, see also Family life
Wood's Creek, Calif., 107
Workmen's compensation, 191
World War, First, 259
Wreckers, see Smugglers and wreckers
Wrestling, 57, 222–3, 241, 272–3, 274, 276
Wyoming, 218, 283

Yale, B.C., 139, 140
Yale, Conn., 279
Yavapai County, Ariz., 219
Yellow fever, 102
Yellowstone, 252
Yorkshire, 24
Young, Brigham, 212, 214
Yuba County, Calif., 111, 125
Yuba River, 123

Zennor, 2
Zinc mining, 169
Zulick, Conrad M., Governor of Arizona, 224